The Demilitarized Zone (DMZ) of Korea

Kwi-Gon Kim

The Demilitarized Zone (DMZ) of Korea

Protection, Conservation and Restoration of a Unique Ecosystem

Springer

Kwi-Gon Kim
Seoul National University
Seoul
Republic of South Korea

ISBN 978-3-662-52060-4 ISBN 978-3-642-38463-9 (eBook)
DOI 10.1007/978-3-642-38463-9
Springer Heidelberg New York Dordrecht London

Printed on acid-free paper

Springer is part of Springer Science+Business Media (www.springer.com)

Foreword

With rapid urbanization and globalization, the balancing act between environmental conservation and socio-economic development in ways that may not endanger the natural resources and stifle opportunities to pursue growth, has become an important concern for policymakers and citizens alike.

As the UN Habitat Agenda indicates, sustainable development is essential for human settlements' development, and thus we need to commit ourselves to the goal of sustainable human settlements in an ever-urbanizing country such as Korea. This can be made possible by developing societies that will make efficient use of resources within the carrying capacity of ecosystems, thereby contributing to the achievement of national sustainable development goals.

The UNDP, through its office in Seoul, has long been a close partner to Korea's governmental and non-governmental endeavors in meeting such national sustainable development goals, as well as in fulfilling the common commitment of the international community to environmental protection. Seoul National University (SNU) and UNDP successfully completed their joint project in conducting a study on an "Environmentally Sound and Sustainable Development (ESSD)" in the Demilitarized Zone (DMZ). The Study was first initiated in 1996, and has since provided valuable assessments and recommendations for the development of the DMZ, which have been critically documented in this book.

The study has especially been important in providing a systematic and comprehensive analysis on the environment in the DMZ, including the Civilian Controlled Zone (CCZ) and, in raising awareness of the value and significance of conserving these areas. The 155 mile (248 km) long DMZ is rich in wildlife and despite political and security issues, the DMZ has become a haven for rare and endangered species of animals and plants. The project successfully compiled extensive records of these ecological features. The protection and permanent preservation of the DMZ ecosystems in South and North Korea hopefully can make a significant contribution to the establishment of a zone of peace between the two Koreas, since the DMZ environment has naturally evolved from the remnants of the Korean War and preserving it is of utmost mutual concern.

Albeit seemingly, environmental issues may be the least considerable concern that would probably break the ice. Nevertheless, this joint effort could gradually build confidence and eliminate the distrust between the two countries, leading to further collaboration in other fields of common interest like promoting peace in this part of the world. Such concerted effort in biodiversity conservation and in other possible common interests will not only enhance environmental security but ultimately strengthen human security in the Korean Peninsula. It is my sincere hope that this book will contribute not only to the sustainable management of the DMZ and in building inter-Korean cooperative relations, but also in sharing valuable experiences with other countries.

I would like to thank Professor Kwi-Gon Kim of SNU and DMZ National Project Director, and as well as all the contributing experts in Korea and abroad for their significant roles in this project, including the sincere assistance of the military authority, all of whom provided support, guidance, and valuable information about the sustainable management of the DMZ eco-systems, most especially for their selfless contributions to the strengthening of inter-Korean relations.

Anne-Isabelle Degryse-Blateau
Director, UNDP Seoul Policy Center

Preface

The Civilian Controlled Zone (CCZ) and the Demilitarized Zone (DMZ) in Korea are the only one of its kind found throughout the world. These areas provide an opportunity for the conservation of the ecosystem and the stimulation of cooperation between disputing regions. The area has been virtually uninhabited by humans since the Korean War, and its inaccessibility allowed damaged ecosystems to rehabilitate and return to its natural state.

Recently, however, the western part of DMZ has been put under serious developmental pressure. There had been rapid changes in land use and such was the conversion of reed fens and wetlands into agricultural land, particularly in CCZ. If this was allowed to continue, the aftermath would have had tremendous devastating impact on the natural resources of the affected areas. In light of the situation, it was by then imperative to develop mid- and long-term land-use strategies that incorporate Environmentally Sound and Sustainable Development (ESSD) principles.

Based on international recommendations and sustainability criteria, the UNDP, Ministry of Science and Technology (MOST), Seoul National University (SNU), and Korea Land Corporation (KOLAND) formed a partnership to develop an ESSD model for the regions affected by agricultural development. The developed model, in turn, was applied to the western CCZ (Paju) and DMZ area since these areas are the most developed before the division of the Korean Peninsula and, that are still under greatest pressure for development today.

The outcome of this twenty-year study is a long-term plan with an implementation strategy that will allow the central government, Kyonggi province and Paju city to manage the area based on sound local ecosystem principles.

The long-term plan cannot be simply put into practice without support from relevant policies, both domestic and international. Therefore, the strategy offered in this study should be understood and aligned with domestic and international efforts for sustainable development. Those who are involved in natural resources management should participate in setting priority action plans like that of what is being laid out in this study to bring them into action. I firmly believe that the long-term sustainability of CCZ and DMZ will be ensured when new challenges are met with creative solutions from all participants.

To the research staff of the Environment & Ecology Planning Laboratory (EEPL) and to the study team of SNU, in particular, Miss Je-Soon Moon, Mr. Dong-Hyun Lee, Mr. Hoon Lee, Dr. Mi-Yong Park, Professor Chang-Hwan Kim and Mr. Grant Jones, I would like to offer my deepest acknowledgement for being instrumental in preparing this book. I wish to offer special thanks to Miss Tessa Eduave, Dr. Stella Gonzales and Dr. Yong-Hoon Kim for their help by editing and reviewing the manuscript. Equal gratitude is also extended to Miss Anne-Isabel Degryse-Blateau who also made significant contributions to the final outcome. I would like to express my sincerest gratitude to UNDP, UNESCO, MOST, KOLAND, Paju city authorities and the Eco-advisory Group for their cooperation and support. My special thanks must go to the Korean government including Ministry of Environment, Ministry of National Land and Transport and Ministry of Defense. I am indebted to the members of the DMZ Environment & Ecology Research Group commissioned by the Korean government. Finally, I wish to thank all my family for support given throughout the writing process. Without the support from the above personnel and institutions, this book would not have been possible.

The Demilitarized Zone, July 2013 Kwi-Gon Kim

Contents

1 History and Evolution of the DMZ Ecosystem 1
1.1 Overview of the Past, Present and Future of DMZ 1
1.1.1 The Past of DMZ 1
1.1.2 The Present of DMZ 2
1.1.3 The Future of DMZ 6
1.2 Environmental Description of DMZ 11
1.2.1 Introduction 11
1.2.2 Topography 11
1.2.3 Functions of DMZ Ecosystem 12
1.2.4 Ecosystem in the Western DMZ. 12
1.2.5 Ecosystem in the Middle DMZ 13
1.2.6 Ecosystem in the Eastern DMZ 13
1.3 Land Use Change in the DMZ. 14
1.3.1 Research Method of Land Use Change 14
1.3.2 Comparison of Land Use between 1918, the Present and the Expected Future 15
1.3.3 Past and Present of the Selected DMZ Sites Based on Old Maps (1918), Landsat Imagery and Photographs. 16
1.4 Evolution of the Wetlands 22
1.5 Evolution of Fire Resistance 24
1.6 Twenty (20) Years of DMZ History in Photos 24

2 Landscape Ecology of the DMZ Area 29
2.1 Introduction 29
2.2 Landscape Ecological Approach. 29
2.2.1 Landscape Ecology at Regional Scale 29
2.2.2 Bioregional Approach Model 32
2.3 The Application 34
2.3.1 Classification of DMZ Ecosystem 34
2.3.2 Pattern Classification of DMZ Based on Forman's Pattern Classification 37

2.3.3 Landscape Ecological Analysis Based on Landscape Physiognomy and Composition 49
2.3.4 Landscape Ecological Analysis Based on Ecological Network (International Biosphere Belt (IBB)) 51

3 Staus and Ecological Resource Value of the DMZ Area 55
3.1 Background of Ecological Survey 56
3.2 Ecological Survey Methods 56
3.2.1 Site Description 56
3.2.2 Consolidation of Field Survey Outcomes 57
3.2.3 Value Assessment Methods 57
3.3 Status of Habitats 58
3.3.1 Status of Wetland Ecosystems 58
3.3.2 Status of Forest Ecosystems 60
3.3.3 Status of Grassland Ecosystems 62
3.4 Status of Species in the DMZ and CCZ 62
3.4.1 Assessing the Value of Ecological Resources of DMZ 63
3.5 Summary 70

4 Habitats in the DMZ Area 73
4.1 Analyzing the Habitat 73
4.1.1 The Concept of Habitat 73
4.1.2 Significance of Habitat Investigation 74
4.1.3 The Process of Analyzing the Habitat 74
4.1.4 Habitat Classification 74
4.1.5 Habitat Investigation by Type 79
4.1.6 Relevant Information Collection 80
4.1.7 Preliminary Determination 80
4.1.8 Site Visitation 81
4.1.9 Determination of Habitat 81
4.1.10 Mapping Process of Habitat Types 81
4.2 The Application 82
4.2.1 Site Selection 82
4.2.2 Analysis by Habitat Type 83
4.2.3 Type, Function, and Value of wetlands in Photos of the DMZ 134

5 Species in the DMZ Area 145
5.1 Overview of Survey and Analysis Methods of Species 145
5.1.1 The Flora and Vegetation Survey and Analysis 145
5.1.2 Fauna Survey 147

5.2 The Application 160
5.2.1 Methods Used in Analyzing Data 160
5.2.2 Tool Development 161
5.2.3 Survey Results 161
5.2.4 Total Species by Western-Central Area of the DMZ 161
5.2.5 Detailed Description by Species 190

6 Biodiversity 299
6.1 Components of Biodiversity 299
6.1.1 Classification of Biodiversity 299
6.1.2 Landscape Biodiversity 299
6.2 Biodiversity and Ecosystem Values 299
6.2.1 DMZ Biodiversity as a Wildlife Habitat 300
6.2.2 Historical Value 301
6.2.3 Recreational Opportunities 301
6.2.4 Education 301
6.3 Biodiversity Strategy 301
6.3.1 An International Tendency 301
6.4 Cases 303
6.4.1 Biodiversity Strategy in Cheorwon 303
6.4.2 DMZ Biodiversity Hotspot 310

7 Wetland-Type Classification 399
7.1 The Need for Wetland-Type Classification 399
7.2 How to Classify Wetland Types 400
7.3 New Classification System: Summary of Wetland Types 404
7.3.1 Overall Study Process 405
7.3.2 Detailed Methods of Each Phase 405
7.4 Application for the DMZ: Overview of the Study Site Wetland Classification 409
7.4.1 Overview of Study Site: The DMZ Paju Phanmun Area 409
7.4.2 Wetland Identification 411
7.4.3 Wetland Delineation 411
7.4.4 Wetland Survey 414
7.4.5 Pollutant Sources and Land Use Status 416
7.4.6 Threatening Factors 416
7.4.7 Wetland Classification 416
7.5 Summary of Wetland Types 419
7.5.1 General Wetlands 422

8 Biological Integrity: Assessing the Health of Wetland Ecosystems 429
8.1 Concept and Objectives of Bio-assessment 429
8.2 Wetland Integrity Assessment Process 430
8.3 Methodology and Design for Assessing the Health of DMZ 430
8.4 Case Presentation 434
8.4.1 Site Selection 434
8.4.2 The Process 435
8.4.3 Results 435

9 Threatening Factors in the DMZ Area 445
9.1 Homo Sapiens, an Endangered Species 445
9.2 SWOT Analysis of the DMZ Ecosystem 446
9.2.1 Strengths 446
9.2.2 Weaknesses 446
9.2.3 Opportunities 447
9.2.4 Threats 447
9.3 Development Pressure 447
9.3.1 Threats by Farming and Fishing 449
9.3.2 Threats to Military Activities 450
9.3.3 Forest Fire and Natural Disaster 452
9.3.4 Use of River Levees 455
9.3.5 Streams Transformed into Muddy Water 458
9.3.6 Conversion of Wetlands into Farmlands 461
9.4 Disturbance of the Ecosystem 462
9.4.1 Invasion of Exotic Species 462
9.4.2 Disappearance of Indigenous Species 472
9.5 Environmental Pollution 474
9.5.1 Soil Contamination 474
9.5.2 Water Contamination 476

10 Conservation and Management Strategies 479
10.1 Land Use Strategies 479
10.1.1 Site Selection 479
10.1.2 Process 479
10.1.3 Application of UNESCO/MAB Model 480
10.1.4 Application of Nature Environmental Audit 481
10.1.5 Application of Bioregional Approach Model 487
10.1.6 Application of Land Suitability and Carrying Capacity 491
10.1.7 Total Assessment and Zonation 494

10.2 DMZ Ecotourism and Wetland Tour 496
10.2.1 Site Selection 496
10.2.2 Plan for Ecotourism Route 497
10.2.3 Ecotourism Resources of DMZ 498
10.2.4 Development of DMZ Ecotourism Program 506
10.2.5 Planning for DMZ Ecotourism 506
10.3 Ecological Restoration 520
10.3.1 Construction of Replacement Wetlands 521
10.3.2 The Replacement Wetland Created in DMZ and CCZ 521
10.3.3 Eco-Bridge 528
10.4 Trans-boundary Cooperation 535
10.4.1 One Ecological Unit: Biodiversity Management and Restoration 535
10.4.2 Operational Strategy to Develop the DMZ TBR 541
10.4.3 A Proposed Joint Project for North and South Korea 544
10.4.4 Implementation Plan 546
10.4.5 A Strategy for Regional Cooperation for Northeast Asian Migratory Species Sustainability: A Proposal for Northeast Asian ECONET 548

References 573

Index 577

Introduction

Demilitarized Zone is Life

Ecosystem in the DMZ has its characteristics in various aspects.

First, farming culture has created a large scale of wetland ecosystem. The DMZ was once a cultivated farmland until the Korean War in 1950 and eventually became a vast wetland with high conservation value after its non-cultivation within a period of more than 60 years

Second, DMZ and CCZ areas combined, is four times larger than that of the city of Seoul. Half of the area is wetland. The scale of the wetland in the area is considered as one of the largest conservation areas next to the Amazon forest and the tropical forest of Indonesia.

Third, due to constant forest fires, the area developed its own unique plant succession. In a normal plant succession, coniferous trees would first appear. In the DMZ and CCZ areas, however, young broad-leaved trees appeared first. The broad leaves are well spread and are better tolerant than other forest trees.

Fourth, although materials on genetic diversity are lacking, the area provides a rich repository and living laboratory of natural ecosystems.

Fifth, the DMZ connects two Koreas into one ecosystem.

Approximately 2,000 species of wildlife animals and plants have been identified in the area, and many species have not yet been identified due to landmines. Among the identified species, there are 81 endangered species and protected species. In particular, 14 species which are registered in the IUCN Red List of Threatened Species, including Lutralutra, Hydropotesinermis, Grus japonensis, Eurystomus orientalis, Aix galericulata, Aegypiusmonachus, Platalea minor, Kaloula borealis, and Nannophyapygmaea rambur.

Forest fires have caused the development of mosaic habitats or patches in the river basin.

The area is valuable for providing the following ecological goods and services: Ecological value, Eco-tourism value, and Economic value (as it provides raw materials for medicine and medical supplies). Sovereignty over natural resources and the opportunity for the participation of international programs (the Convention on Biological Diversity, the Climatic Change Convention) has also been provided.

Geographic Information System (GIS) tool was used for spatial data regarding high-conservative-value sites that can be the basis for decision making in land and seascapes. Monitoring and evaluation of the implemented restoration of damaged ecosystem and the initiated replacement ecosystem are based on the principles of restoration ecology. Knowledge regarding diversity of DMZ life and ecosystem services are yet insufficient. Only around two-thirds of DMZ biodiversity and life are known due to the presence of landmines and security issues and, only around 10 % of known species and habitats, status, and trends are documented.

The author once suggested the possibility to apply existing and available information system to DMZ in the international conference regarding Global Biodiversity Information System (GBIS) held in Korea in 2010. Limitation has been identified on the credibility of existing database regarding the protected areas in the DMZ since understanding on ecological process and structure of DMZ ecosystem has yet been in an early stage. The need for acquiring scientifically based knowledge for the establishment and implementation of policies became an important task henceforth. Based on this point of view, DMZ became a "living laboratory" for experts to critically investigate environmental conditions and accumulate enough knowledge on the diversity of life as well as ecosystem services to be used as a strong basis for policymakers to provide a politically friendly environment for DMZ protection, conservation, and sustainable development projects.

One of the uniqueness of the DMZ ecosystem is in the process and formation of its wetland. As a living laboratory, it harbors an ecosystem that triggered the intellectual curiosity of concerned scientists to pilot "ecosystem approaches and wetland sciences." Many wetland studies have been performed in controlled laboratories (en-vitro) but DMZ is a "live" experimental wetlands (en-vivo). In this wetland, ecological studies can be conducted on a spatial and temporal scale that includes most if not all processes of that ecosystem. Studies on the ecosystem of wetlands had been considerably insufficient so far. Thorough and continuous studies are imperative to fully understand and manage wetland ecosystem and ecology. Investigation on its nutrients, function, and role of plants and animals that supplements a fully functional wetland ecosystem must be sought by researchers. Cultural landscapes are also intriguing areas and objects of study. This is again because the ecosystem that developed and flourished in the DMZ had been formed by past agriculture.

Self-organization, self-design, and ecosystem engineering theories and practices such as ecological engineering, adaptive management, and ecosystem modeling are not fully explored in the DMZ. The chapter on wetland contained in this book should be supplemented by scientifically proven concepts and principles, methods and strategies to support the process of data analysis and utilization. It is important to consciously note that the true values of the DMZ wetland are to be continuously explored and understood as the DMZ constantly evolves itself. The uniqueness of DMZ can only be described through long periods of investigation according to scientists and experts on Ecology. Yet even so, from time to time, it continues to amaze and trigger critical curiosity.

As the recent UNEP GEO-5 report had advised, sufficient data should be provided in order to establish and implement policies that are premised on scientifically proven theories and methods. Specifically, time-series (quasi experimental studies) data should be provided to establish policies. Discussions regarding policies, plans, and management for DMZ to date have been done and implemented without having been supported by sufficient data. Science-driven priorities for DMZ data are established in this book and the data utilization is based on various types of ecologies. Intergovernmental Science-Policy Platform on Biodiversity and Ecosystem Services (IPBES) was recently established and created four functions namely, Environmental Assessment, Knowledge Generation, Policy Support, and Capacity Building. The environmental evaluation of DMZ and the acquisition of knowledge that can contribute to the development of DMZ Resources Management Policy is the primary focus of this book. Diversity of DMZ life includes: landscape diversity, habitat diversity, species and genes' diversity. Further, it includes other evaluations and measurements of environment such as status, trends and co-relations, the use of theories and methodologies of Conservation and Invasion Ecology, types of habitats, evaluation on interactions that necessitated a balanced policy establishment and performance on DMZ biomass. All are based on theories and methodologies of Landscape Ecology and Life.

Societal responses on the changes of biodiversity, ecosystem services, land use, performances in Environmental Ecology, and in Ecological Engineering are considered. The evaluation and recommendations of researchers and experts, including the author's critical judgment, are significant to the development of programs and initiatives.

Thus, this book was formulated and sought to describe holistically the environmental status of the DMZ and to explore scientific methods and implementing strategies to achieve an ESSD in the Korean Peninsula. As humbling and enabling this DMZ endeavor can be, this book specifically attempted to:

1. Provide science-based assessments and community-based technologies for the management of the DMZ;
2. Foster an environmentally friendly cultural and socio-economic disposition among the many publics the DMZ serves;
3. Provide appropriate methods and approaches for societies to make efficient land use and developmental plans within the carrying capacity of the ecosystems and natural resources;
4. Provide basis for policymaking and priority setting of action plans;
5. Align domestic initiatives with international efforts for sustainable development;
6. Contribute to the improvement of the quality of natural resource management;
7. Build inter-Korean cooperative relations;
8. Contribute to the establishment of peace between the two Koreas;

9. Share valuable experiences with other countries that are having concerns of the same nature; and
10. Enhance human and ecological security.

Significance of the DMZ

Pertinent data continuously documented reveals significant aspects of the DMZ and its implication to human and ecological security not only in Korea but also to the world. Preserving, protecting, and conserving the forest lands and other life-sustaining natural resources in all parts of the world, is an international concern. This common vision does not just apprehend the effects of natural catastrophes brought about by global climate change, but most significantly improves, prevails, and maintains the quality of life of humans as well.

The DMZ corridor has some distinguished features of its own. The land once completely devastated by war, has now become a natural treasure where rare birds and animals inhabit and where succession of plant species occur and thrive. Almost all the ecological features are intact without being shredded by human interference. Here we find a lofty ideal that peace and life can be rediscovered even on the ashes of war. For the Koreans, it is also a place where human agonies and jubilations, heroic stories and sagas remain alive. At the advent of the new millennium, the world starts to shed new light on the DMZ. It is transforming itself from a legacy of the cold war to a symbol of new reconciliation and harmony.

The DMZ Ecosystem regulates Flood and Erosion. Intertidal Habitats such as salt marsh and mudflats provide a buffer system preventing the tidal energy of the ocean to hit the coast. This geological process of the beach plays an important function in protecting the coast. Five rivers running through the DMZ play critical roles in controlling inland floods as well. Grass species along the river tributaries control silts from clogging the rivers while its roots strengthen river banks, thereby preventing erosion.

Water Quality

Wetlands, as natural water purifiers, process water by filtering sediments and absorbing many pollutants in the surface waters. They function as a natural medium for purifying not only surface waters, but also groundwater reserves. Some wetland systems were found to have mechanisms that enhance the quality of groundwater resources.

Habitat for Wildlife

The DMZ ecosystem provides a vital habitat to endangered species. It serves as a flyway for Northeast Asian birds, for many wading species, and waterfowls. It is a stopover for these birds to rest, feed, and build up energy before traveling further.

Global Concern

The DMZ forest is second to the Amazon in value. The Amazon is considered the world's foremost stronghold in controlling climate change. So must DMZ be. The world must realize this, and should help in sustaining this valuable ecosystem because DMZ is a buffering ecology that combats the dreaded effects of global warming.

The collaboration of North and South Korea in the DMZ conservation project is indeed humbling and enabling. It is a clear evident indication that there is a possibility of peace between the two Koreas. This is only the beginning that can open up many doors for partnership, starting with programs of mutual concerns and interests like perhaps in preserving their precious ancestral culture and heritage. The DMZ is humbling because it builds a natural bridge to unify the two countries. Even as they should be independently functioning as two different countries, they can also share their material and human resources, just like friendly neighbors do. The natural healing of the DMZ is an important enabling lesson. It taught the world to allow its wounds to heal naturally and not to nurture the pains caused by them. And to move forward without fear of committing the same mistake that caused such wounds.

Research Value

Environmental Researchers specify their areas for study. It must remain untouched and free from human influence, high impact on human and ecological security, and require collaboration and partnership with international research institutions and agencies. The DMZ is a valuable research that provides a never ending curiosity to the critical researchers and a never ending satisfaction on the results of their advocacy.

Chapter 1
History and Evolution of the DMZ Ecosystem

1.1 Overview of the Past, Present and Future of DMZ

After the Korean War broke out on 25th June 1950, an armistice was established at a deadlock, and North Korea occupied the north of the 38th parallel. Military presence, weapons development and installation of military facilities had been banned in the zone by the armistice pact. In addition, the CCZ was set up to limit civilian entrance in the armistice line areas. Thus, the DMZ and CCZ have formed a boundary belt between the North and South Korea, expanding from the east coast to the west coast, not including the seas.

The length of DMZ from the end of the west to the end of the east is about 248 km (approximately 155 miles), while the width from the south to north is 4 km, covering a total of 907 km^2 as a whole. In addition, the CCZ—the northern area of the civilian controlled line was set with the width of 5–20 km from south boundary line of the DMZ and the total area is 1,369 km^2. As the Borderland Promotion Act took effect in 2000, some parts of the southern area of the CCZ were set as borderlands. As a result, the total area became 7,678 km^2. If borderlands are not counted, the total area of the ecosystem treasure which crosses the Korean Peninsula is 2,276 km^2 as shown in Fig. 1.1.

1.1.1 The Past of DMZ

DMZ and CCZ, hereto referred as "DMZ", are the political products from the past. Many areas in the DMZ, including Panmunjeom and Cheorwon fields in particular, contain a large farmland where Korean people had farmed and resided in the past. The remains of such farming and residing activities were observed during the survey. One example was the Jangdanmyeon town office in the Paju DMZ area which indicates that the surrounding area was a large district with the town office serving as a central zone. Also, the area was an important transport terminal of the Gyeonggi railroad. The Fig. 1.2 shows some of the remaining evidence of the past.

K.-G. Kim, *The Demilitarized Zone (DMZ) of Korea*,
DOI: 10.1007/978-3-642-38463-9_1,

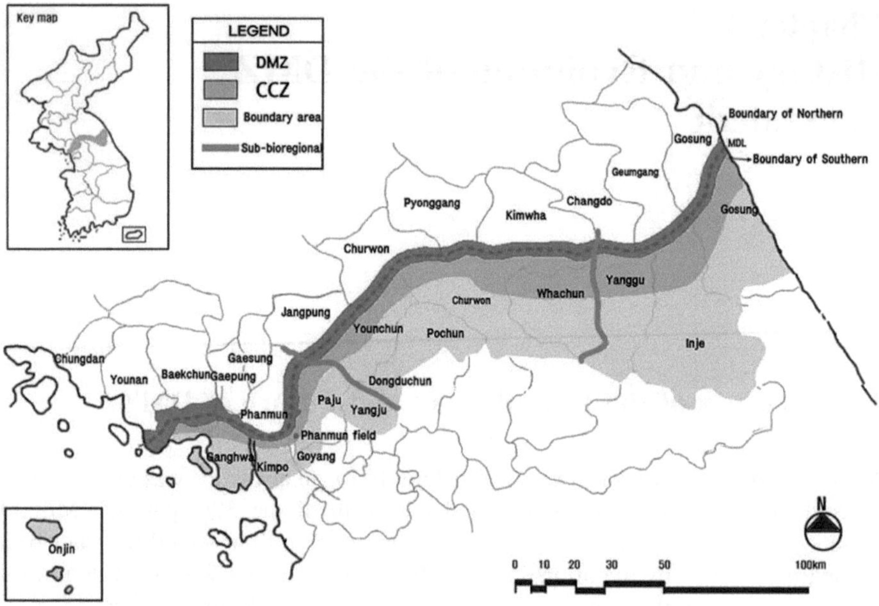

Fig. 1.1 Location of the DMZ and CCZ

Further, Cheorwon DMZ area is a flatland which was mostly used as farmland. In this particular area lies the places where Goguryeo ruled in year 829–936 and Gung Ye who ruled in 901–918, with the remains of their residences indicated the glory of their times. Some of the remains in the Cheorwon DMZ area are shown in Fig. 1.3.

Having different environmental features from previously described Paju and Cheorwon, the DMZ of Goseong was an area where people inhabited. *Pseudosasa japonica*, poplar (*Populus deltoids*) and tombs, are unique characteristics of villages formed in those areas that were observed during the on-site survey. Specific views of the area are presented in Fig. 1.4.

1.1.2 The Present of DMZ

Being a residential and agricultural area in the past, DMZ area has been changed to wildlife habitat today. After the Korean War in 1950 and ceasefire line on the 38th parallel as the armistice pact was signed, the DMZ has remained 'a no-man's land' for the past 60 years. The military left mine fields in many areas while the border is known to be the most heavily armed in the world. Under such circumstances, the farmlands and villages left by its inhabitants had time to naturally vegetate and undergo ecological succession, a process of natural change. The farmlands, in particular, became wetlands and wildlife habitats. Many abandoned rice paddies

Fig. 1.2 Remains of past residence in Paju DMZ area. **a** Janndanmyeon town office. **b** Remain of Gyeonggi railroad

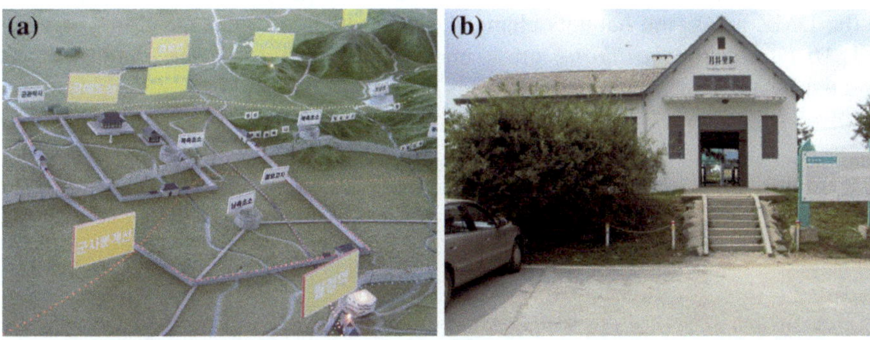

Fig. 1.3 Remains of human residences in the Cheorwon DMZ area. **a** Model of capital in Cheorwon DMZ area. **b** Woljeong-ri station in Cheorwon DMZ area

Fig. 1.4 Villages of the DMZ of Goseong area as a ground of life in the past. **a** Poplars in the area of Goseong DMZ. **b** Tombs in the area of Goseong DMZ

and farmlands in Panmunjom—a joint Security Area then, became a sanctuary of birds including the white egret, cattle egret and, the common heron and white heron. This is shown in Figs. 1.5, 1.6 and Table 1.1.

The DMZ in Cheorwon has also been transformed into habitats for birds as the farmland areas in the past was abandoned and transformed into wetlands. Particularly in Cheorwon DMZ area, virtually no farmland could be seen anymore and abundant reeds developed instead, covering the wetlands (Fig. 1.7).

Costal dune was widely developed in Goseong DMZ area brought about by the sands swept away from the coastline. As the DMZ area has been conserved, widely spread alluvial land used as farmlands in the past have been transformed into wetlands where willows and alder trees thrived (Fig. 1.8).

The peaceful utilization of the DMZ areas has been raised in accordance with South–North reconciliation mood since the South–North joint Declaration has been signed on June 15, 2000. Construction works started and completed in 2002 at country level. In addition, as a number of development plans were implemented in the DMZ areas, the habitats changed. The wetlands where meadows of reeds and *Miscanthussacchariflorus* grew and widely thriving, were damaged and affected by the construction works (Figs. 1.9, 1.10 and 1.11).

Fig. 1.5 Abandoned rice paddies of Panmun field in Paju transformed into forested wetlands

Fig. 1.6 Birds found at abandoned rice paddies that transformed into forested wetland in the Panmun field. **a** Some white herons and their breeding. **b** Egrets' multiplication

Table 1.1 The current status of specially protected species in the DMZ

Classification	Endangered species designated by Ministry of Environment	Red data	Natural monument species	Indigenous species	Total
Vegetation	–	–	–	–	0
Mammals	Small-eared cat	Chinese water deer, small-eared cat	–	–	2
Birds	Golden eagle, Red-crowned crane, white-naped crane, marsh harrier, swan goose, einereous vulture, common buzzard, goshawk, bean goose	Red-crowned crane, white-naped crane, hooded crane, einereous vulture, swan goose	Red-crowned crane, white-naped crane, golden eagle, goshawk, marsh harrier, kestrel, swan goose	–	10
Amphibians and reptiles	–	–	–	–	0
Insects	–	–	–	–	0
Fishery	Common Korean bitterling	–	–	Korean rose bitterling, Squalidusgracilismajimae, Abbottinaspringeri Banarescu et Nalbant, Korean spotted sleeper	5
Total	–	–	–	–	17

Fig. 1.7 Abandoned rice paddy field transformed into wetlands in Cheorwon DMZ area. **a** Wetland area transformed from abandoned rice paddy field in Panmun field in Paju. **b** Wetland in the DMZ area at the inside of Weoljeongri station. **c** Wetlands in Cheorwon DMZ area viewed from the victory observatory

1.1.3 The Future of DMZ

Ocular surveys revealed that the wetlands nearby DMZ were damaged and succinctly affected by development. Efforts for restoration and management had been made to minimize the effect. A joint survey team for ecological environment has been established to minimize the devastating effect on the ecological system of the highly sensitive ecological areas that arises from the construction of Gyoenggi Line, Donghae Line and North–South connecting road inside the DMZ. The joint survey team for ecological environment played an important role in these areas. The team identifies specific ecological changes by periodically monitoring the area and, provides appropriate conservation strategy for the damaged habitats by enforcing the restoration and maintenance policy. They accelerated and prompted the execution of the conservation and restoration plan by introducing a 'Fast Track' system as a first priority of the national government. The activities of the joint survey team for ecological system are shown in Fig. 1.12.

The reduction and restoration for the damaged areas nearby DMZ had been accomplished through the decisions and actions made by the joint survey team for ecological environment. Their epitomic accomplishments are the construction of ecological bridge and the replacement wetlands. The features after the habitat restoration are shown in Figs. 1.13, 1.14, 1.15. They tried the ecological links by

Fig. 1.8 Estuarine wetlands and abandoned rice paddy field transformed into wetlands in Goseong DMZ area. **a** Marsh wetlands developed in Goseong DMZ. **b** Widely developed coastal dune wetlands at the inside of barbed wire in Goseong DMZ area. **c** Alluvial valley wetlands developed in Goseong DMZ area

Fig. 1.9 Changes in habitats due to the construction of the road connection of South–North Korea in Paju. **a** Before the road construction. **b** Habitats affected by the road construction

Fig. 1.10 Changes in habitats due to the construction of the Gyeonggi Railroad line, Paju. **a** Before the construction of the railroad. **b** Damaged state after the construction of the railroad

Fig. 1.11 Changes in habitats due to the construction of Donghae Railroad line and the road that connects South Korea to North in Goseong. **a** Before the construction of the railroad. **b** Habitats affected by the construction of the railroad

Fig. 1.12 Photos on the activities of a joint survey team for ecological environment led by the author. **a** Scene for installation of survey plots. **b** A view for ecological survey

means of connecting the mountainous habitats through the building of an ecological bridge in the isolated habitats and damaged areas. The functions of nature can be maintained by nurturing the replacement wetlands which perform the roles and forms the structures similar to the damaged wetlands.

The wetlands, forest and other habitats are restored through the ecological restoration process at the DMZ area. Such features help explore ways to save the DMZ natural ecology from constant development pressure. A model for continuous land use with attached natural conservation plan should be applied in DMZ area. Guaranteeing future environmental sustainability and maintenance of the

Fig. 1.13 The view of the constructed ecological bridge at the DMZ in Paju. **a** The damaged habitats in road construction. **b** The restoration habitat after the construction of the ecological bridge

Fig. 1.14 The view of the constructed replacement wetland in new Gyoenggi line at Paju. **a** The damaged area after the railroad construction. **b** The restored habitat by construction of a replacement wetland

Fig. 1.15 The effect of the constructed bridge along Goseong line, Donghae line and North–South road. **a** The habitat damaged by the railroad construction. **b** The effect was minimized by construction of a eco-bridge

DMZ must be given sincere consideration and that such responsibility must be inculcated to the present generation and the generations to come.

The DMZ is currently under the pressure of development. The discussion on the joint cooperative development project of North and South Korea in utilizing DMZ is gaining force. It was pointed out that a model on the conservation and continued use of DMZ for the cooperation between North and South is urgent and a national task. The Panmun field where the Gyoenggi Line and roads cross is located at the west of DMZ is highly evaluated in many respects as the areas of the model suggested for North and South cooperation with the opportunity and potentiality to draw out the future vision of the proximities at North–South border line. In biological and geographic points of view, the Panmun field is a link to the ecological system developed in a wider range at the estuary of Sacheon River which originated from Kaesung city and flowing into Imjin river. However, it is identified as the most feasible area for the integrated management of estuary and land use by the North–South. The author would like to suggest a few proposals which would help the development of a model for North–South cooperation in their joint project at the Panmun field of DMZ.

First, provide a model for North–South environmental cooperation. It may include a model for the management of wetlands and rivers, model for its designation as a nature reserve, and the model for information generation. North Korea released a survey report on the ecological system after declaring the Panmun field as a nature reserve area. According to the report, 15 orders, 38 families, 149 species of birds live in the area. The white-naped and hooded cranes were included. The ecological survey, led by the author, on the southern part of Panmun field has been conducted through the permission of the United Nations Command since 1996. During the survey, four dead bodies of eagles were found at the proximities of JSA inside DMZ. These eagles are designated as Natural Treasure No. 243. These dead eagles were handed over to the College of Veterinary, Seoul National University where the prescribed investigation was performed in order to find out the cause of death. It was presumed then that dead eagles can also be found at the northern part of DMZ since these birds thrive and cover the same environment as to those in the immediate south. However, it is difficult to establish the cause of death of these eagles because information is not shared.

Second, develop a model for the cultural buffer region. The suggestion to develop a model for the recovery of cultural identity with the cooperation of both North–South is feasible in Panmun field even before as well as after the unification. The native towns at the Panmun field areas can be fostered as the ecological peace zone of North–South and as the cultural buffer area. Designating Kaesung city as the world cultural heritage complex of UNESCO will be helpful for the recovery of cultural identity.

Third, develop the model of special region for North–South eco-tourism. It will be the best of both worlds if the water path from Yeomhwa river of Incheon, through the estuary of Han river, to the Sacheon river and to the proximities of Panmun field is developed for eco-tourism. The Gyoenggi Railroad and the roads near the DMZ can be launched as a sightseeing course for ecology enthusiasts. This opportunity for eco-tourism at the Panmun field of DMZ will suggest a term-based, comprehensive integrated model that boasts the features of the DMZ to the world can eventually boost the collaborative economic future of North–South. The models to be developed here cover the total area of Northern and Southern parts of the DMZ, hence can initiate a significant road map for North–South partnership and hopefully contributes to the eventual peace in the Korean peninsula.

1.2 Environmental Description of DMZ

1.2.1 Introduction

There will be no peace and life in the DMZ without the strong conviction and common efforts of government and non-government agencies to conserve its natural resources—a unique biosphere in the world, and paralleled with the collaborative efforts of the public.

1.2.1.1 DMZ

The DMZ is a strip of land running across the Korean Peninsula, which served as a buffer zone since the ceasefire between North and South Korea in 1953. It is 155 mile long and 2.5 mile wide along the military demarcation line (MDL) where civilians' pass is strictly restricted. It has been untouched ever since and is now considered as a precious natural habitat, a rich source of flora nowhere else found in other parts of the country. Lots of efforts are being made to preserve its biodiversity and to transform this zone into a sanctuary of nature, life and peace.

1.2.1.2 Scars of Tragic History

Since the ceasefire between the two Koreas in 1953, there had been various provocative actions from North Korea. One critical provocation was the four underground tunnels the North Korean army built. It was evidently designed for an invasion of the south. When discovered by the south, North Korea insisted that it was designed by South Korea for a surprise attack on the north. North Korea's allegation was proven untrue as evidenced by the direction of blasting inside the tunnels that were obviously going towards the south. Since the thawing tide of the cold war however, attention and great interest from various sectors had been concentrated in the area.

1.2.2 Topography

Steep mountains cover the eastern part of the DMZ, which made it a very good forest ecosystem and a favorable habitat for birds and mammals. Vast plains cover the central part of the DMZ with lush forests where rare migrant birds, including cranes, refuge here in winter time. The western DMZ features lowlands and various types of wetlands creating high biodiversity. It is inhabited by various rare animals and plants. The conservation value of the DMZ is high with its excellent ecosystem. For the same reason, people's access to the area has been controlled for the past six decades.

1.2.3 Functions of DMZ Ecosystem

The DMZ ecosystem has many functions even though there are needs for more studies to describe functional response.

Habitat for wildlife: The DMZ ecosystem provides vital habitat to endangered species. It serves as flyways of North-east Asian birds, for many wading species and waterfowls, a stopover to rest, feed and build up energy before traveling further.

Protection against flood: Five rivers running through the DMZ play critical roles in controlling flood as well as providing water resources.

Research value: Environmental researchers specify their areas for study. It must remain untouched and free from human influence, high impact on human and ecological security and, require collaboration and partnership with international research institutions and agencies.

1.2.4 Ecosystem in the Western DMZ

In the western area of DMZ, a large area of wetland expands around Jangdan peninsula, Imjin and Sacheon river, where an estuary is running into the West Sea then conjoined with Imjin and Han river. These areas serve as intermediate landing spots for migratory birds that are flying over the western coastal area and, provides a winter home for many water birds as well.

A variety of wetland ecosystems can be seen conjoining with the estuarine and coastal wetlands. That is why winter migratory birds designated as Korean natural treasures like the Swan Goose and Brent Goose, the Red-crowned Crane among other rare birds are found here. Natural wetlands provide wildlife habitats not only for birds but also for insects such as *Nannophyapygmaea*, reptilia and as well as *Elapheschrenckii* species.

Necessity for protecting and restoring the Western DMZ ecosystem
The DMZ ecosystem is expected to be subjected to high potential environmental degradation arising from high demand of development. These conditions attract tremendous concerns especially when the Gyoenggi railroad was restored. It was an area that required regional and international partnership for its restoration and conversation. We should conserve the DMZ wetlands along with the equally important Donghae (East Sea) wetlands, as they are part of the critical flyway for water birds.

1.2.5 Ecosystem in the Middle DMZ

The Cheorwon plain is the winter home of the world-renowned cranes and eagles. These migratory birds need habitats where they can find sufficient food, shelter and a relatively warm environment. The plain is always ice-free even in winter and provides spacious wetlands with enough food from its broad fields. Many other winter migratory birds and ducks make use of this area for over-winter residence or utilize it as an intermediate landing spot en-route to other destinations. The plain takes a peculiar terrestrial landscape which was formed by a volcanic explosion in the distant past. The plain shows natural wetlands formed by the idle farmland. The CCZ rice fields and wetlands inside the DMZ assume a critical function in providing food, shelter and suitable habitat for these migratory birds.

Necessity for protecting and restoring the Middle DMZ ecosystem
Conservation activities for cranes on the part of local residents, military servicemen, local government and civil organization are very active. However, more concerted efforts are still needed for the conservation of ecosystem and habitat environment where these birds harbor. The ecosystem is particularly at risk due to reduction of food resources and destruction of habitat due to the increasing development activities such as the construction of buildings and roads.

1.2.6 Ecosystem in the Eastern DMZ

Most of this area consists of forested wetlands and Gamho lagoon wetlands, facing the East Sea. It also displays magnificent valley wetlands and river wetlands formed mostly along the Nam river inside the forest stretching from Mt. Geumgang to Mt. Seorak. Here, the Yong Swamp is situated, a unique highland wetland in Korea, that is currently under protection along with other wetlands including the Upo Swamp after these swamps were designated as the Ramsar site. It also serves as the landing spots for the rare red goral and water birds, egrets and other diminishing bird species flying over the East Sea.

Necessity for protecting and restoring the Eastern DMZ ecosystem
This area urgently needs conversation activities as well as research for methods and effective strategies for the conservation of rare wildlife like the red goral. This area is also threatened by development projects like the Mt. Geumgang tourism road construction. This area is dynamic and with pristine and primitive ecosystems where rare species like *Hanabusayaasiatica Nakai* and the sundew plant are found in abundance. Thus, this area is of utmost importance in efforts associated with the conversation of biodiversity.

1.3 Land Use Change in the DMZ

1.3.1 Research Method of Land Use Change

This section illustrates the changes of land use in the DMZ area based on the author's recent research.

1.3.1.1 Importance of Historical Records

Documentary review of historical records is the most important element in natural environment restoration planning. The focus of analysis is on the processes that occurred during such changes which have contributed to either the regeneration of nature or to the causes of the concurrent damages in the target area. Analysis of environmental conditions, including landform and hydrologic structures, as well as the status of the species that ever inhabited in the area, are most helpful information in planning appropriate measures. Therefore, sufficient data should be available for decision making and in order to determine the scale, objective and methods to be used for the present and continuing conservation and restoration initiatives in the future.

1.3.1.2 Analysis Methods of Historical Records

Analysis based on old maps
Old maps provide important clues even though the quality of these maps cannot be compared to the accuracy of information provided by modern technology. The maps in particular, are more useful in regional scaling. It cannot scale minute details of information. The regional scale information can be valuable in the identification of areas needed to be restored and in the conduct of network planning for areas with restoration potential.

Analysis based on topographic maps
A topographic map contains many different types of geographic information. The map contains detailed information on landform, which can be useful in identifying the previous stage of landform in the restoration area that may have experienced severe landform changes. The systematic survey for the creation of topographic maps has begun rather recently in South Korea. Topographic maps (50,000:1) from the Japanese colonial period (1910–1945) and from the 1960s can be easily accessible from the National Geography Institute, however, the maps from the previous earlier years are not easily accessible.

Analysis based on aerial photographs
The advantage of an aerial photograph is that it provides an easier understanding of the pre-damage period of a specific area compared to the old maps and topographic maps. Furthermore, the aerial photograph allows the identification of the regenerative state of vegetation in the target restoration area although the information does not allow scientific or quantifiable outcomes.

Analysis based on satellite images
Survey and analysis based on satellite images have been widely used in many areas and served an important role in ecological restoration.

The Pollen survey
Pollen is a fine to coarse powder containing the microgametophytes of seed plants, which differs depending on types of plants. Pollen is chemically very stable, and is not easily changed against the chemical. After the changes of geological features and climates in a long period of time, the pollen would stay close to the original form as in fossil form. Therefore, pollen provides important information on the vegetation of the past. The pollen analysis of plants found in a target restoration site would give scientific and accurate understanding of its past vegetation.

Analysis based on the perceptions of local residents
Key Informant Interviews can be conducted among local residents in order to acquire a general view of the past conditions of the target area, especially on their observations in the process of its transformation from being a farmland to becoming a battle field and then to its becoming a valuable ecosystem. The respondents selected are the local residents who have lived near the target site for a long period of time. Knowledge on past land use are obtained through interviews with land owners. They can provide some clues on the previous state of the site, including the identification of species thriving in the area in the past. The knowledge received is useful for planning vegetation design, method and strategy. As vegetation undergoes changes in time, as seen in invasion theory or succession theory, it provides assistance especially in selecting suitable management directions. It is an effective method in identifying life species, and comparing past and present life species. Information on life species are hard to be figured out in topographic maps or in aerial photographs, and even in satellite Images.

1.3.2 Comparison of Land Use between 1918, the Present and the Expected Future

In this section, the findings of the research regarding changes of land use in DMZ which was conducted by the author will be discussed. The study used the Land Use Change Method Landsat which was earlier mentioned in the previous paragraph. Figure 1.16 shows the relationship between ecological process, structure and functions affected by natural phenomena and threats. Figure 1.17 explains the spectrum of the past and present of the DMZ based on the relationship shown in Fig. 1.16.

1.3.3 Past and Present of the Selected DMZ Sites Based on Old Maps (1918), Landsat Imagery and Photographs

In this section, land use changes of spots selected from major natural areas of the DMZ shown in Fig. 1.18 are addressed. They were examined using topographic maps, satellite images and aerial photographs among other methods used in this study.

1.3.3.1 Location of Selected Areas

Looking Fig. 1.18 itself, anyone could say that the DMZ has many assets and opportunities for peaceful and significant other uses for human existence.

Four areas were selected as examples of natural areas within the DMZ to review land use changes. They include Panmunjeom·Sacheon Area, Naepo·Gangsan Area, and Goseong Area.

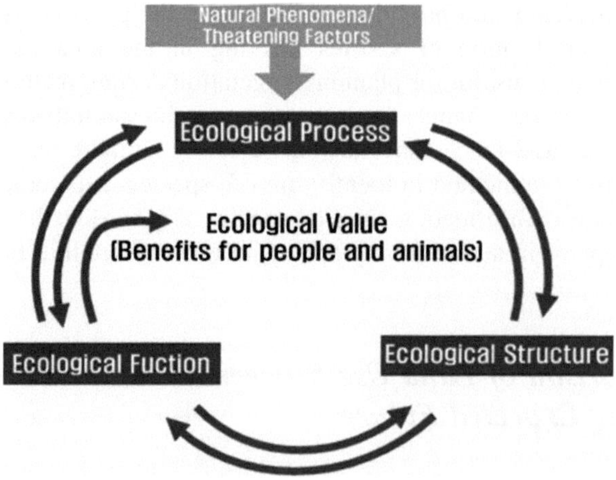

Fig. 1.16 Relationships between ecological process, structure and function affected by the drivers such as natural phenomena and threats

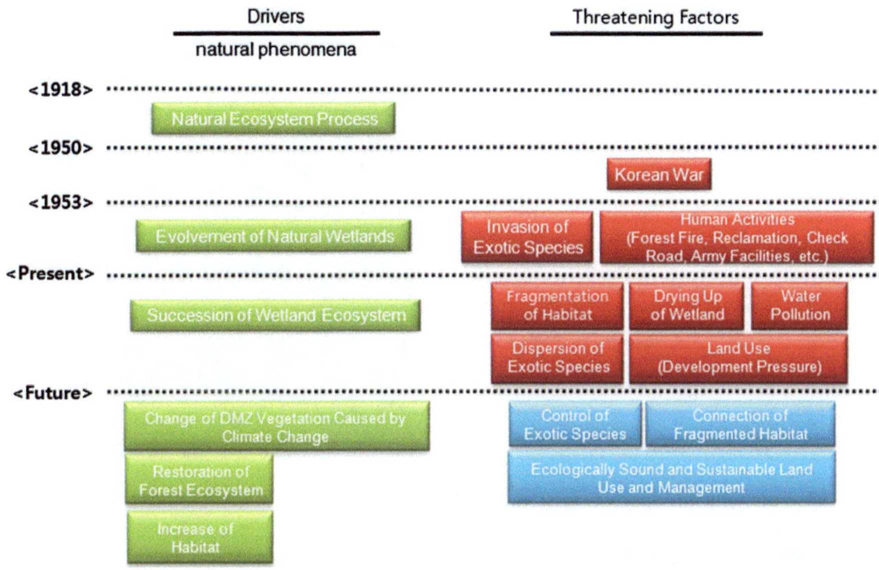

Fig. 1.17 Spectrum of past and present of the DMZ

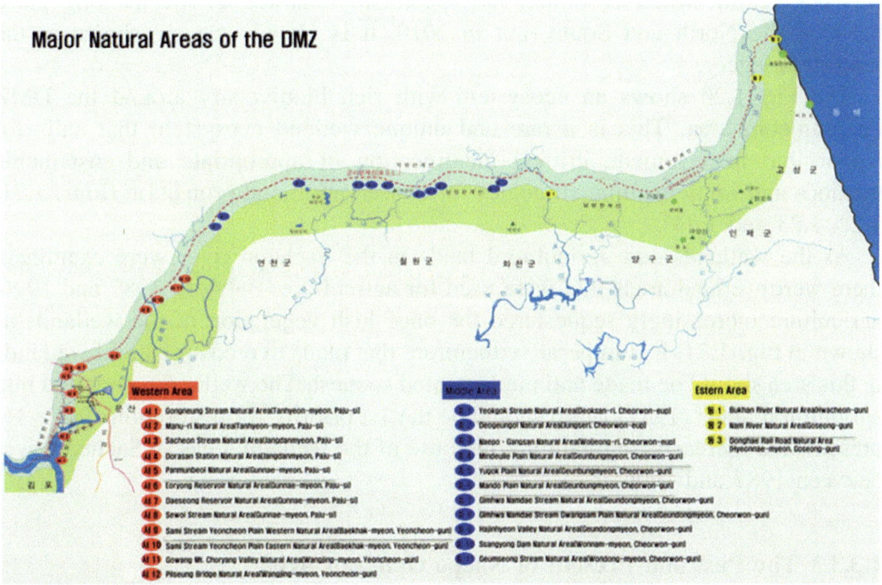

Fig. 1.18 Major natural areas of the DMZ

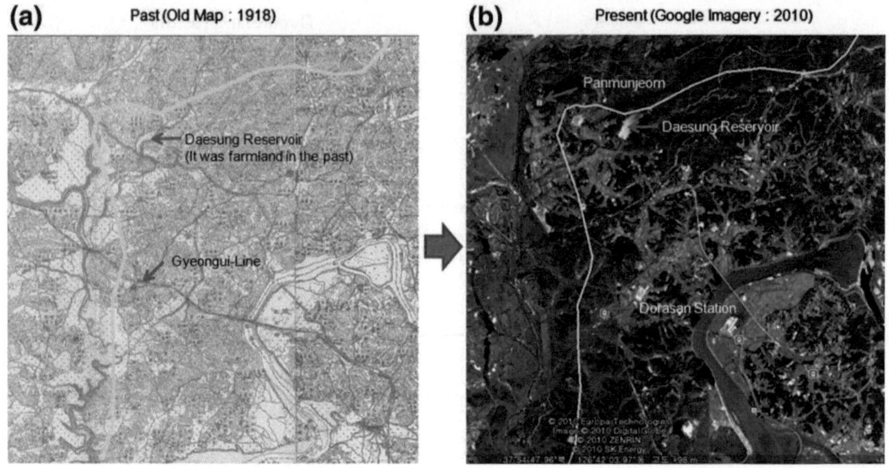

Fig. 1.19 Panmunjeom·Sacheon area of the DMZ

1.3.3.2 The Past and Present of Panmunjeom·Sacheon Area

Panmunjeom·Sacheon Area within the DMZ as shown in Fig. 1.19, compares the old map published in 1918 to the Google imagery taken in 2010. In 1918, the area was not yet identified or rather, not specifically marked as the meeting place between the North and South, but in 2010, it is already clearly shown in the satellite imagery.

The Fig. 1.20 shows an ecosystem with rich biodiversity around the DMZ Panmunjeom area. This is a rare and unique wetland ecosystem that calls for responsible management, critical identification of appropriate and sustainable methods and implementing strategies for conservation and protection (Figs. 1.21, 1.22, 1.23 and 1.24).

As the wetlands and agricultural lands in the Sacheon river were examined, there were wetland areas that were used for agriculture. Between 1987 and 1996, agriculture increasingly sequestered the once lush vegetation of the wetlands as shown in Fig. 1.21. It is imperative therefore that plans to recover natural wetlands in this area should be made and implemented soonest. The wetlands converted into agricultural lands (1987–1996) should be the 1st priority for restoration. Table 1.2 presents the increasing agricultural land use of the wetland areas in Sacheon river between 1987 and 1996.

1.3.3.3 The Past and Present of Naepo·Gangsan Area

Naepo·Gangsan which is part of Cheorwon is a world-important winter home for cranes. These migratory birds need habitats where they can find sufficient food, shelter and a relatively warm environment. The migratory birds include white-naped crane, red-crowned crane, hooded crane and ducks use this area for over-

Fig. 1.20 Photographs of Panmunjeom·Sacheon area of the DMZ

winter residence or utilize it as intermediate landing spot en-route to other locales. It takes a peculiar terrestrial landscape which was formed by a volcanic explosion in the distant past. The CCZ rice fields and wetlands and grasslands inside the DMZ assume respectively a critical function of providing food, shelter and suitable habitat for these migratory birds.

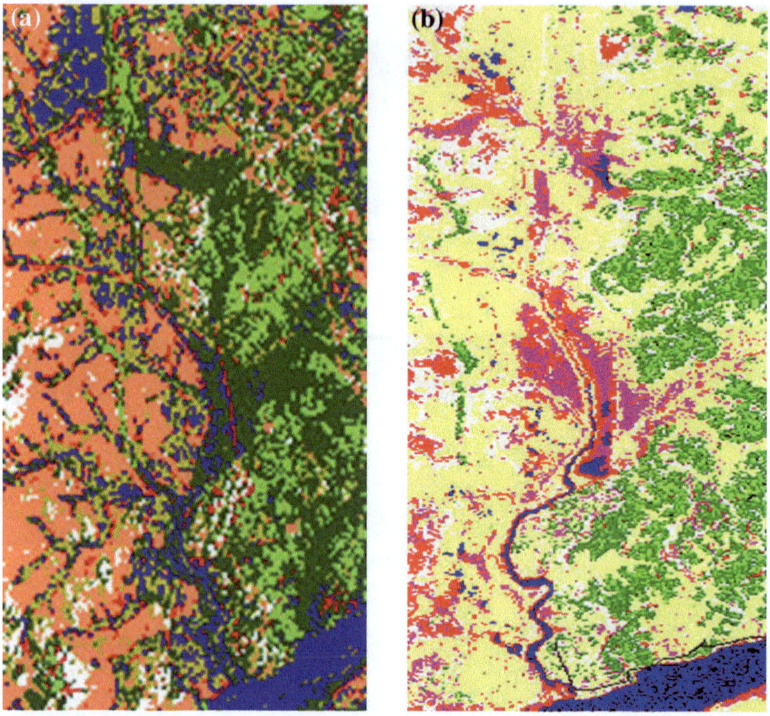

Fig. 1.21 Changing process of wetland areas in the Sacheon river. **a** Land use summary in western DMZ, Sacheon (Satellite NDVI 1987). **b** Land use summary in western DMZ, Sacheon (Satellite NDVI 1997)

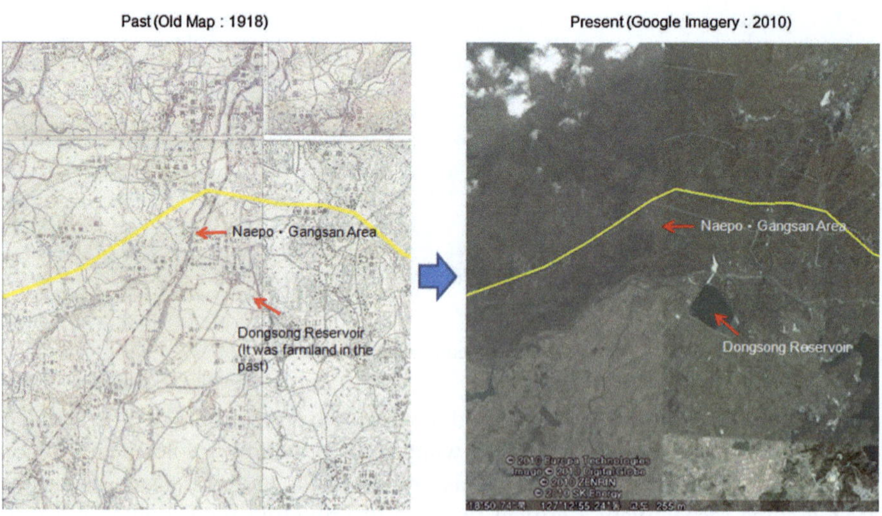

Fig. 1.22 Naepo·Gangsan area of the DMZ

Fig. 1.23 Photographs of Naepo·Gangsan Area of the DMZ

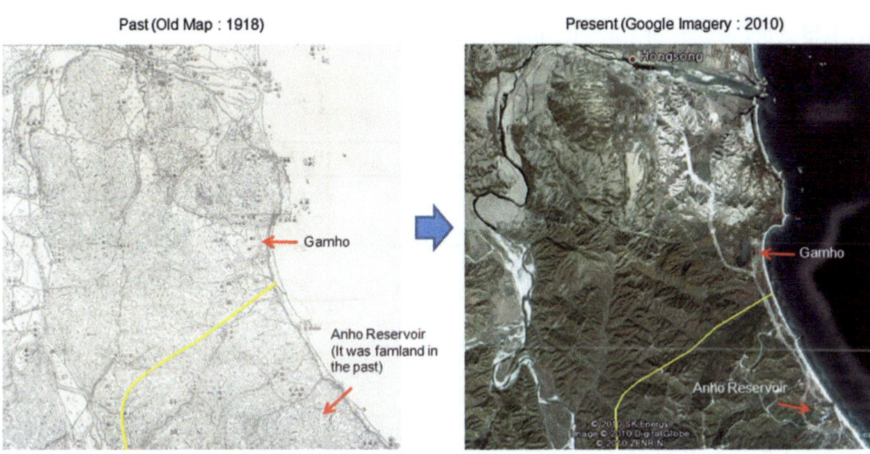

Fig. 1.24 Goseong area of the DMZ

1.3.3.4 The Past and Present of Goseong Area

Most of this area consists of forested wetlands and lagoon Gamho lagoon wetlands, facing East Sea. It displays magnificent valley wetlands and riverine wetlands formed mostly along the Nam river inside the forest stretching from Mt. Geumgang to Mt. Seorak. It also serves as landing spots of the migratory birds including egret and water birds on their flyway. It provides habitats for rare species like red goral.

Fig. 1.25 Photographs of Goseong area

Table 1.2 Change of the wetland areas in Sacheon river between 1987 and 1996

	1987 (%)	1996 (%)
Wetland	25.6	21.7
Farmland	19.3	22.3

1.4 Evolution of the Wetlands

The metamorphosis of the wetland formation in idle abandoned lands was examined by visual judgment through systematic observation during the conduct of ocular survey. The wetlands were classified according to its form and topography as follows; mountain valley form, plain land form, riverside form and, further classified according to other different types of wetland characteristics.

Details of wetland-type classification can be found in Chap. 7 of this book.

The ecological value of DMZ has been enhanced as it had naturally transformed itself from being a rice paddy into a forest wetland at present. A picture of the typology of wetland and its examined characteristics are presented in below.

Plate 1
Plain type

The previously wide plain of rice paddies were converted into a lush wetland and vegetation had naturally grown since it was made free of human interference for 60 years. Various native and rare animals, birds and insects have begun to multiply and played an important role in forming a succession of plant community in these areas which eventually became conducive habitats for their refuge of migrant birds in winter.

Plate 2
Mountain valley type

Areas near mountains and valleys were replaced by wetland vegetation, favored by flying-over birds and by the nesting permanent/resident birds.

Plate 3
Riverine type

The once agricultural lands near rivers and streams were naturally transformed into wetland vegetation, and became an effective habitat for native and rare ducks as well as perennial fresh water fish, whose life cycle is dependent on each others' survival.

Fig. 1.26 Natural restoration process of woodland patches formed by forest fire

1.5 Evolution of Fire Resistance

Forest fires during military operations in the DMZ affected the structure of its ecosystem. It was subjected to frequent burning by the army people then, perhaps one of the reasons why resident farmers evacuate the area and forced to leave their fields. Albeit seemingly, the fire contributed to the DMZ's magnificent structures and landscapes ranging from shape of its tropical savanna to its boreal forests. As it is seen in Fig. 1.26, wetland ecosystems along the sides of rivers and streams also support a vegetation of stable grasses, bushes and medicinal plants used traditionally by the locals. Riverside wetland ecosystem has evidently maintained patches of herby and woody wetland vegetations which were naturally formed as a result of periodic forest fires. The Evolving Process of Ecology in its natural adaptation to local cultures (philosophy and behavioral practices) is clearly illustrated by the series of landscape elements and classes of dominant vegetations in the vicinity.

1.6 Twenty (20) Years of DMZ History in Photos

This section contains pictures documented during the past 20 years of DMZ history. The photos illustrated how nature could find its way back to its natural form and contents. From the panoramic landscapes, rare birds are now freely flying and reproducing. When the plants complete themselves into lush vegetations, other living things follow supporting the propagation, like the insects that facilitate pollination and the birds also help in widely spreading its seeds. Korean animal treasures began to appear in the vicinity and joined by rare birds never to have known and seen in this area. More animals are attracted to the habitats including migrant birds, where they stop over in winter time. In 20 years, DMZ transcended into an environment with spiritual fervor and peace, a "dream come true" for humans and animals alike.

A Bird gliding over Mindeulle plain

A group of eagles in Cheorwon plain

Estuarine wetland in the East Coastal Area

(continued)

Estuarine wetland of the Han river

Wetland of the flood plains and its rich biodiversity in the Yeonchon field

Vegetation

Birds

Bird gliding over Yeoncheon plain

Bird on rice paddies in Daeseongdong

Vegetation inhabited in eastern DMZ

Naemorhedus goral found in the eastern DMZ

Naemorhedus goral

(continued)

Fish

Mammal

Invasive plants of DMZ

Amorphafruticosa

Aster pilosusWilld

Ambrosia artemisiifolia

Phytolacca americana L.

Chapter 2
Landscape Ecology of the DMZ Area

2.1 Introduction

This chapter encompasses the landscape ecological point of view of the DMZ. In the first part of this chapter, the concepts, scientific theories, principles, and models in the characterization of the ecological landscapes are presented. Moreover, in the latter parts of this chapter, the results of analysis and evaluation of expert investigation follow.

2.2 Landscape Ecological Approach

2.2.1 Landscape Ecology at Regional Scale

In every region on Earth and at every stage in history, human survival has required a basic understanding of landscapes. In contemporary ecology, a landscape is a heterogeneous area consisting of distinctive patches, which landscape ecologists refer to as landscape elements organized into a mosaic-like pattern. The elements of a mountain landscape may include forests, meadows, bogs, and streams, while those in urban landscape include parks, industrial districts, and residential areas (Molles Jr. 2010). Landscape Ecology is defined as the study of the relationship between spatial pattern and ecological process over a range of scales.

2.2.1.1 Concepts of Landscape Ecology

Earl N. Troll (1939), a geologist, coined the term 'Landscape Ecology' at the time he was studying the issues in land development and use in Eastern Africa. As he was taking aerial photographs and consequently analyzing the landscapes, he found out that there is an intimate connectivity between geology and ecology and declared landscape ecology as "Ecoscience", distinguishing it from geoscience.

K.-G. Kim, *The Demilitarized Zone (DMZ) of Korea*,
DOI: 10.1007/978-3-642-38463-9_2, © Springer-Verlag Berlin Heidelberg 2013

Landscape ecology is a science which explains the phenomenon of composite interaction under comprehensive and certain rules existing between life community and its surrounding environmental conditions. This science has been developing and is based on the existing geology and ecology since 1960s up to date.

Ecology is a study which explains vertical relations arising between water, wind, energy flow, animals, and plants. Landscape ecology is a horizontal study which explains phenomenon arising between spaces. Serious attention has been given to planning and designing areas, specifically focusing on the phenomenon arising between geographic spaces (Landscape) considering the existing ecological characteristics which evolved in these spaces. As a result, landscape ecology can be briefly defined as ecology objecting to landscapes. However, the important matter is the fact that landscape ecology studies prioritize the special distributions and how these develop in specific landscapes and, emphasize the role of disturbances, including human influences in the subject area.

2.2.1.2 Objectives of Landscape Ecology

Landscape ecology intensively studies the structures, functions, and changes of landscapes. The specific tool employed by researchers is the geographical information system (GIS). Data from terrestrial and aerial photographs including satellite imagery are also pertinent to the assessment and evaluation. These studies concentrate on the planning process utilizing all the gathered information.

Landscape ecology is classified into four kinds: (1) patches, (2) edges and boundaries, (3) corridors and connectivity, and (4) mosaics: Pattern and scale.

Wherefore, landscape ecology data encompass specific variables in describing, analyzing, and measuring patches, edges, boundaries, corridors, and mosaics. It is quantified according to metrics, connectivity, pattern and scale of the distribution, and migration of life species. These studies significantly contributed to the development of landscape ecological models and methods regarding migration of life species within mosaics, diversification of life species in patches, and corridors. In case of plants, landscape ecology determines the elements, moisture, soil, nutrition, and identification of primary species thriving in the landscape under the study's interest. The characteristics based on landscape ecology are described in Table 2.1. On the other hand, Table 2.2 shows Forman's patch types.

Meanwhile, principles of landscape ecology are more concretely suggested as:

First, sizes and forms of patches are ecologically suggestive of its status in terms of productivity, diversity of species, soils, and waters.

Second, habitats decide the size of patches. A patch is a wide area which is relatively homogenous but different from surroundings considering the inhabited radius of major life species.

Third, natural vegetation in wide patches are necessary in order to preserve and protect aquifers that ensure the quality of lake water.

Fourth, several wide patches should be needed to preserve life diversification of ecosystem (pine trees patches in different size, vine yard patches, and so on).

Table 2.1 Characteristics based on landscape ecology

Elements	Characteristics based on landscape ecology
Patches	• Distinguished and analyzed in the size, numbers, and location's point of views • Size of patches: Core parts' habitats and species, internal habitats and species, areal extinction possibility, distinctions, diversification of habitats, barrier against disturbance, large-scale patches, and small-scale patches • Number of patches: loss of habitats, activities of distorted entities, number of large-scale patches, number of small-scale patches, and collectivized patches into habitats • Locations of patches: extinction, selection of patches for preservation
Edges and boundaries	• Importance of edges becomes lots of opportunities in order to consider ecological succession • Structure of edges: diversification, width, ecological boundaries, edges as filtering device, and slopes • Boundaries (lineal or curve type): natural and artificial Edges, lineal and curve types of boundaries, rough boundaries and smooth boundaries, curve degree and width of edges, tail, and spur • Types of patches: edges and internal species, interactions on surrounding environments, ecologically optimized types of patches, features, and directions
Corridors and connectivity	• Habitats' networks • Species' corridor: control of corridor function, GAP's effectiveness in corridors, and structural similarity versus plant's similarity • Stepping stone: connectivity, distance, loss, and concentration • Road and baffle wall: roads and corridors, wind erosion and its control • Stream corridors and dissolved substances, widths of stream corridors in the main streams, widths of stream corridors, and connectivity of stream corridors
Mosaics: pattern and scale	• Segregation of habitats due to transformation of land use occurred by human activities • It itself can emphasizes functions of landscapes • Networks: connectivity and circularity of networks, circulation and other alternatives, density of species and size of nets inside of corridors, effects of cross-points, species in the connected small patches, distribution of species, and small connected patches • Types of fragmentation: Internal versus entire loss of habitats, fragmented patches, suburbanization, and protection areas for foreign species • Barometer: the lowest scale of mosaic, identification of barometers for animal fragmentation, special species, and general species, types of mosaics for multiple habitats

Fifth, Islands close to continents preserve more species than isolated islands far from continents. Likewise, wetlands close to forests preserve more species than wetlands distant from them.

Sixth, corridors should be made wider than 60–90 m before it becomes a "Disturbance Patch" for birds roaming in open farm areas according to a study regarding electric pole corridors in temperate deciduous forests.

Table 2.2 Forman's patch types

Disturbance patch	Recently created by some form of disturbance, e.g., forest fire
Regenerated patch	Patch recovering from disturbance, e.g., forest clearance in the process of regrowth
Environmental patch	Some sharp discontinuity, e.g., change in soil type that cause a distinct change in the plant community
Remnant patch	Area remaining of the original habitat isolates by an encroaching matrix, e.g., Woodland patch amongst agricultural fields
Introduced patch	A habitat created by human activity, e.g., a Woodland clearance

Seventh, encountering points of waterways should be accounted for large vegetated nodes of large-scale vegetation.

Eighth, form of patches should be curve and irregular.

Ninth, mosaics should be made of combining patches, corridors, and metrics by arranging the principles of land use through synthesizing.

In the landscape ecological point of view, the analysis of present conditions is based on the four kinds of landscape ecological principles (edges and boundaries, corridors and connectivity, mosaics, pattern and scale). This analytical method is used in the planning of the restoration of streams and landscapes. A data-based restoration plan and environmentally friendly land use plan influence successful evaluations. Simple methods of qualifying the ecological properties of a landscape are shown in Fig. 2.2. Figure 2.1 shows the patch work of traditional DMZ agriculture in terms of land use.

2.2.2 Bioregional Approach Model

Bioregional approach is an approach based on plan, preservation, and development. It is an approach combining each unit of analyzed landscape setting high

Fig. 2.1 The patch work of traditional DMZ agriculture

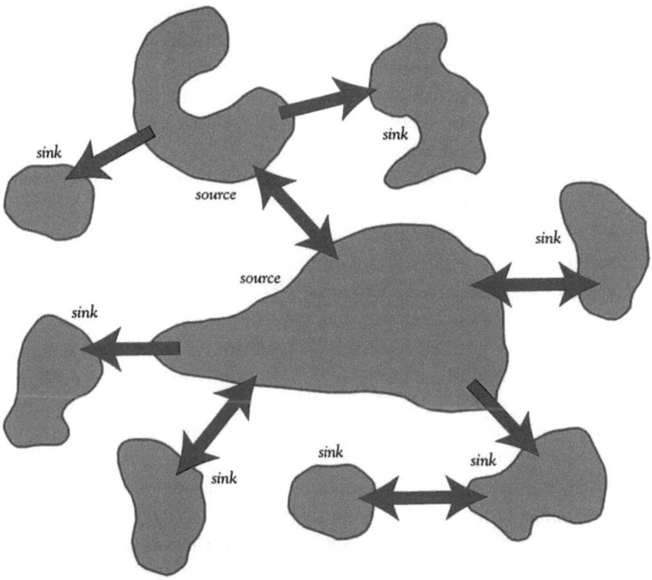

Fig. 2.2 Simple methods of quantifying the ecological properties of a landscape. Landscape Physiognomy is the physical layout of an ecosystem in the landscape. The habitats of interest are the ponds and two measures are indicated here namely; the length of the boundary (*red*) of each patch and the distance between pairs of patches (*green*)

values on continuity of landscapes apart from the exiting method which divides dissected parts of landscapes (Jones and Jones and SNU 1998). This is for establishing place unit as a basic unit based on geographical characteristics in areas and analyzing landscape resources through examining ecological elements (natural, cultural resources) which compose specific characteristics in the each area (UNDP 2000). That is, units of bioregional areas are in the most significant level and are divided as landscape district in lower significant levels and, place unit in the lowest significant level. Table 2.3 arranged the contents of the concepts and features regarding life bioregion, landscape district, and place unit using the bioregional approach. Characteristics of bioregional approach can be found in the course of this process as its striking feature. Each classified place unit is to establish a management plan targeting areas by combining them to a bigger area (e.g., targeting these areas by combining landscape district, subregion, and bioregion) of homogenous units after the evaluation process.

Table 2.4 suggests the contents according to each level as it undergoes the process to establish preservation plans of landscape resources through the bioregional areas. Featured contents in this process can be mentioned as second and third steps. In the second step, characteristics of geological areas are stipulated according to physical elements, such as water system and land form (biological characteristics) and the fundamental resources that compose the characteristics of these areas. In the third step, characteristics and features of each area are stipulated

Table 2.3 Concepts and characteristics of bioregion, landscape district and place unit

Bioregion	• Bioregional area, the largest unit of approaching unit • Classification by distinctive climate in areas, Landform, vegetation, river basin, and types of land use (in classification, a process was made through to boundaries and features recognized by area residents through contact and participation of area residents) • Concepts of evaluation of preservation values, plan, and management are provided in landscapes of object areas
Landscape district	• Lower unit in Bioregion • Classified by waters and mountains and range to recognize observers • There are many cases where area residents have intimate names on the areas and it is often recognized as specific region • Primarily classified focusing on landform, animal/plant habitat conditions, waters, land use patterns and additionally classified by adding culture, and life styles through site visitation
Place unit	• It is lower unit of landscape district and the lowest unit in area classification of bioregion areas • "Surrounding Space" as unique visual featured river basin • Possible to immediate visual distinction • Decision by vegetation, landform (Landform), ridge (Ridge) in boundaries

as the area's map sectioned by life species' area and divided into place units. The mapping is based on the observation made from photographs of each sectional area and the testimonies of area residents. The relation between the resources of each area composing the landscape to that of each place unit will be better observed in the bioregional approach. This reveals the natural, cultural essence, and the process in the bigger bioregional position (macroscopic point of view).

However, the most important matter is the comprehension of values and attitudes of the people living in the area through site visitation. That is, classification by physical aspects should be simultaneously performed along with the consideration of the cognition of the residents, and the classification of predetermined spaces.

As this area classification is performed, consideration includes the academic, social, and physical aspects of the area. The classification, therefore, should necessarily be based on sufficient basic knowledge, data review, and field experiences to be precise in the classification.

2.3 The Application

2.3.1 Classification of DMZ Ecosystem

The bioregion of the DMZ ecosystem can be classified into the following: the Western (Paju bioregion area), Central (Cheorwon bioregion area), and the Eastern (Goseong bioregion area). This is shown in Fig. 2.3.

Table 2.4 Process of bioregional area approach

Steps	Contents
First step project preparation	• Development of organization and targets in the task (composition of performing group, unofficial discussion on landscapes, development of targets) • Object areas (selection of candidate areas for the study) • Comprehension on data (decision of necessary information, decision of information process method) • Process plan and schedule (decision of process method between work group, adjustment of work period)
Second step understanding of landscape of target area	• Mapping process of bioregion (natural area in candidate areas and regulations and mapping process of bioregion) • Mapping process of Landscape district(Landscape District) (Bioregional Division into landscape, utilization of natural Boundaries, such as water system) • Mapping process of place unit(Place Unit) (division of landscape districts into place units, title imposition including areal feature) • Recording of characteristics and documentation for concerning matters of the community (deduction of concerning matter through discussions with area residents, photographing on major areas showing features of place unit)
Third step identification of essential resources	• Review and modification of landscape district, place unit boundaries, and names • Final mapping process and examination of resources (actualization of essential resources through collective description on objects and features and drawing of the final map)
Fourth step determination of landscape quality	• Development of evaluation standard and ratio (selection of landscape quality and determination of ratio) • Evaluation of place unit (Application of standards to essential resources in each place unit) • Combination with preference results (combination of other evaluation methods and preference study deducted as results) • Analysis of the entire results (determination of each Landscape district and landscape quality per unit) • Composition of the final report
Fifth step development of implementation plan	• Preservation of areas' landscape resources and establishment of preservation plan

The identifiable composition of the species in the areas is the diversified characteristics of ever green needle-leaf trees, the broadleaf trees, and bamboos.

The Korean Peninsula is divided into eight plant geographical regions namely: (1) northern high mountain area, (2) northern and southern high mountain area, (3) central mountain area, (4) southern mountain area, (5) central and western island area, (6) southern island area, (7) West Sea, South Sea, and East Sea island area and relating inland area, and (8) segregation area between South and North.

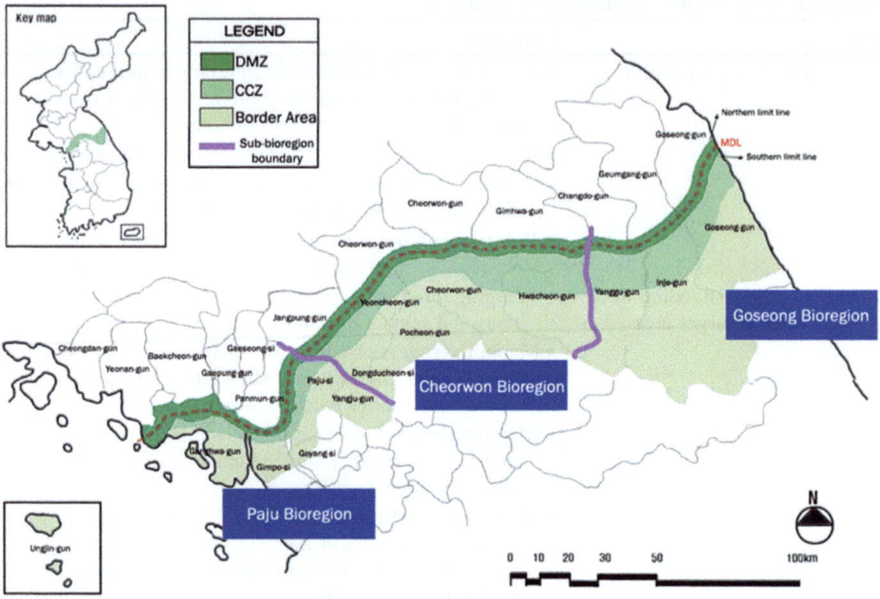

Fig. 2.3 Classification of DMZ ecosystem biogeography

In this book, characteristics are analyzed per each regional area and divided into west, central, and eastern region, specifically considering the plant geological classification, current diversification, and water system networks in the DMZ.

As the characteristics per bioregions are examined, the Goseong area maintains a generally excellent life species in its forest ecosystem, forest wetland, and coastal ecosystem. In the case of the Cheorwon bioregion area, it became the wintering area of the world's rare birds due to the existence of farmlands and inland wetlands. The Cheorwon plain, on the other hand, is widely spread with many developed reservoirs such as the Sanmyung Lake, Togyo Reservoir, Hak Reservoir, and among other smaller reservoirs. There are also naturally formed swamps in the Cheorwon-like Saemtong and the abandoned paddy field are also well-developed marshes. Lastly, the Paju Bioregion area shows high diversification of life species and well-developed wetlands and mud flats that help prevent soil erosion and floods.

There are specific areas in the bioregions that display significant biosystems. The Hyangrobong area (part of the Goseong bioregion area) preserves an ecosystem close to pristine forest, with very diversified plants and serves as the connecting axis that links to Mt. Kumkang and Mt. Seorak. The only high peatland in Korea is located on the top of Dream Mountain and designated as wetland protection area by the Ramsar Convention and the Ministry of Environment.

Hwajinpo Lake, located nearby East Sea area, is also highly evaluated as a leading wetland with sea water flows. The Dutayeon area is also known as the best habitat for lenok in South Korea. In the Cheorwon plain, rare birds such as the

Grusjaponensis and the white-naped crane (among other types of rare cranes) spend their winter in its wetlands.

The Imjin river estuary of Paju Bioregion area, there are specific world rare birds that frequently inhabit in the area namely; the blackfaced spoonbills and the Grus japonensis. In the Sacheon river basin, near Panmunjeom area, blackfaced spoonbills are also observed to inhabit there. In these rivers, a rare fish called Mugilogobius Abei is also seen thriving here.

Various life species are observed in Eoryong Reservoir and Daeseong Reservoir, such as rare fish, reptiles, and amphibians besides the birds and animals mentioned earlier to have inhabited there.

2.3.2 Pattern Classification of DMZ Based on Forman's Pattern Classification

This section expands the discussion on the ecological pattern analysis method applied to DMZ previously mentioned in the preceding sections. This method is the Forman's Pattern Classification and applied to the following major areas of the DMZ.

Estuary of the Han River Area

As the whole sections of Han river flow through Gangwon province, Chungcheongbuk province, Gyeonggi province, and Seoul city, only part of estuary of the Han river, of which flows into Yellow Sea among sections of the entire 281.7 km stream, is accounted for DMZ and CCZ. The estuary of the Han river has plain form due to characteristics of landform in Korean Peninsular and shows a form of mud flat as a form of estuarine wetland.

Figure 2.4 shows an image of the estuary of the Han river with wetland or meadow wetland form in most sections. It is not an area with evident mountains or developed trees and does not show figures of noticeable patches except for corridors due to roads or waterways. This area, like other DMZ and CCZ, also shows disconnection between habitats and ecosystem due to cease-fire line. This is not desirable in the landscape ecological point of view but is rather an irony in a sense that the present natural environment has been preserved.

Fig. 2.4 Landscape ecological pattern analysis on the estuary of the Han river

Fig. 2.5 Landscape ecological pattern of surrounding farmlands in the Imjin river, Paju

Paju Imjin River Area

Figure 2.5 shows an image of the surrounding areas of the Imjin river that flows through Paju area, Gyeonggi province. Presently, this area belongs to CCZ and it shows an area with widely developed wetlands along with the Imjin river. But cultivated areas into rice paddies appear in all over the places for the purpose of residents living in CCZ.

This area is in nature composed of Woodland patch and surrounding forests encircling farmlands which can be considered to be area's mosaics. The main stream of the Imjin river and its tributary which flows from the Imjin river become corridors, and it is divided into patches in wetland areas, patches in farmland areas, and patches in grassland areas centered as tributary of the Imjin river. The cease-fire line passes through between farmland areas and the Imjin river. Yet, habitats appear to be satisfactory in overall landscape ecological point of views, even though it shows forms of habitats' disconnections in part as farmlands are located in the center of the area.

Fig. 2.6 Landscape ecological pattern near the Imjin river, Paju

Fig. 2.7 Landscape ecological pattern of Chopyeong island, Paju

The area shown in the Fig. 2.6 is also a place near stream of the Imjin river, Paju and this place can be considered to be the representative image of habitats' segregation due to political reason which DMZ and CCZ have in the country. Well-developed wetlands along with the Imjin river show figures of habitats' fragmentation with the cease-fire line as boundaries. Habitat fragmentation is considered to cause serious problems in DMZ and CCZ due to the cease-fire line according to landscape ecological point of view, considering sections of DMZ and CCZ that are composed of soils without vegetation due to the establishment of the cease-fire line.

The area composing Woodland patches in Fig. 2.7 is an island area called Chopyong island located in the Imjin river. It is natural area with no human residence and where a village called Chopyeong Dong once existed and currently human access is controlled due to security issues.

The center area of Paju Chopyeong island area shows the regenerated Woodland patches, such as Morusalba, Populus deltoides, and surrounding shows wetland patches, such as Salix koreensis Andersson, Phragmites australis. The Imjin river flows between vicinity inland area and the Chopyong island, and the surroundings of the Imjin river show grassland edges.

The area in Fig. 2.8 is called Jangdan peninsular. The Imjin river flows from the east to the south and joins with Munsan river in the Southeastern direction. It is used to be wetland area with developed Phragmite saustralis community in the past, but presently many areas are cultivated as farmlands. In the partially

Fig. 2.8 Landscape ecological pattern of Jangdan peninsula, Paju

access control area, Phragmites japonica Steud community or Miscanthus sacchariflorus Benth communities are developed. Water flow gate is blocked between the Imjin river and Jangdan peninsula due to the bank constructed there.

Most of Jangdan peninsula wetland area is rice paddy matrix and wetland patch is located in the center. Stream corridor flows in the middle, which connects rice paddies and the wetland. But habitats in Jangdan peninsula wetland are disconnected as a bank is constructed between rice paddies and the Imjin river. Woodland patch (Woodland Patch) and the Imjin river (River Corridor) are located in these surroundings.

The Imjin river, often quoted as "expressing sorrows of division", as it flows through South and North, has many existing cultural heritages along with the river surroundings and it is one of the interesting focuses in DMZ and CCZ due to its clear waters and beautiful sceneries. As the stream, shown in Fig. 2.9, is also in the area, especially in the first area where the Imjin river flows into the South and a bridge called Pilseung bridge exists.

One side is composed of Woodland patch and the other side has developed flood plain wetlands and centered by the mainstream of the Imjin river. One side also has the cease-fire line and it is a cause of habitats' fragmentation. It functions as a Road Corridor since migration along the cease-fire line is possible. However, this area also preserves beautiful sceneries as well as very clean natural environments due to fragmentation from the outside for a long time, and therefore patches have not occurred due to the fragmentation.

Sacheon River Area

Sachoen, Paju area, was originally farmland area but it became an abandoned paddy field as human access has been impossible. At the present time, it is being changed into a feature of wetlands. Therefore, the wetland area is the matrix of this area and the Woodland patch is located in the center of the wetland. The forest axis of the Dora Mountain and Baekhak Mountain connected along with DMZ area is a Corridor Area.

Fig. 2.9 Landscape ecological pattern of Pilseung bridge area in the estuary of the Imjin river, Yeonchoen

Fig. 2.10 Landscape ecological pattern on surroundings of Sacheon river, Paju

The area shown in Fig. 2.10 is the rear wetland area of Sacheon stream in DMZ near Sagok Li area, Paju Gyeonggi province. It is changing into a natural wetland since human access is prohibited. Woodland patches are located in some places as well. It becomes a good habitat for wild animals and plants since wetland habitats and forest habitats are located in one area.

In the rear wetland area of Sachoen stream located near Dora Mountain Station, beautiful sceneries are shown as human access have not been possible. Therefore, the wetland area is the matrix of this area and the Woodland patches are also located in the center of the wetland. The forest axis of the Deongmul Mountain connected along with northern DMZ area to the Sacheon stream which flows along with DMZ is a Corridor (Fig. 2.11).

Figure 2.12 shows the image of Gyeonggi Rail Road in Paju, where the road ran across the dimensions of the formed patches and habitats' disconnection is inevitable in the left side. The reservoir located in the center forms a patch which functions as a wetland with a reservoir in the center. Forests are shown as forms of patches not as consecutive forest patches and new types of patches are shown as meadow wetland which appears between the wetland and the forest. In this area, ecological bridges can be considered in order to link the disconnected habitats. The

Fig. 2.11 Landscape ecological pattern on surroundings of the Dora Mountain station

Fig. 2.12 Landscape ecological pattern on surroundings of Gyoenggi rail road, Paju

Fig. 2.13 Landscape ecological pattern on the reservoir in Daeseong–dong area, Paju

habitats' disconnection poses as the most serious problem in the conservation efforts.

This area is a village situated in the Daeseong-dong area (Paju city, Gyeonggi province) of the DMZ and is the only area of the DMZ where human reside. This area shows typical farmland scenery patterns with simultaneous appearance of paddy patches, lake patches, and Woodland patches, since most of the residents are farmers. Farmlands and forests are distributed and the Daeseong reservoir provides the water needed for agriculture. Daeseong reservoir also provides water to the habitats where migratory birds and wild animals dwell. The area is shown in Fig. 2.13.

Paju Sewol Stream Area

This area cannot be seen in general maps and topographic maps in the country but can be seen in military maps. As shown in Fig. 2.14, the Paju Sewol stream area is entirely forest based, but patches are formed as roads pass through the forest. Stream corridors are formed in this area along where the Sewel stream flows.

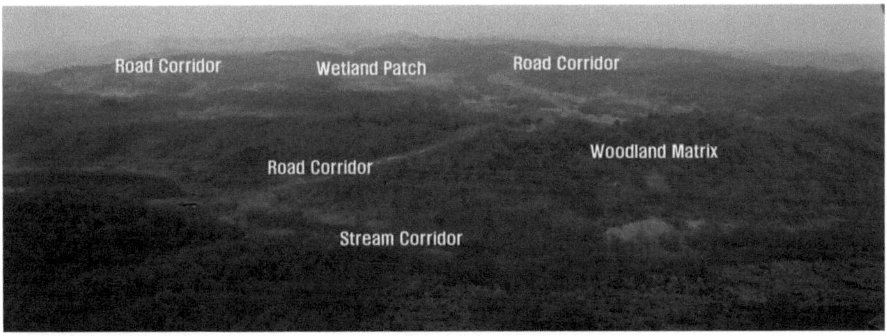

Fig. 2.14 Landscape ecological pattern of Sewol stream area, Paju

Fig. 2.15 Landscape ecological pattern in Sami stream area, Yeoncheon

Wetland is not widely developed since the stream is not large enough to form as patch. It only formed an edge wetland. The forest matrix is disconnected by roads that branched out up to the rear portion of the Paju Sewol stream area. In the rear portion, there are wide spaces of wetlands and formed as patches as shown in Fig. 2.14.

Yeoncheon Sami Stream Area

Figure 2.15 shows the Sami stream in the Yeoncheon county of Gyeonggi province, which emanates from the Sami river basin. This stream naturally curves across the dimensions of the area and the wetland surrounding it propagated a very well-developed Phragmites australis community. The forest ecosystem observed near the Sami stream enjoins with the stream ecosystem and the wetland ecosystem. Thus, it is a salient need to preserve this area and management efforts should be geared toward protecting it from threatening elements. This area did not show human-related damage as evidenced by the absence of disconnected habitats. The stream forms corridors of wildlife animals and the forest acts like a natural Ridge that encircles it. Wide and small-scale wetlands patches appear in this area influenced by the flow of the Sami stream but, in general, the various landscape ecological patterns are not yet established.

Fig. 2.16 Landscape ecological pattern of Sami stream river basin

Fig. 2.17 Landscape ecological pattern in wetland area in the rear part of Sami stream

Figure 2.16 presents the Sami stream river basin. The picture does not succinctly illustrate the stream as a Corridor because of the wide spread wetland formed in this specific area. Trees formed as patches in some portions, but the meadow wetland formed between the wetland and the forest demarcated the boundaries (Edge). A road located at the center of the wetland divided the wetland into two sections nevertheless, both the separated wetlands still maintain similar forms of vegetations.

The wetland area at the rear part of Sami stream is situated at the Yeoncheon county, still within the province of Gyeonggi. There used to be villages and farmlands in this area in the past. Presently, the area transformed into natural landscape patterns since human access has not been possible. A matrix wetland formed in the rear part of Sami stream and Woodland patches formed in the middle portion. Sami stream flows from south to the north corridors of the DMZ. The bushes in the neighboring area acted as buffer zones between the wetland and forest ecosystems. This area has high landscape ecological value. It is a favorite heaven for migratory birds and wild animals (Fig. 2.17).

The picture shown in Fig. 2.18 is the Gowang mountain area, still within the Yeoncheon county. It has typical mountain landform forming a forest-based Matrix.

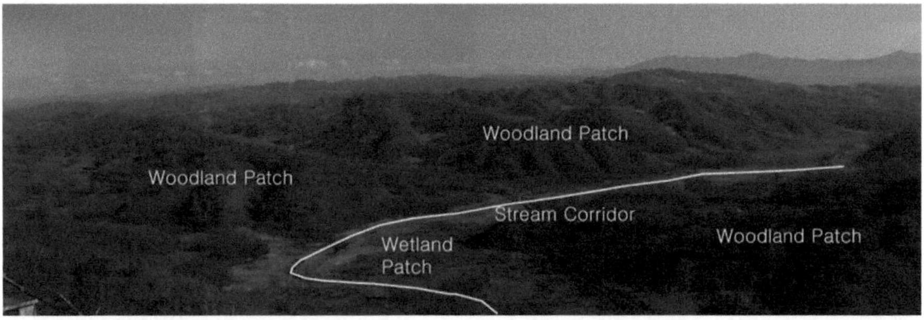

Fig. 2.18 Landscape ecological pattern in Gowang mountain area

Fig. 2.19 Landscape ecological pattern in the Yeokgok stream area, Cheorwon

The Sami stream which flows in this forest becomes stream corridors of wildlife. There are wetland patches observed between the stream and the forest.

Cheorwon Yeokgok Stream Area

This area of the Cheorwon is where the Yeokgok stream flows. It is a fairly large stream that flows mostly in the entire DMZ among other streams flowing here. Actually, Yeokgok water flows across the entire country. This area of the DMZ is also well preserved since it enjoys almost no interference or disturbance from the outside civilization for a long time. A wetland vegetation is widely developed along the Yeokgok stream.

This Yeokgok stream forms corridors and landscape ecological patterns with diverse wetland vegetation of various bushes and trees. The wetland is wide spread and forest Matrix eventually formed patches near Yeokgok stream (Fig. 2.19).

Figure 2.20 presents an image of the Yeokgok stream expanding down to Pilseung observation point (OP) of the Cheorwon. What is significant about the Yeokgok stream is that, it is located deep inside of DMZ, and it is almost the sole

Fig. 2.20 Landscape ecological pattern in Pilseung OP area, Cheorwon

source of the surrounding very well-developed vegetation. It forms major corridors and adjacent to it, there are patches of wetland formations and in the plains' Landform are grasslands and forest Patches. In a landscape ecological aspect, diversified patterns are formed of relatively various types of patches.

As it is shown in most of the landscapes in the DMZ, roads are not visible but it shows the cease-fire line passing across the middle as shown in Fig. 2.20. As a result, these cease-fire lines in many sections of the DMZ actually disconnects the habitats. This resultant disconnection made by the cease-fire lines, including the roads constructed across the conserved landscape, are considered as 'disturbance' elements, an interrupting factor for the migration of wildlife and the succession of plant ecosystem, which required access to interrelating processes of plant gene propagation. The DMZ's ever developing environmental ecology and genealogy must be given high value by this Peninsula, and perhaps by the whole world. Hence, alternative management actions should be installed through joint

Fig. 2.21 Landscape ecological pattern in the upper area of the Northern Han river

governance and convergence of resources of both North and South Korea for the main purpose of protecting and sustaining the DMZ.

Northern Han River Area

The Northern Han river is the primary tributary that flows from Gangwon province through the Gyeonggi provincial areas and then finally joining with the Southern Han river in Yansoo ri. Figure 2.21 presents the upper river area within the Northern Han river sections, which is surrounded by dense forests in both sides of the stream. Wetland is not widely developed since water resource is tantamount to the needs of the density of the forest and the landscape does not permit such storage of surface waters. The composition of this area are mostly Woodland patches that protect the area from floods, small meadow wetlands are also formed, while lush shrubs and bushes buffer the boundaries (Edge) between the stream corridors and the forest Matrix thereby protecting the area from silt and soil erosion.

Cheorwon Hantan River Area

The Hantan river originated from Gangwon province, passing through the Gimhwa, Cheorwon, Pocheon, Yeoncheon, and joins the Imjin river that flows across the long and narrow Chugaryeong valley. This valley was formed by volcanic materials spurted by the volcanic explosion in the remote past. Subsequent cliffs and valleys were developed allowing the river basin to be formed. The steep forest patches that are directly next to the corridors of the streams of Hantan river can be clearly seen in Fig. 2.22 as well. Shrubs and bushes also form the boundaries between the stream Corridor and the forest patches in some sections. The wetland that functions as a riparian buffer area near the stream is hardly seen in this picture.

Cheorwon Namdae Stream Area

The Namdae area nurtures a stream that originated from Oseong mountain of North Korea flowing down to Hantan river after passing a 50 km radius of land to the south, then finally to the DMZ. The stream passes across the farmlands that spread out in Namdae and Gimhwa-town down to the Gimhwa plain. Figure 2.23 features the

Fig. 2.22 Landscape ecological pattern of Cheorwon Hantan river and surrounding wetland

Fig. 2.23 Landscape ecological pattern in Namdae stream area, Cheorwon

stream that branches out to the Gimhwa, Namdae, and Gimhwa plain. The forest formed along the stream is also vividly shown dominating the whole area. The forest patches formed in the mid portion act as a corridor of the stream and the road.

Goseong Anho Reservoir Area

The Anho Reservoir is located in Goseong county, Gangwon province. This is the area where the most astounding metamorphosis of the wetland ecology has been displayed. It is a conglomeration of plain, estuary, coastline, and valley wetlands evenly distributed and that have highly developed ecosystems. These features, therefore, are equally and highly distinguished to be maintained.

The picture shown in Fig. 2.24 is the actual documentaries of the Anho Reservoir. The coastal wetland corridor is the most prominent feature that is beyond

Fig. 2.24 Landscape ecological pattern near Anho Reservoir, Goseong

comparison and cannot be seen in the other landscapes of the DMZ areas. Vegetation near the Anho Reservoir may have incomplete settlement but the wetland vegetation in the boundaries (Edge) is distinctive. Patches are also observed along the wire fences surrounding and crisscrossing inside and out of the reservoir. This was made possible because human access is restricted.

2.3.3 Landscape Ecological Analysis Based on Landscape Physiognomy and Composition

This section presents the analysis of findings based on Landscape Ecology theories and methods introduced and applied in one area of the DMZ utilizing satellite imaging and actual ocular survey.

Figure 2.25 shows different mosaics of habitats. It was known that the variance in geomorphological process in this landscape is attributed to the degree of human interference and the area was exposed to.

Dandelion Fields in the Hantan River Basin

Mindeulle plain is the first pilot study area for the landscape physiognomy and composition.

The agricultural area formed a very uniform homogeneous landscape with short (straight-line) boundaries relative to the geographic outlines of the areas. As it is located in the central part of DMZ, the farmland having temperate climate has been converted into wetland due to 60 years of noncultivation and filled up by the Han river that connects the south to the north (Fig. 2.25).

The Mindeulle plain in DMZ Hantan river basin is a large-scale process landscape. The plants and animals migrate together by genes, energy, and nutrients that have been transported through wind and water. Networks of corridor rivers, also known as roadside verges provide routes for some species to move between habitat patches to another. The biological transfer is dependent on the climate within the valley, and the movement of soils and water down to the valley (Fig. 2.26).

This ecological properties of landscapes can be measured in terms of physiognomy and composition point of view (Fig. 2.27).

The Landscape Physiognomy of Mindeulle plain can be calculated using the length of patches' boundaries and the distance between patches, ponds, and wetlands habitats.

Through systematic observation during ocular surveys, the Landscape composition of Mindeulle plain noted to be highly heterogeneous apart from the mosaic patches but associated to the Hantan river and the tributary corridor network. Ridges are traces between rice paddies and houses and, within the military check route which is believed to have contributed to the formation of these various types of patches. Patch types discovered in this landscape are: disturbance patches,

Fig. 2.25 Two landscapes in the DMZ. **a** DMZ Hantan river basin: Mindeulle plain area. **b** DMZ Sacheon river basin: Panmun field area. (**a**) with low degree of human interference and (**b**) high degree of human interference. These examples contain a number of different ecosystems

Fig. 2.26 Landscape physiognomy and composition of DMZ Hantan river basin: Mindeulle plain

Fig. 2.27 Patch types of Hantan river landscape

regenerated patches, environmental patches, remnant patches, and introduced patches classified by Forman.

2.3.4 Landscape Ecological Analysis Based on Ecological Network (International Biosphere Belt (IBB))

The DMZ ecosystem is very significant in itself, but it may also play a crucial role in eco-network in the Korean peninsula and even in the entire Eurasia. With this end in view, a conceptual plan that valued the study area as a starting point, and the establishment of a core base of Northeast Asia and Eurasia ecologic-network was developed (Fig. 2.28).

2.3.4.1 North-East Ecological Network

The study area is located in the central part of the Korean peninsula and is connected to the tidal flat in the western coast, wetland, and the Han river estuary. Expanding in a 155 mile-long DMZ and CCZ, it is linked to the central area of the Korean peninsula and forms an ecologic belt ranging from the western coast to the eastern coast. Such geographic features allow the formation of a natural Ecosystem Belt stretching from the tidal flat to the valuable natural resources in the western coast up to the Paikdoo main range in the eastern area. This ecosystem network arrangement continues to link Paikdoo main range to Mt. Kumkang, to Mt. Paikdoo, up far to the large wetland in Korea–China–Russia border area. The network now expands and forms the Northeast Asia ecological network. The proposal is described more in detail in Chap. 10 of this book (Figs. 2.29 and 2.30).

Fig. 2.28 DMZ biosphere concept

Fig. 2.30 Connecting to Mt. Kumkang and western wetland belt

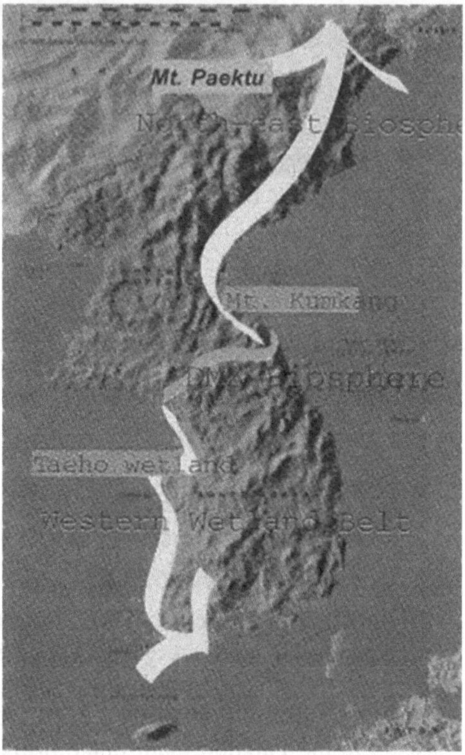

Fig. 2.29 Connecting the west to the east of DMZ

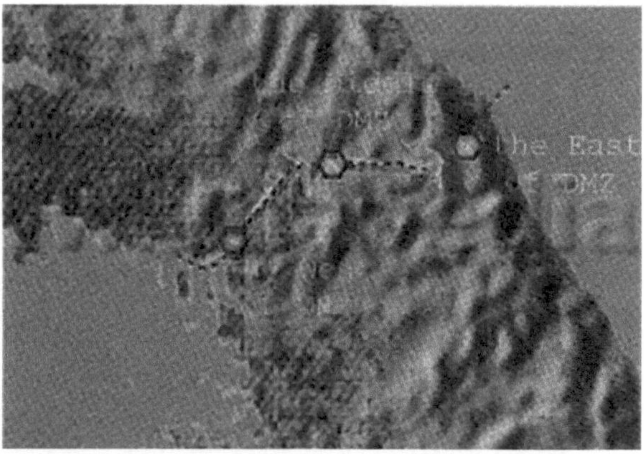

2.3.4.2 Global Ecological Network

This North-east Asia eco-network may develop further toward Central Asia and to EECONET (the eco-network in Europe). Such a network will provide a good opportunity to promote mutual understanding and management on ecosystem in Europe and Asia. This network opens door for academic exchanges that can strengthen ecological functions of all nations.

Chapter 3
Staus and Ecological Resource Value of the DMZ Area

This chapter explores the status and value of ecological resources, in terms of habitats and species found in the DMZ, the area separating North and South Korea. To maximize the knowledge on the ecological resources within the DMZ, this chapter is not delimited to the results of the study conducted by the environmental and ecological group led by the author of this book. Outcomes of surveys conducted by various organizations were also consolidated in this presentation. This chapter seeks to be used as a material for the conservation and sustainable use of the DMZ. Finally, the chapter aims to strengthen the advocacy on the designation of the DMZ as a World Natural Heritage site through the identification and suggestion of key or prime biodiversity areas (PBAs) within the DMZ.

There had been few surveys of ecological resources in the DMZ due to the presence of land mines and security issues that confronted researchers in the conduct of their investigation. As such, there are habitats and its respective species that are not possibly investigated in some critical areas of the DMZ. Contrary to the species-driven approach, the investigating team was very particular to develop an ecological map for the DMZ in order to predict the possible species that may be thriving in less investigated areas and, to estimate its corresponding population based on the estimated ecological space.

Conservation value was assessed by using the priorities identified by a number of Korean and foreign institutions. The conservation value of habitats, particularly of wetlands including peatlands, is based on the Ramsar Site criteria, the International Peat Society criteria, and the designation criteria for the UNESCO World Natural Heritage and Biosphere Reserve. Species value is based on the IUCN's Red Data Book (1997). This is completed by using a model suggested by English Nature (UK) and based on the aforementioned criteria. In-depth studies on the functions of DMZ habitats are not available due to the constraints mentioned above. The habitat assessment is inevitably tentative.

K.-G. Kim, *The Demilitarized Zone (DMZ) of Korea*,
DOI: 10.1007/978-3-642-38463-9_3, © Springer-Verlag Berlin Heidelberg 2013

3.1 Background of Ecological Survey

The DMZ in Korea is both a tragic symbol of the confrontation and division between South and North Korea, and a global treasure house of ecosystems. The place, a heartbreaking and tragic site to Koreans, is now one of the largest ecologically rich pieces of land on earth divided for political reasons. Accordingly, the area has been completely controlled and restricted from human disturbance since July 1953. However, the DMZ is now threatened by many factors and can no longer be safely treated as a treasure house. Overland routes are opened to promote inter-Korean reconciliation, economic cooperation, cultural, and tourist exchanges, transcending frequent traffic between South and North Korea. The onset of these threats to the DMZ requires the development of an efficient conservation and management plan as soon as possible. To this end, the DMZ 2-km wide south of the Military Demarcation Line (MDL) needs to be surveyed and analyzed in detail. However, a systematic and detailed survey of the DMZ is yet to take place, until such time that the mines are assured totally removed.

A study team led by the author of this book conducted an ecological survey (UNDP•SNU 2000) of the western DMZ and CCZ. The ecological survey was conducted as an impact assessment focused on Paju in the wake of the restoration of the Gyeonggi Line (Gyeongeui inter-Korea Road Joint Environment Survey Team 2001) and the Donghae Line and the construction of inter-Korea roads (Korea Rail Network Authority 2004). In addition, intermittent ecological surveys took place over the iron fence. A document to organically synthesize these data in order to suggest habitats and species prevalent in the DMZ is still unavailable.

Every idea and plan for development should be preceded by precise, scientific surveys and systematic analyses of a target area. As a scholar specializing in environment and ecology planning, the author regrets that many discussions on development in the DMZ are taking place with no support from comprehensive and systematic surveys and analyses of the ecological resources of the DMZ and their functions and values.

3.2 Ecological Survey Methods

3.2.1 Site Description

The DMZ is 248 km long from west to east and 4 km wide, covers an area of 907 km^2, and lies across the central part of the Korean Peninsula. The CCZ is in the northern part of the Civilian Controlled Line (CCL). Its width ranges from 5 to 20 km from the Southern Limitation Line and its area is 1,369 km^2. Under the Trans-boundary Area Support Law, which was approved in 2000, the southern

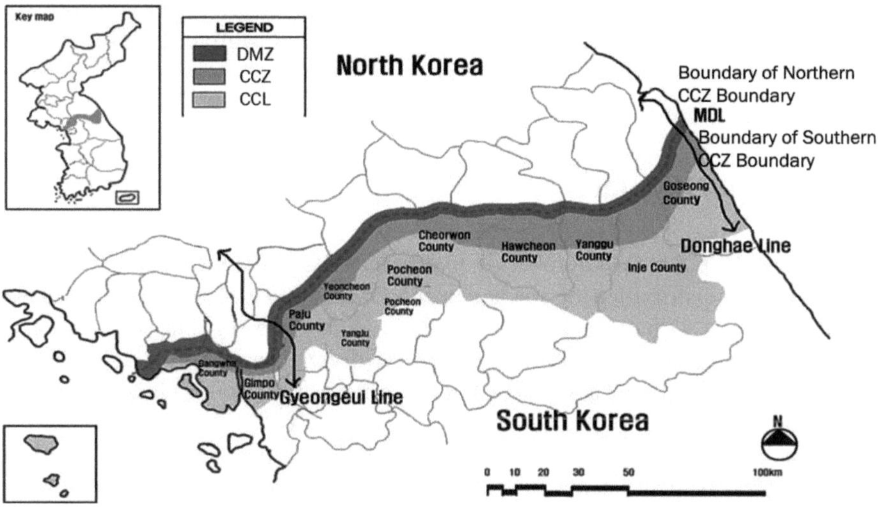

Fig. 3.1 DMZ and CCZ geography

portion of the CCZ has been designated as a trans-boundary area. Its area is 7,678 km^2. Excluding the trans-boundary area, the ecosystem treasure house running across the Korean Peninsula covers about 2,276 km^2 (see Fig. 3.1).

3.2.2 Consolidation of Field Survey Outcomes

The status of ecological resources of the DMZ presented here are based on the outcome of a UNDP Project (1996–1999) led by the author (UNDP/SNU, 2000), field visits facilitated by the Joong-ang Daily News (2000) and the Seoul Daily News (2004), survey of natural ecosystems in an area adjacent to the DMZ by Goseong county (2002), and based on surveys completed in the wake of the construction of the Gyeonggi Line, the Donghae Line, and inter-Korean roads.

3.2.3 Value Assessment Methods

The value of ecological resources of the DMZ is assessed based on surveys and analyses conducted so far. Although the DMZ has always been quoted as a treasure house of ecosystems, it has not been well examined according to specific criteria. Therefore, the conservation value of the DMZ can only be assessed according to the priorities of PBAs and other internationally recognized criteria, including criteria put forth by the IUCN (1997), RAMSAR and UNESCO (2000). However, all the results of these assessments are to be considered tentative since the surveys conducted are not sufficient enough to provide details.

3.3 Status of Habitats

The DMZ and CCZ have a number of diverse habitats. Studies to date have mostly focused on the distribution of species in the DMZ and CCZ with little emphasis placed on the systematic review of habitats. Wherefore, in this compilation, a comprehensive presentation of the status of the diverse habitats is sought. Habitat details are described in Table 3.1 using the criteria for habitat classifications of Sutherland and Hill (2002).

As for ecosystems based on topography, the identified habitats are wetland ecosystems, forest ecosystems, grassland ecosystems, farmland ecosystems, water bodies, and coastal ecosystems.

3.3.1 Status of Wetland Ecosystems

Wetlands in the DMZ and CCZ are much more diverse and extensive than the common expectation. Due to a higher concentration of lowlands, wetlands are more diverse in the western portion of the DMZ than in the eastern portion. Valley wetlands and lacustrine wetlands are mainly observed in the eastern area. Tidal flats are well developed in the Gangwha and Gimpo areas since it is adjacent to the western coast. There are 11 inhabited islands and 17 uninhabited islands in this area. The latter maintain habitats conducive to the coastal bird populations that they sustain at present. An area presumed to be a peatland was found in the DMZ near the Gyeonggi Line railroad path in Paju. In addition, riverine wetlands created by the Imjin river and the Sacheon river were distributed extensively in the western area of the DMZ.

Meanwhile, in the Yeoncheon area, riverine wetlands and palustrine wetlands are located extensively around the Sami stream. In particular, the Sami stream which originates from North Korea, is an important site offering an image of undestroyed natural habitat within the DMZ. Abandoned rice paddy wetlands and forested wetlands are well developed around the Paju Story Shooting Range. Wetlands of various sizes are interspersed around the Eoryong reservoir, the Han river estuary, the Myeolgong stream, and the Sewol stream. Reservoirs, lakes, and farmlands occupy Cheorwon county in the central part of the DMZ. Hydrologic resources include Togyo reservoir, Hak reservoir, and Sanmyeong lake. Cheon-tong is a wetland biotope with a high conservation value inhabited by *Hydropotes inermis*. In addition, wetlands created in a process of a succession of being an agricultural land in the past, then to abandoned rice paddies, and then finally into

Table 3.1 Overview of DMZ Habitat

Habitat classification	Target area within the DMZ
Coastal habitats	
The open sea and subtidal habitats	Gimpo tidal flats, Gangwha tidal flats, etc.
Rocky shores	Yudo, etc.
Sandy shores and estuaries	Hangang estuary, Gimpo Wolgot, eastern coast estuary, etc.
Coastal lagoons	
Coastal lagoons	Gamho, Hwajinpo, Yeongrang lake, etc.
Sand dunes	Sweetbrier communities along the eastern coastline, etc.
Salt marshes	Wetlands along the Han river estuary, etc.
Sea cliffs	Eastern coast cliffs, etc.
	NA
River habitats	Imjin river, Hantan river, Sacheon river, Sami stream, Yeokgok stream, Nam river, Bukhan river, Sewol stream, Myeolgong stream, Seongnae stream, Suip stream, etc.
Water bodies	Eoryong reservoir, Hak reservoir, Togyo reservoir, Sanmyeong lake, an lake, Paro lake, etc.
Forested wetlands	Gojindong valley in Goseong county, forest valley in Dutayeon, Bangchuk-dong in Paju, forested wetlands from Baekhak, Paju, to Pilseung bridge in Cheorwon, every riverside, etc.
Reedbeds	Samtong in Cheorwon, Paju story shooting range, abandoned rice paddies, rivers, etc.
Peatlands	Peatlands (estimated) in Gyeonggi line, Yong Neup, etc.
Grassland	Between Ojak bridge and Andongpo bridge, Paju story shooting range, etc.
Acid grasslands	
Mesotrophic grasslands	NA
Calcareous grasslands	NA
Farmland	Gimpo plain, Gangwha agricultural land, Paju agricultural land, Cheorwon plain, Yeoncheon agricultural land, and Goseong agricultural land, etc.
Natural woodland and scrub	Coniferous Geonbong mountain, Hyangro peak, Daeam mountain, Baekam mountain, etc.
Tree communities	
Broad-leaved	Gachil peak, Dosol mountain, Daewu mountain, Gajeon-ri, Daeseong mountain, Dora mountain, Baekhak mountain, and forest around Baekma height
Coppice	Goseong and Yanggu forest fire areas, etc.
Scrub succession	Goseong and Yanggu forest fire areas, etc.
Urban areas	Villages scattered in Daesong-dong and CCZ in Paju

becoming natural wetlands distributed extensively that eventually served as a home to various wild animals. The Hantan river and the Namdaecheon stream wetlands also support important species (Fig. 3.2).

Yong neub on the other hand is a high peatland located in Daeam Mountain which is situated in the boundary between Yanggu county and Inje county. Due to its unique biota and ecology, Yong neub has been designated as an ecosystem

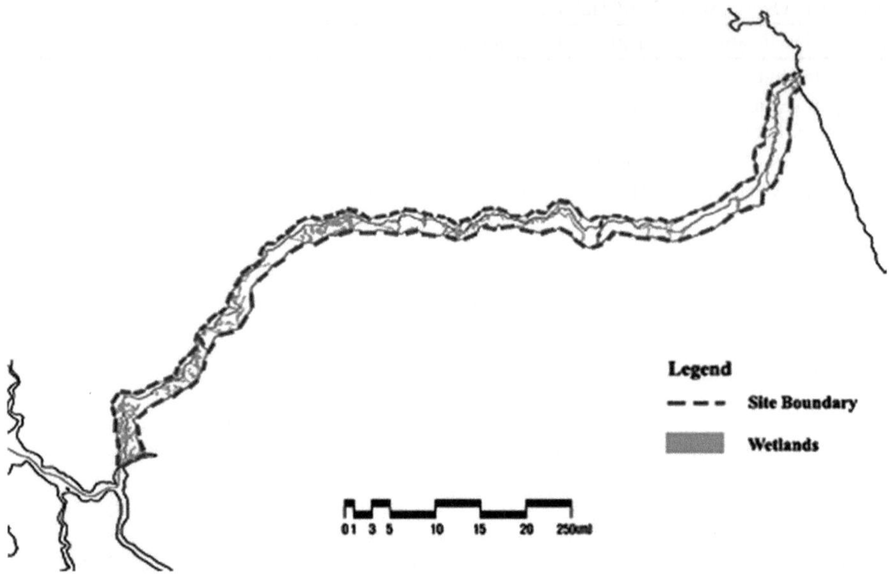

Fig. 3.2 Wetland ecosystems in the DMZ (Kim 2003a, b; Ku and Kim 2001)

conservation area, a wetland protection district, and a natural monument by the Ministry of Environment (2003). It is also registered as a Ramsar Site. In Hwacheon county, the lowlands near Daeseong mountain, the wetlands developed in the valleys form various plant communities. Natural riverine wetlands are also well preserved along the Naerin stream in Inje county. Diverse wetlands including forested wetlands, riverine wetlands, coastal wetlands, lacustrine wetlands, and palustrine wetlands are observed in Goseong county (2002). Coastal wetlands are particularly well developed in Goseong county (2002) and they include what is estimated to be the largest Rosa rugosa community in Korea. This community extends from the front of the Unification Observation Deck, across the Military Demarcation Line, on to the sand dunes that continue to the Guseon Peak. Further, Hwajinpo and Gamho embody the largest lagoons in the eastern coast which are additional habitat resources of migratory birds.

3.3.2 Status of Forest Ecosystems

Topographically, forest ecosystems are most prominent in steep terrains. Geographically, they range from the eastern DMZ to the Cheorwon plain. Based on prevalent vegetation, forests can be classified into four classes namely coniferous forest, broad-leaved forest, mixed forest, and scrubs. Major forest ecosystems observed in the DMZ and CCZ are mapped in Fig. 3.3.

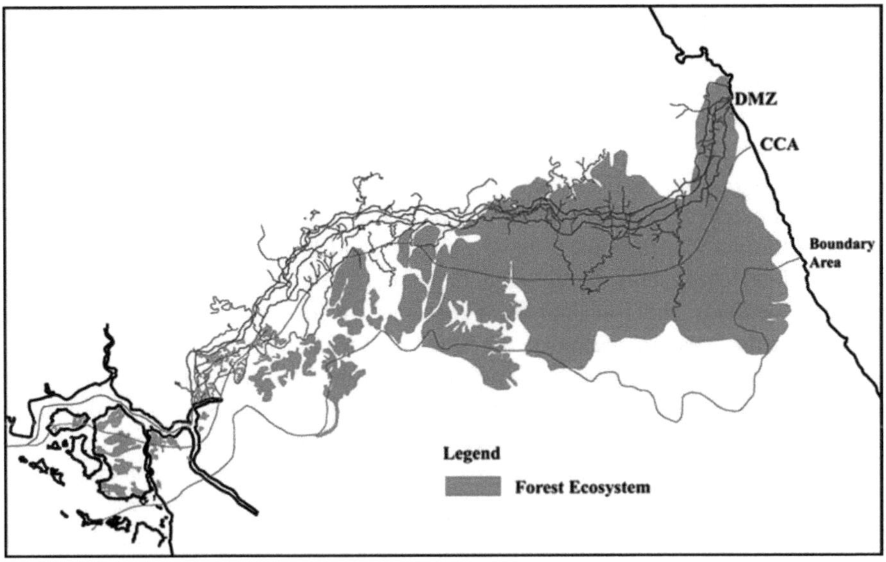

Fig. 3.3 Major forest ecosystems in the DMZ and CCZ

The western portion of the DMZ and CCZ maintains an attractive forest environment that was formed in relatively low hills and islands. Forest ecosystems near Panmunjeom and the forests on the low hills around Dora mountain and Baekhak mountain maintain connectivity with water resources, wetlands, and grasslands, which serve as a habitat for a number of species. Forest vegetation characteristic of the post-disturbance secondary succession is observed in the hilly areas of Cheorwon and in an area around the Baekma Heights Observation Deck. The area is broadly dominated by *Quercus mongolica*. Some species that are presumed to have been intentionally planted are found sporadically which include *Castanea crenata* var. dulcis, *Pinus densiflora*, and *Pinus koraiensis*.

Military exercises, including forest fires, intentionally done to improve visibility, frequently disturb vegetation inside the DMZ and to those environments adjacent to the DMZ specifically in Yanggu and Goseong counties. These continued forest disturbances resulted in a poor quality forest biota of shrubs and stunted forest development. In Gachil Peak, Dosol mountain, Daeu mountain, and Gajeon-ri area, the vegetation is dominated by broad-leaved shrubs. But the forest that stood in Dutayeon, Geonbong mountain, and in the Hyangro Peak areas are well developed. The Hyangro Peak area is particularly mountainous and serves as a watershed for the east to west division habitats. Steep topography links ecosystems in Chiljeol Peak, Hyangro Peak, and Geonbong mountain. These areas are assessed to be preserving its natural state.

3.3.3 Status of Grassland Ecosystems

Some grassland habitats within the DMZ are also restricted from successional progression due to frequent disturbances such as intentional and natural forest fires. Grassland communities are dominated by *Potentilla fragarioides* var. major, *Erigeron Canadensis* or *E. annuus*, and *Miscanthus sinensis* var. purpurascens that are easily observed on the levees of rice paddies, in fields and on mountains of the DMZ. Although it is difficult to determine how much grassland exists in the area, the grassland ecosystem is regarded to be important in promoting biodiversity between wetlands and forest areas. Grasslands in low, flat areas are distributed throughout the western region of the DMZ and CCZ. Alpine grasslands are also present at high elevations of the DMZ while bottomland-grasslands are present in the central and eastern regions. In Cheorwon county, vast grasslands are present in the slopes of lowlands and at the margins of nearby agricultural lands. These areas are inhabited by *Physalis alkekengi* var. francheti, *Commelina communis*, *Geranium sibiricum*, and *Chelidonium majus* var. asiaticum. The alpine grasslands are more common in the eastern portion of the DMZ which maintain a rather weak community of grasses dominated by *P. fragarioides* var. major and Viola mandshurica. *Caltha palustris* var. membranacea, and *Duchesnea* chrysantha communities are observed on the summit of Hyangro Peak. Highly dispersed *M. sinensis* var. purpurascens communities are also present in the DMZ.

3.4 Status of Species in the DMZ and CCZ

The status of species has been determined by synthcsizing all data available on species confirmed to be living in the DMZ and CCZ. This is based on existing literatures, records, and the results of surveys actually conducted by the author and his team. This is summarized in Table 3.2. Field surveys were limited due to circumstantial constraints such as land mines and restricted access. A follow-up detailed ecological survey is highly recommended, hopefully in the near future if conditions are more enabling. If conducted, it is expected that more species can be identified than those described here. Moreover, the number of reported species would likely increase if future surveys include portions of the DMZ located in North Korea. The incomplete nature of the results presented in Table 3.2 implies that it is very important for South and North Korea to jointly conduct a detailed and well-structured ecological survey on the DMZ and CCZ. The distribution of major species is shown in Fig. 3.4.

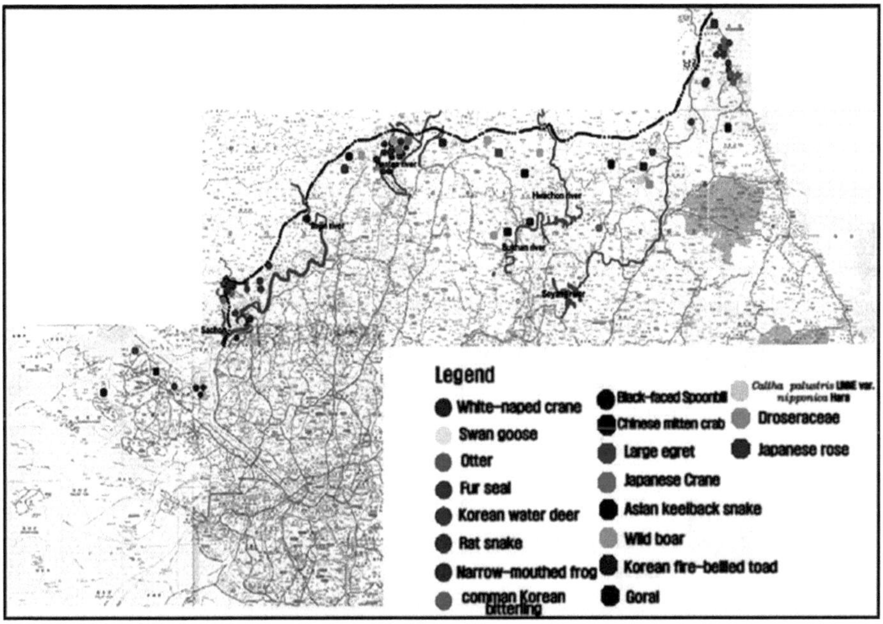

Fig. 3.4 Distribution of major species in the DMZ and CCZ

3.4.1 Assessing the Value of Ecological Resources of DMZ

Assessing conservation value according to PBAs

Conservation Value Based on Prime Biodiversity Area(PBA) Priorities. The initial step of the ecological resource value assessment is the identification of Key PBAs. This is to prioritize salient areas that needed immediate attention and have a multiplier effect on other prospected ecological areas. Key PBAs refer to sites with especially high measures of biodiversity. In general, PBAs exhibit a distribution of wildlife and biogeographic features representative of natural conditions. High species diversity and distribution within the DMZ enable assessment of conservation value through the identification of PBAs in the region. The general process and details regarding conservation value assessment in accordance with key PBAs is shown in Fig. 3.5.

The criteria for the assessment of PBAs including size and diversity are defined at a macro-level. Other criteria which define sub-levels include habitat size, species richness, diversity, distinctive habitat formats, and unique habitats. The results of the assessment when these criteria were applied to a few sites where surveys were possible are shown in Table 3.2.

As described in Table 3.2, when conservation value was assessed based on surveys conducted by the team and on all data presently available, the Chopyeong Island in Paju, the Gyeonggi Line section, the Sami stream, the Sewol stream, the

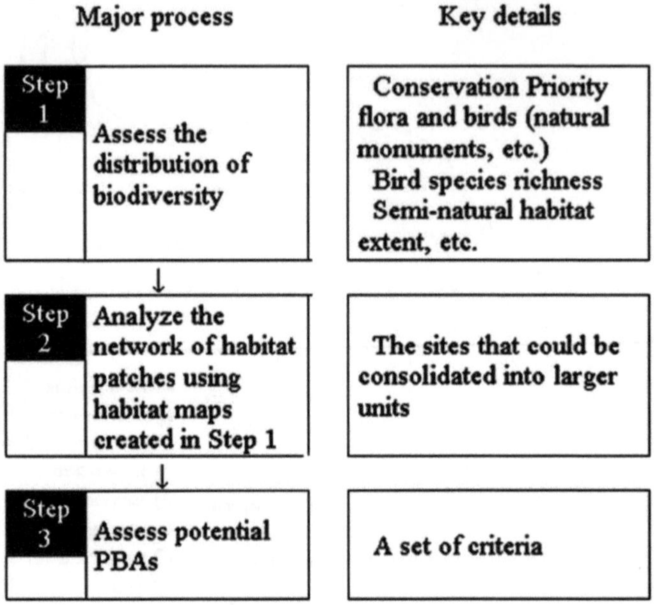

Fig. 3.5 Process of assessing PBAs

Yeokgok stream in Cheorwon, the Seongnae stream in Hwacheon, the area around Dutayeon, Hyangro Peak, and the Donghae Line section are identified as key PBAs. These areas widely maintain the natural wetlands, forests, and grasslands that sustain the existing resident and migrating wildlife.

Assessing conservation value according to criteria suggested by international organizations

The purpose of utilizing the criteria suggested by international organizations in assessing the ecological value of the DMZ and CCZ is to find solid evidences that the natural ecology of the DMZ deserves to be protected and should be designated or at least recognized internationally as a treasure of ecosystems. In Fig. 3.5, the PBAs in the DMZ were assessed using the criteria used in the IUCN (1997) Red Data Book, criteria in identifying Ramsar Sites and Trans-boundary Biosphere Reserves, and the criteria suggested by the International Peat Society and by the UNESCO (2000) Natural Heritage program.

Assessing value according to the IUCN (1997) Red Data Book

The Red Data Book of the IUCN (1997) provides the basic data on the status of endangered species by identifying the worldwide distribution of important wildlife for the purpose of preventing species extinguishment (http://www.iucn.org). In addition, it aims to provide information for sustainable conservation through the

Table 3.2 Assessment table for PBAs on major sites that may be surveyed

ry sites	Assessment criteria																																			
	Railway & road const-ruction project area	Area where resto-ration is raised as an issue (Under threat)	Areas with unique habitats: Rivers	Wetlands: Estuary	Palustrine	Lacust-rine	Riverine	Forested	Peat land	Forest	Gras-sland	Estuary	Sand dune	Reser-voirs	Areas with unique species: Slender bitter-ling	Chinese mitten crab	Goral	Wild boar	Korea water deer	Ardeidae: Large egret	Little egret	Gray heron	Black-faced spoon-bill	Diamond bluebell	Drose-raceae	Japanese rose	Kaet-meklot	Area with natural scenic beauty	Areas with unique biological process: Sand dune	Peat-land	Geo-logy	Forest fires	Original riverine wetland areas not disturbed by human-kind	Other: Winter migra-tory bird arrival sites	Tidal flats	Islands
ıpo Wolgot				□			□					□				□				□	□	□						□						□	□	□
hon Stream							□												□									□		□			□			
ıutaries in Paju																																				
›ongeui Line in Paju	□	□							□		□			□				□	□	□	□	□						□								
›pyeong Island Paju					□		□				□				□			□	□	□	□	□						□								
ıi Stream Yeoncheon			□																□	□	□	□						□					□			
ol Stream Yeoncheon			□																□	□	□	□						□					□			
untainous wetland					□			□		□									□									□								
nmunities from Baikhak u to Pilseuing Bridge Cheorwon																																				
a Pilseung Bridge Cheorwon					□			□		□								□	□									□								
·kgok Stream, Cheorwon			□																	□	□	□						□					□	□		
·orwon Plain wetland	□																		□									□						□		
myeong Lake, Cheorwon	□					□								□														□						□		
yo (Samtong) Cheorwon	□																		□									□						□		
k Bridge, Hwacheon	□						□												□	□	□	□						□								
ıdongpo Bridge ↔ ak Bridge)																																				
acheon 504OP					□			□		□								□										□								
ıg Swamp, Daeam ıuntain, Yanggu	□							□		□													□	□				□			□					
› Stream, Yanggu			□																	□	□	□						□					□			
ngnae Stream, Hwacheon			□												□					□	□	□						□					□			
;ok mineral fountain, nggu					□																							□								
ıghae Line	□	□			□								□				□	□	□	□	□	□			□	□	□	□								
und Dutayeon			□		□			□		□							□	□		□	□	□						□				□		□		
ıngro Peak	□				□												□	□										□			□					
n River (Gojin-dong valley)			□		□			□		□																		□								
n River (Oso-dong valley)			□		□			□		□																		□								
ers																																				
ra Mountain ›servatory deck, Paju	□																											□								
olsoe Observatory :ck, Yeoncheon																												□								
:pung Observatory :ck, Yeoncheon																												□								
›ljeong Observatory eck, Cheorwon	□																											□								
ification Observatory :ck, Goseong	□																											□								
ıce Dam	□																											□								
inhabited islands																												□						□		□

Table 3.3 Result of value assessment according to the Red Data Book

Items	Species found in the DMZ
Mammals	*Hydropotes inermis*
	Felis bengalensis manchurica
	Moschus moschiferus parvipes
	Nemorhaedus goral raddeanus
	Sciurus vulgaris coreae
Birds	*Grus vipio*
	Grus monacha
	Aegypius monachus
	Anser cygnoid
	Platalea minor
	Haliaeetus albicilla
	Tringa guttifer
	Numenius madagascariensis
	Chinese black-headed gull
	Aquila heliaca
	Eurynorhynchus pygmeus
	Haliaeetus pelagicus
	Locustella pleskei
	Megalurus pryeri
Amphibians and reptiles	DD species
Insects	DD species
Plants	DD species

DD: data deficient

restriction of destructive activities. About 80 important wild animal species found in Korea are included in the Red Data Book. The species observed in the DMZ that are listed in the Red Data Book are shown in Table 3.3.

Assessing value according to criteria for Ramsar Site

The Ramsar Convention held in Ramsar, Iran in 1971, signed an international agreement to prevent the loss of wetlands with high protection value recognized at an international level. It designates and manages wetlands that have a high global conservation value (http://www.ramsar.org). The conservation value of the DMZ was assessed according to the Ramsar Convention criteria. Results of this assessment are presented in Table 3.4. According to this assessment, the wetlands in the DMZ and CCZ qualify as Ramsar sites since basically they are located in the transboundary area of South and North Korea. These wetlands support over 1 % of the global *Grus japonensis* population and are visited by over 20,000 water birds regularly. Therefore, the wetlands should be officially recognized as an area of international ecological importance, and should be protected accordingly. A strong proposal based on extensive assessment of DMZ and CCZ should be submitted for global conservation efforts. Inhibiting factors for detail investigations and official assessment should be eradicated to enable further studies. Full advocacy support from governments and global publics should be initiated.

Table 3.4 DMZ value assessment results according to Ramsar site criteria

Criteria	Potential sites within the DMZ	Remarks
Group A		
Representative or unique wetlands	Wetlands in the entire DMZ Yong neub in Daeam mountain	Designated as a Ramsar site
Group B		
Criteria based on species and ecological communities	Wetlands in Gangwha, Paju, and Cheorwon	
Specific criteria based on waterbirds	Wetlands in Paju and Cheorwon inhabited by the Japanese	
	Crane, Gyodongdo in Gangwha, and Yudo in Gimpo	
Specific criteria based on fish	Inhabited by the black-faced spoonbill	
	Tidal flats and estuary wetlands in Gangwha	

Assessing value according to criteria put forth by the International Peat Society

Peatlands or mires are wetland ecosystems that are characterized by the accumulation of organic matter that is produced and deposited at a greater rate than is decomposed, leading to the formation of peat. Peatlands play an important role in the biosphere. They interact with fundamental life-support processes, involving biogeochemical cycling, food-chain support, hydrological dynamics and water quality, and provide habitats for many characteristic and highly adaptive plant and animal species. Peatlands within the DMZ and CCZ have distinct features. Peatlands are considered as an internationally important wetland ecosystem. The assessment of the peatlands in the DMZ according to these criteria is presented in Table 3.5. Assessment of peatlands requires continuous monitoring, field surveys, and analyses. Therefore, there should be future surveys of the peatlands in the DMZ and conservation programs must be implemented.

Table 3.5 The results of a value assessment according to the international peat society criteria (tentative)

Criteria	Key potential sites in the DMZ	Remarks
Where peat accumulates over hundreds, or thousands, of years	Yong neub in Daeam Mountain	
Where depths are at least 45 cm and where depths range from 2 m to 12 m		The depth of a peat layer cannot be actually measured in many sites within the DMZ
Undrained peat consisting of 95 % water and 5 % solid material	Natural wetlands created through succession of abandoned rice paddy wetlands	
Types of peatlands are fens or bogs	Yong neub in Daeam Mountain and wetlands next to Gyeonggi Railroad	

Assessing value according to criteria for designation as a UNESCO World Natural Heritage site

The World Heritage Committee was launched at the UNESCO General Conference in 1972 in order to restore, protect, and preserve the world's natural and cultural heritage sites that, despite their internationally high value, have been deserted and are on the verge of ecological deterioration, destruction, or extinction (http://www.unesco.org). Natural Heritage refers to sites with high global conservation value that are designated by UNESCO (2000) according to the "World Heritage Convention." Sites qualified for designation as Natural Heritage include those showing a stage of the earth's evolutionary processes, or unique kinds of physical and geologic formations, or are representative of biological evolution, or contain the natural habitats of endangered animals. It may be a scene of exceptional beauty or a reserve for large numbers of wild animals. Based on these criteria, 149 Natural Heritage sites have been designated to date. The results of a DMZ assessment completed using UNESCO (2000) Natural Heritage designation criteria are shown in Table 3.6.

Table 3.6 Results of the DMZ value assessment according to the UNESCO (2000) world nature heritage criteria

Criteria	Potential Sites in the DMZ	Remarks
Earth's evolutionary processes (or representative of biological evolution)	Rice paddy wetlands (the entire DMZ) Peatlands (Daeam Mountain and Gyeonggi Line) Burned forests (Goseong, Cheorwon, etc.)	
Natural habitat	Wetlands (entire area) Quality forest (in particular, quality forests in the eastern part of the DMZ)	
Endangered animals	*N. goral raddeanus* *G. japonensis* *M. moschiferus parvipes*, etc.	
A scene of exceptional beauty (or a spectacular view)	Hyangro Peak Geonbong Mountain Daeseong Mountain Daeam Mountain Baekam Mountain, etc.	
A reserve for large numbers of wild animals	Wetlands (entire area) Major rivers and streams (Imjin river, Nam river, Sacheon river, Sewol stream, Myeolgong stream, etc.) Major forests in the eastern area such as Hyangro Peak and Geonbong Mountain	Sites considered to be PBAs in Table 3.3
Physical and geologic formations	Columnal joints (Hantan riverside, Imjin riverside) Limestone caves (Sangmeori in Cheorwon)	

Table 3.7 Results of a value assessment of the DMZ according to the UNESCO TBR criteria

Criteria	Approach
Span the territories of two or more nations while maintaining spatial continuity	As the DMZ is situated traversing South and North Korea, this criterion may be applied to the entire area. As a result, the entire DMZ may be approached for designation as a transboundary biosphere reserve
Meet the requirements for designation as a biosphere reserve	The DMZ may be designated as a biosphere reserve because it maintains high species diversity. Sites where particularly important species are located may be approached individually
Economic, tourism, cultural, and personnel exchanges take place between transboundary nations	In areas where the Gyeonggi Line and Donghae Line pass through the DMZ, economic, tourism, and cultural exchanges take place between South and North Korea. These areas may be considered for designation first
Develop a joint management plan and a mechanism to adjust the regional functions of each nation	In the future, a joint management plan with North Korea needs to be developed. Designation of TBRs in relevant areas may be approached

Most areas of the DMZ possess unique kinds of physical and geologic formations that are products of biological evolutions that sustain natural habitats of endangered animals. The DMZ has been an object of eco-tourism because of the beauty of its evolving landscape, the wild flowers that adorn the scenery, and especially the colorful birds that live and migrate there. Definitely, the DMZ meets the criteria for designating it as a World Natural Heritage site. More studies should be conducted by various organizations and agencies for thorough verification in order to provide stronger basis and pressure. Therefore, information campaigns, massive education and advocacy movements for its international designation and recognition as a world natural heritage should take place soon before modern developments jeopardize its existence.

Assessing value according to criteria for designation as a UNESCO Transboundary Biosphere Reserve

Major habitats in the DMZ can also be designated as UNESCO's Transboundary Biosphere Reserves (TBR). With this status, the management of the preservation and protection of the DMZ will be raised to international levels. TBRs refer to areas designated by UNESCO (2000) for the purposes of conserving biodiversity, developing communities, and preserving cultural values. Results from an assessment of the DMZ using TBR criteria are shown in Table 3.7. The requirements for designation as a biosphere reserve include (1) an ecosystem representing a biogeographic region, (2) an area containing a variety of species and habitats to be conserved, (3) an area where a concept of sustainable development can be applied, and (4) an area where public institutions, regional communities, and private sectors may participate. The DMZ and CCZ meet all criteria and are in fact, rich in ecological diversity and functions. Therefore, South Korea and North Korea should make joint efforts to have DMZ and CCZ designated as a TBR. By doing so, international knowledge, material and human resources, as well as funds may be infused in their conservation programs and projects.

3.5 Summary

In terms of ecosystems and habitats, the investigation confirmed that diverse forest vegetation, water bodies, grasslands, farmlands, and wetland types comprise the landscape from the eastern mountainous area to the western low wetland area. It further attested that ecological succession continues in the ecosystems of the DMZ and CCZ despite continued disturbances by humans and natural forest fires.

In terms of species, the Comprehensive Report on the Ecological Survey of the DMZ released by the Ministry of Environment (2003) revealed that the identified biota in the DMZ and adjacent areas include 1,597 plant species (34 % of all plants of South Korea), 106 fish species (12 % of fish species of South Korea), 29 amphibian and reptile species (71 % of all amphibians and reptiles of South Korea), 201 bird species (51 % of all birds of South Korea), and 52 mammalian species (52 % of all mammals of South Korea).

The conservation value assessments of the DMZ conducted using criteria from the IUCN (1997) Red Data Book, Ramsar Site criteria, International Peat Society criteria, UNESCO (2000) Natural Heritage, and TBR critera and PBA criteria, strongly provide the basis that the DMZ and CCZ are ecosystem treasure houses that should be recognized and designated internationally as having high conservation value that impacts global environmental conditions. In terms of habitats and species, the conservation values are found high specifically; the tidal flats in Gyodongdo, Gangwha, Yudo in Gimpo, Chopyeong island, Eoryong reservoir, the Sacheon river, areas adjacent to the Gyeonggi Line, the Sewol stream, the Myeongong stream, the Choelwon plain, Togyo reservoir, the Yeokgok stream, Geonbong mountain, Hyangro Peak, Baekam mountain, Daeseong mountain, and the sand dunes in the eastern coast. Although this study may be considered a tentative assessment which was completed without a DB or without the support of comprehensive and detailed studies, the investigation utilized various scientific methods and approaches of research like systematic observations, ocular surveys, and validations through record and documentary reviews of research abstracts and reports. The DMZ and neighboring CCZ was assessed using internationally recognized criteria and found to be maintaining ecosystems significantly suggesting and calling out for international recognition. With just a handful of data, the DMZ and CCZ had already met the criteria. How much more if comprehensive studies are conducted with research locale extended to include the mine-torn fields and restricted areas?

Although the economic welfare and the quality of life of Koreans are heavily dependent on natural resources including wetlands and forests, it has been observed that the DMZ and CCZ have been damaged to the extent that significant wetlands and forested areas have been lost. Therefore, habitats that do remain are even more valuable.

From this point forward, in order to preserve wetlands and forests for the sake of all species and that of future human generations. South and North Korea should cooperate proactively and take actions protect its natural ecosystems. To this end, it is necessary to decide area-specific conservation values and to develop a comprehensive plan for environmental management based on sound scientific and systematic ecological surveys.

In order to realize this, a DMZ South–North Korean Environment Cooperation Committee should be created. A joint study of ecological resources within the DMZ should be conducted with the support of international organizations. Efforts for the designation of the DMZ as a UNESCO TBR and a World Natural-Cultural Heritage site as proposed by President Kim Dae-joong, and as a Ramsar Site should be exerted. In particular, an environmental impact assessment of the northern part of Gyeonggi inter-Korean road, Mt. Geumgang road, and inter-Korean railroad restoration should be conducted. The two Koreas should work together in setting alternatives that will minimize negative impacts on ecosystems and seeking actions to reduce the impact. Through such efforts, natural ecosystems in the DMZ and CCZ may be seen as a benchmark where an environmentally sound and sustainable international management model is applied.

Based on details and key conclusions presented in this chapter, a few ideas that may be helpful for the efficient future conservation and management of the DMZ are suggested.

First, further comprehensive studies and surveys on wetlands should be conducted since surveys on the DMZ are just focused on forest ecosystems. Wetland classification and mapping, wetland species surveys, and assessments of their functions and values should be conducted in a systematic and scientific manner. To this end, it would be reasonable to include the DMZ in the Nationwide Wetland Survey Project initiated by the Ministry of Environment (2003) as a UNDP/GEF project.

Second, a DMZ habitat management plan should be developed. In advanced nations such as the UK and the USA, habitat management plans are implemented for environmentally sensitive areas. Therefore, a habitat management plan should be developed, albeit in phases, for the DMZ. The plan should go beyond the level of simple conservation by including wetland restoration with compensatory mitigation policies.

Lastly, a World Heritage Application needs to be prepared. Through Focus Group Discussions with the concerned ministries, agencies, and specialists, the application for the designation of DMZ as a conservation area should be in accordance with the criteria set by the convention. Then, it would be reasonable to provide the details of the listed variables and indices through surveys and studies.

Chapter 4
Habitats in the DMZ Area

There is much variation in the way DMZ surveys have been carried out, ranging from what has been recorded to how the communities were identified. With regard to community identification, there are two principal techniques used namely; identification by eye in the field and identification using quadrats.

This chapter attempts to address how to analyze the habitat and its application for the DMZ.

4.1 Analyzing the Habitat

4.1.1 The Concept of Habitat

Habitat is a place where the food, water, shelter, etc., are located for creatures to exist and to grow and multiply (Adams 1994). Therefore, the components of habitat are; feeding, shelter, water, space, and other things needed for the existence and survival of creatures. Shaw (1985) said that feeding is the most obvious component of habitats. In particular, feeding changes depending on temperature, area, and season. Further, the food capacity of the habitat is necessary for predators. This food carrying capacity process is dependent on the relationship of the food chain, and basically, the vegetation should be present. The shelter capacity of habitats means that it can provide protection or shield from threatening components, i.e., predators and harsh weather. These elements play important roles especially to many kinds of birds. Water is an indirect element that changes habitats. Wetlands are formed where water is present. Wetlands are important habitats to waterfowls and aquatic plants. The small swamp type of wetlands developed along with rivers and streams play an important role in connecting corridors for the interaction of habitats with those in other areas. Space should be sufficient for creatures to dwell, to move, to take in clean water, and to breed. However, it is desirable to determine the size of the space according to the size of communities requiring specific space or habitat area.

K.-G. Kim, *The Demilitarized Zone (DMZ) of Korea*,

DOI: 10.1007/978-3-642-38463-9_4,

4.1.2 Significance of Habitat Investigation

Habitat survey and analysis had been normally done based on the species observed in a specific area. Although habitat surveys should include the description of the area and environment where organisms live, systematic analysis had not been fully conducted. But the ultimate goal in the ecological restoration is the reconstruction of sound habitats and not the restoration of life species such as the endangered and rare species. Therefore, habitats have significant implications in the restoration of the natural environment. In the same view, the need for the analysis of habitats and finding measures for its preservation or restoration is significant. The proceeding aspects are proposed to specify the appropriate investigation process or method relative to this endeavor.

4.1.3 The Process of Analyzing the Habitat

There are four steps considered in the process of habitat survey and analysis (Fig. 4.1) namely; selection of target areas, habitat type classification, investigation per classified types, and map preparation based on the analysis of data. In many cases, the target area for investigation is preselected since primarily, it has to meet the criteria in determining areas for restoration or conservation. If the coverage of the survey is vast and difficult to define, the following prioritization are helpful in planning for the process of investigation; First, areas recognized as most important habitats; Second, areas with public issues; Third, areas where important life species inhabit; fourth, areas with the existence of local characteristics, and Fifth, areas with development plans.

4.1.4 Habitat Classification

Types of habitats can be very diverse depending on its level and how far it should be. Habitat classification is more complicated in urban areas. However, in case of Korea, the description of the habitat is limited on rivers and parks like; neighborhood parks, nature parks, etc. On the other hand, structured classification is presented in detail for foreign cases (Kim et al. 1999). A classification system is used in London, England and efforts are based on scientific standards which can be understood by anyone. Simple descriptions are being used in naming a specific title and code per habitat type.

On the other hand, Kiemstedt (1994) emphasized the importance of habitat classification and mapping processes in survey contents especially in establishing landscape planning. He classified each habitat type in detail after classifying 14

Process		Contents & Methods of Survey & Analysis
Selection of target area	:	Selection of target area for habitat survey & analysis
Type classification of habitat	:	Preparation of standards for habitat classification
Survey per habitat classification	:	Collection of related information Pre-judgment Modification & supplement through site visitation Final judgment on habitat classification Concrete survey & analysis per habit classification
Mapping process of habitat classification	:	Preparation of habitat classification map based on survey & analysis results

Fig. 4.1 Process of habitat survey & analysis

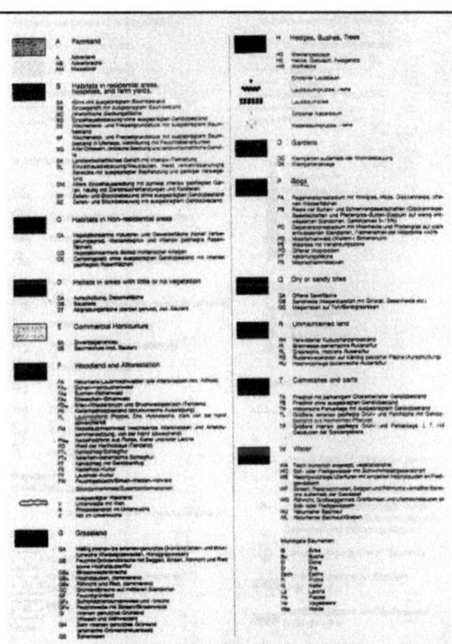

Farmland
Habitat in residence, hospitals, farms
Habitat in nonresidential areas
Habitat with little vegetation little or no vegetation
Commercial horticulture
Forests and forest plantations
Grassland
Fences, trees, shrubs,
Garden
Bogs
Dry or sandy areas
Wasteland
Cemeteries
Water

Fig. 4.2 Habitat classification by Hans Kiemstedt. *Source* Kiemstedt (1994)

habitat types by combining land use and vegetation as shown in Fig. 4.2 (Table 4.1).

In Finland, the habitats are classified into 24 types considering the characteristics of the region and the habitat (Virolainen 1996) as shown in Table 4.2.

Table 4.1 Classification and definition of wildlife animal's habitat in England

Code	Name	Definition
01/02/03	Woodland	Place where 76 % is covered by shrub-sized forest and shrub
01	Broad-leaved indigenous forest	Forest consists of broad-leaved indigenous species with at least 75 % of canopy
02	Broad-leaved exotics forest	Forest consists of broad-leaved exotics species with at least 75 % of canopy
03	Coniferous forest	Forest consists of Coniferous species with at least 75 % of canopy
37	Scattered trees (scattered trees)	Forest formed with less than 75 % canopy
05	Recent logging forest	Bushes not included
06	Scrub	At least about 75 % of the shrubs (less than 5 m) dominated
38	Planted bush	Dominated by shrubs, but usually not indigenous species, plated neatly except hedges
25	Hedges of Indigenous species	Hedges composed of indigenous species
34	Hedges of non-indigenous species	Hedges composed on non-indigenous species
31	Orchard	Fruit trees forming at least 50 % of canopy
36	Wall green belt	Area with vegetation which can detect ruins, fence, and other man-made structures (but artificial riparian vegetation is classified into different types of habitat)
26	Barren soil and rock	Areas planted with least tress except farmland
27	Sterile artificial habitat	Least planted area near concrete, railroad, cobbled streets, buildings, and artifacts of the riparian habitat surrounding
08	Acidic grassland	Areas with less than 25 % covered with grassland
09	Neutral grassland, semi-improved	
35	Neutral grassland, herb-rich	
10	Alkaline grassland	
11	Improved or re-seeded agricultural grassland	
07	Pleasant grassland	Areas with often trims and non-diversified genes such as park etc.

(continued)

Table 4.1 (continued)

Code	Name	Definition
12	Ruderal or ephemeral	Community composed of invasive species when transitions occur in heavily changed areas
13	*Arachniodes aristata* (G.Forst.) Tindale forest(Bracken)	*Arachniodes aristata* (G.Forst.) Tindale forest-dominant area
14	Tall herbaceous herbs	Tall, non-grass herbaceous areas
33	Roughland	
15	Heathland	
39	Alloumenis(active)	
28	Farmland	Cultivated area, Cog, flowerbeds, etc.
16	Bog	Where the dominant Sphagnum paper
17	wetland(reed swamp)	Areas with at least 75 percent of reed as coastal areas
18	Humid vegetation	
19	Fen carr	Forests or copsy covered by herbaceous vegetation nearly year-round with water
20	Standing water	Areas such as lakes, detention pond, ponds, pools (pools)
21	Ditches(Water filled)	
22	Flowing water (rivers and streams)	
23	Intertidal mud sand, shingle, etc.	
24	Saltmarsh	
30	Area where habitat information is unknown	Areas where observation and investigation are not possible or inaccessible areas
29	Other	
32	Paper rich herbaceous wetland	Areas that is distinct from humid vegetation with nearly year-round water and less than about 75 % of the herbaceous vegetation. Very rare in London and habitat type with no further investigations

Table 4.2 Types of habitats classified in Finland

1. Herb-rich delicious forests	13. Springs
2. Esker forests(pine)	14. The shore of the Baltic sea
3. Dry upland forests(pine)	15. Riparian habitats
4. Moist upland forest(spruce)	16. Flooded areas
5. Rich fens	17. Beaches
6. Open mires(poor fens)	18. Rocks
7. Pine mires	19. Calcareous rocks and quarries
8. Spruce mires	20. Herb-poor dry meadows
9. The Baltic sea and its outer islands	21. Wet meadows
10. Oligotrophic lakes	22. Arable lands
11. Eutrophic lakes	23. Parks and gardens
12. Stream and rivers	24. Industrial/urban habitats

Park (2000) in Korea presented examples of classifiable habitat types for the purpose of assessing management strategies and habitat evaluation as shown in Table 4.3.

Though diversification and details are important in habitat classification, habitat type should be fundamentally based on the overall ecological environment of the areas that should be investigated. The classification system used in the natural environment cannot be equated to the habitat investigation of the urban ecological

Table 4.3 Habitat type classification in domestic natural areas

Types	Category	Subcategory	Ratio (%)
Water	Rivers and streams	Artificial	
	Reservoir	Natural	
	Pool		
	Low wetland		
Farming	Paddy	Farmland	
		Abandoned Paddy Field	
	Field	Arable land (field ginseng, beans, etc.)	
		Weed-grown fallow	
Forest	Softwood		
	Hardwoods		
	Mixed forest	Quercus communities	
	Area with lots of dead trees,		
	Bush		
Grassland			
Town	Residential		
	Road		
	Other nine military facilities, etc.)		
Others	Derelict land		
	Edges		
	Cliff		

Source Park (2000)

environment having severe deformations. The characteristics of the sites for planning should be considered in classifying habitats.

4.1.5 Habitat Investigation by Type

Research on the habitat is to be performed in the actual target area after the habitat has been classified as mentioned earlier. The process of the investigation is focused in this section.

Habitats are generally classified as; wetland, grassland, scrub, forest, and farmland. But there had been many types of classifications proposed pertaining to the very specifics or details of each of the general types identified. The status of the habitat and the conditions of wild plants and animals can be analyzed by habitat survey format. An example of the format is shown in Fig. 4.3 used in the habitat investigation conducted in Leicester city, England.

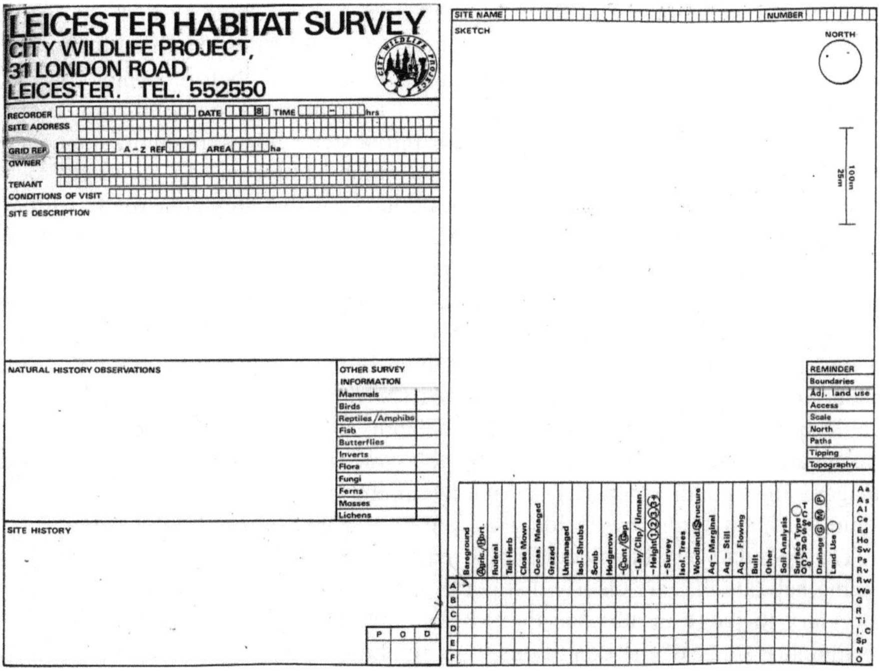

Fig. 4.3 Leicester city habitat survey format

4.1.6 Relevant Information Collection

In order to prepare a habitat map of the site, related information should be completely recorded. The researcher should be an expert in determining the peculiarities of habitats and must have a profound background on the nature of the growth and development of each specific habitat. Related material information should be included such as the following information:

- Habitat status, structure, and characteristics of the soil
- Basic ecological data such as existing plant communities and the animals
- Rare species of plants or animals
- Regional environmental characteristics
- Threats, the situation occurred in recent years, causes of changes, etc.

4.1.7 Preliminary Determination

In the standard process of research, this is referred to as the research locale where the basic profile of the target area has to be predetermined to establish possible variances of situations. Unless a targeted area has not been visited, an ocular survey should be done first in order to draft a sketch of the identified habitats in general. But when the general information is available before conducting the investigation, you can immediately develop a preliminary determination of the habitat. There are so many reasons for this initial step such as; planning the course of the study, identification of the most appropriate strategies and research procedures, the deployment of researchers to specific areas that they are experts with; in identifying prioritized areas; critical areas that require safety gears; special tools and instruments needed for the investigation; preparation of permits in certain areas; in determining how many staffs an expert would need; timeframe, etc. A preliminary habitat mapping needs to be done in advance in order to have a general understanding of characteristics of the habitat in the whole target area before the actual investigation has to be conducted. Nevertheless, the ultimate outcome would be the desired habitat map that contains all the detailed information that specifically defines it. For this purpose, topographic maps, aerial photos, satellite images, among other records are handy in the preparation of the preliminary coding of habitat, and by deciding a scale of 1: 1,000 or 1: 5,000 per area.

The boundary establishment should equally be done in advance for the areas having habitats that are clearly distinct from one another. It is equally and strictly necessary to judge in advance the possible threatening elements of the habitats. Preliminary study of the conditions you are about to put yourself into is actually not just introducing yourself to the locale, but also ensuring the safety of the entire research team and the assurance that the research endeavor is doable and goals attainable.

4.1.8 Site Visitation

The preliminary determined habitat has to be validated through site visitation or in standard term, ocular survey. This is important since habitats are open systems that reacts and adapts continuously to many incidental factors like changes in the soil, climate, shape, space, etc. The habitat is also greatly affected by the sociopolitical, cultural, and economic status of the public that surrounds it. This is the reason why the status of the habitats has to be continually monitored.

In the site investigation, researchers check the veracity of the predetermined map, its anticipated locations, and conditions since it is only based on existing maps, records, literatures, and studies; especially, when these documents are old and the maps outdated. Then there are also parts of the locale that are hard to determine because of inexistent data or in case of incomplete and missing data.

Ocular survey or site visitation of the totality of the environment provides a clear macroscopic view of the research locale especially in establishing the validity and reliability of habitat drawings and ecological network. So to be sure, the researchers must validate through site visitation.

4.1.9 Determination of Habitat

The specificity of the type of Habitat is determined and described based on the interpretation and analysis of data collected. Proposals for restoration and conservation can now be possibly based on systematized knowledge and actual facts of the life existing in the habitat.

4.1.10 Mapping Process of Habitat Types

The information resultant from the investigation and analysis of each type of habitat should be incorporated in the habitat map. Mapping is one the must-do item since at a glance, it provides important data that determine the exact location, landmarks, and boundaries, especially the significant indicators of habitats and the specific wildlife that exist in a specific spot. This is very helpful in planning restoration and conservation projects.

After determining the boundaries clearly including the unclear areas of habitat, it is necessary that there is also a clear demarcation between types of vegetations and types of habitats such as; wetland, forestland, animal habitat, etc. In this step of the process, colors or shapes are marked according to the coding established in the predetermined map.

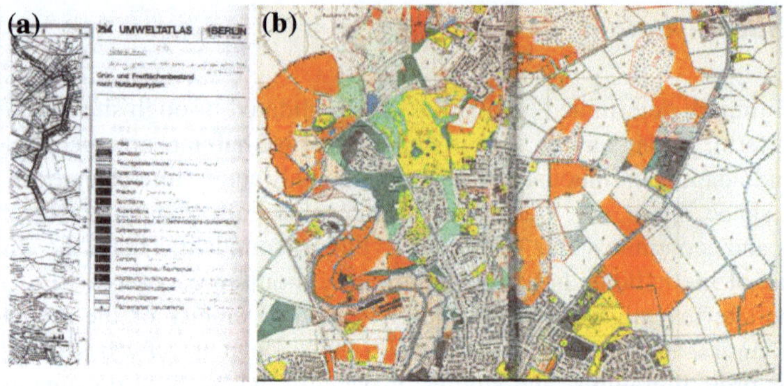

Fig. 4.4 Examples of map drawing in Germany & England. **a** Existing Green and Open Space in Berlin, Germany. **b** Habitat Classification Map of London, England

For existing green fields and open spaces, basic maps are used. A specific example is the Biotope classification used in Germany regarding the method and process of mapping. Figure 4.4a is an example of Germany's Existing Green Open Space map. Whereas in London, maps contain relatively detailed contents based on more detailed habitat classification system as presented in Fig. 4.4b.

Drawing scales 1: 5,000 –1:50,000 are suitable for targets with relatively large areas. For rural areas, 1: 25,000 or 1: 50,000 scales are suitable in executing the mapping process for habitats (Eagles 1984). A scale of 1:25,000 and 1:50,000 existing vegetation map are recommended to be used for the maps showing major flora and fauna. The exact scale regarding major animals or plants is not mentioned (Ministry of Construction and Transportation 1999).

4.2 The Application

Kim et al. (2010) explored the use of the habitat analysis in the DMZ. Wetlands predominate in the DMZ

4.2.1 Site Selection

Table 4.4 shows number of sites and communities recorded in the DMZ for analyzing the habitat.

Table 4.4 Number of sites and communities recorded in the DMZ

Province	Site	Remarks
Gyeonggi Province (8 places)	Jangdan Peninsula, Geogok-ri, Sijangdan-town, Paju city	
	Gunnae-town, Paju city	
	Choypeong Island wetland, Munsan-town, Paju city	
	Bosan island, Soyo-district, Dongducheon city	
	Changsu-town, Pocheon city	
	Youngjung-town, Pocheon city	
	Youngbuk-town, Pocheon city	
	Dongsong-town, Cheorwon county, Gangwon province	
Gangwon Province (20 places)	Woljeong-ri wetland, Cheorwon-town, Cheorwon county	
	Cheorwon Plain wetland, adjacent to Hantan Riverline, Galmal-eup, Cheorwon county	
	Whitehorse Summit valley wetland, Dongsong-town, Cheorwon county	
	Namdaecheon Stream wetland, Galmal-town, Cherowon county	
	Mt. Gyewoong valley wetland, Geundong-town, Cherowon county	
	Mt. Baekam, Hwacheon-town, Hwacheon county	
	Haesanryung-hill, Hwacheon-town, Hwacheon county	
	Peace Dam Wetland, Hwacheon-town, Hwacheon county	
	Punchball Wetland, Haean-town, Yanggu county	
	Eulji Observatory, Haean-town, Yanggu county	
	Buk-myeon/Inje-myeon, Inje county	
	Inje-town, Inje county	
	Kirin-myeon, Inje-town, Inje county	
	Bukcheon stream, Buk-myeon, Inje county	
	Hacheon valley wetland, Seohwa-myeon, Inje county	
	Hyangro summit valley wetland, Sudong-myeon, Goseong county	
	Bukcheon stream wetland, Ganseong-town, Goseong county	
	Unificiation Observatory, Hyunnae-myeon, Goseong county	
	Geojin-town, Hyunnae-myeon, Goseong county	
	Jukwang-myeon/Ganseong-town, Goseong county	

4.2.2 Analysis by Habitat Type

DMZ and CCZ areas range from steep mountains to plains as the area moves from the eastern across to the western parts. The water system flows the same way as the terrain moves from the steep mountains down to the plains through the Imjin river, up to the coastline of the west sea. Hantan river and Namkang river flow on to the East Sea and Soyang river.

The water systems have developed the way the forest ecosystems and wetland ecosystems have developed, a rare characteristic of co-existence and co-evolution peculiar to DMZ–CCZ only. Here, ecosystems develop according to the terrain and topography. Forest grew in the mountains and hills and down on its base, wetland ecosystems formed, then further surrounded by grassland ecosystems and agricultural ecosystems down the plains. While these characteristic habitats develop, reservoirs and lakes develop in places created in the mountains, as well as in the slopes and in the lower than ground levels of the plains and meet coastal ecosystems. It became a perfect water system that provides sustenance wherever the habitats of any kind are situated.

It is very noticeable that these habitats are not separate from each other. They are rather interconnected, as if one does not exist without the other.

Ecosystems of the DMZ are the main focus of the analysis addressed here. It is generally classified as coastal ecosystem, estuarine ecosystems, stream ecosystems, lake and reservoir ecosystems, wetland ecosystems, grassland ecosystems, cropland ecosystems, forest ecosystems, bush grove ecosystems, and the artificial village ecosystem.

Quercus serrata, *Q. mongolica*, Carpinus, pine tree, shrub-scrub ecosystems and Styrax trees are the temperate central standard species that dominate the DMZ and CCZ forest ecosystem. Parts of forests are damaged by the construction of military facilities and roads. They burned down the trees to secure their visibility. Some of the damaged areas turned into grasslands.

The water systems that are directly receiving and supplying the DMZ are: the Imjin river, Hantan river, Han river, Sacheon river and Samichon river, and Namkang river. Five major rivers flow into the Yellow Sea. Only Namkang river flows into the East Sea.

All the main streams of five major rivers are flowing through the DMZ and CCZ before ending into the Yellow sea. Nam river originated from the north, flows across the DMZ and CCZ, flows toward south and flows back to the north toward the East Sea, forming a U-shaped system. Its basin is primarily the main attribute in formation of the wetlands in the DMZ. The eventual ecosystems that developed in the area evolved according to the form of the terrain and topography. The wetland ecosystems are connected at the base of the forest ecosystem. The entire habitat classification per ecosystem in these DMZ areas is shown in Fig. 4.5.

Table 4.5 presents the habitats per ecosystem and selected specific areas. A variety of habitats are found in each ecosystem that exists in the DMZ and CCZ. In this part of the habitat, the distribution map drafted is based on the results of the investigation done by the author's team.

Coastal Habitat

Coastal ecosystems should not be considered as what we usually refer to as foreshores. However, not only foreshores belong to coastal wetlands, but also wetlands formed in beaches and coasts with rocks. Coastal wetlands include areas reached by the ebb and tide of the sea, along with stagnant water areas, where the water salinity content is more than 30 ‰. Fresh water is nearly difficult to detect as it is to salt water. More specifically, it includes the maximum distance that the tide had reached to the inland, even including the water-repellent area.

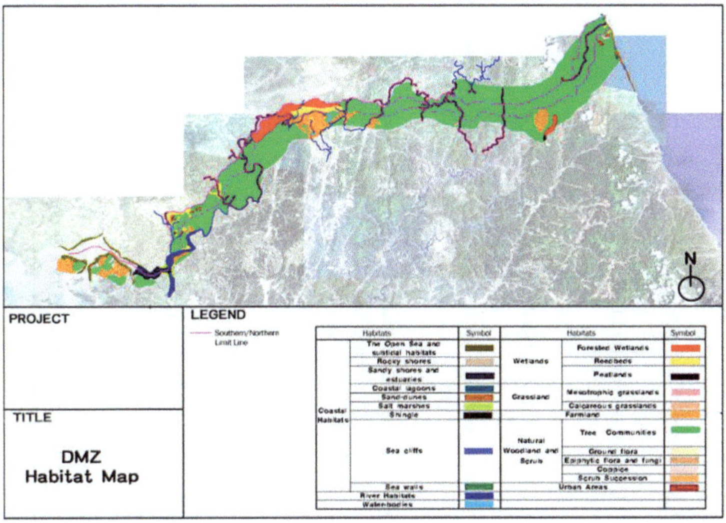

Fig. 4.5 Habitat map of DMZ and CCZ

These coastal ecosystems appear in western DMZ and in the eastern coasts. Though they appear opposite and distant to each other, they are very typical. The mudflat habitat is a characteristic typical to the western region. They are distributed widely in the western region and cannot be found in any other areas. Rosa rugosa inhabits only in coastal habitats of beaches surrounding the East Sea and in the coastal dunes of the eastern coastal areas.

Mudflat

Figure 4.6.

Distribution and habitat characterization

Typically, the tidal flat is a flat terrain arising from prolonged accumulation of fine particles of sand and clay carried by the calm sea water. These areas are immersed in water during high tide and exposed to the air during the low tide. Particularly in the west coast, the tides are very huge. It is attributed to its geographical characteristics. It is a very long bay with indented coast lines. Hence, materials are deposited and increasingly accumulating in time. The vegetation that thrives here is believed to have been brought by the birds.

The investigation revealed that 83 % of the total tidal coastal area in the South Korea is located in the western coast. It is considered as the world's fifth tidal flat, along with the east coast of Canada, eastern United States coast, North Sea coast, and the Amazon river basin.

The tidal flats of the DMZ are the Kanghwa and Kimpo tidal flats. The Kanghwa tidal flat have fresh water in it despite the surrounding sea water. It is supplied by two rivers; Yeseonggang river that passes through as it flows toward the north and the Imjin river and an estuary of the Han river to the south. The

Table 4.5 Summary of habitat in DMZ & CCZ

Ecosystem	Habitat	Targeted area within DMZ	Targeted area within CCZ
Coastal ecosystem	Mud flat		• Kimpo mudflat, Kangwha mudflat etc.
	Rocky island	• Yudo island etc.	
	Beach and estuary	• Estuary of the Han River, Manuri area of the Han River and the Imjin river, Estuary of the East Sea	• Kimpowol Cape etc.
	Lagoons	• Gamho	Hwajinpo, Youngrangho
	Coastal sand dunes	• Sweetbrier community along the east coast	
	Salt marsh	• Wetland in estuary of the Han River wetland	
Stream ecosystem	River & stream	• Samicheon, Yeokgokcheon, Namgang, Sewolcheon, Myeolgongcheon, Sungnaecheon, Sooipcheon etc.	• The Imjin river, Hantan river, The North Han river etc.
Lakes and Reservoir Ecosystem	Reservoirs and lakes	• Eoryong reservoir, Sacheon reservoir, Daeseong reservoir etc.	• Hak reservoir, Togyo reservoir, Sanmyungho, Anho, Paroho, etc.
Wetland Ecosystem	Forest wetland	• Forest wetland within the DMZ viewed from Cheorwon Observatory	• Gojindong valley (Goseong), Forest valley in Dootayeon, Bangchukdong in Paju, Forest wetland from Baekhak Paju to Pilseung bridge Cheorwon
	Rice paddy field wetland	• Paju Panmunbeol Abandoned Paddy Fieldwetland, Cheorwon DMZ Abandoned Paddy Fieldwetland, Yeoncheon Abandoned Paddy Fieldwetland, Sami Stream Abandoned Paddy Fieldwetland, Sewol Stream Abandoned Paddy Fieldwetland	• Samtong in Cheorwon, Story shooting place in Paju
	Peatland	• peat wetland in Gyeonggi railroad	• Dragon swamp in Mt. Daeam etc.
Grassland ecosystem	Grasslands	• Mt. Gyeung in Cheorwon DMZ	• Paju Story shooting range
Farmland ecosystem	Farmland	• Paju Panmunbeol Abandoned Paddy Fieldwetland	• Gimpo plain, Ganghwa plain, Paju plain, Cheorwon plain, Yeoncheon plain, Goseong plain

(continued)

Table 4.5 (continued)

Ecosystem	Habitat		Targeted area within DMZ	Targeted area within CCZ
Forest, shrubs and scrub ecosystem	Forest	Coniferous forest		• Mt. Gunbong, Mt. Daeam, Mt. Baekam
		Deciduous forest	• Mt. Gyeung in Cheorwon DMZ	• Mt. Dosol, Mt. Daeu, Mt. Deasung, Mt. Baekhak
	Shrubs and scrubs		• Forest fire region in Goseong, Yanggu	• Forest fire region in Goseong, Yanggu
Village ecosystem	Artificial towns		• Paju Panmunbeol	• Village in Paju, Cheorwon and Goseong

Fig. 4.6 Distribution map of mudflat located in DMZ &CCZ

Fig. 4.7 Kanghwa Uepori tidal flat

Kanghwa tidal flats, on the other hand, are supplied by the estuary of the Han river. In the southern part of Kanghwa, uninhabited islands such as Seongmodo and Buleumdo became important habitat for migratory birds (Figs. 4.7 and 4.8).

Biota in mud flats

The creatures living in the tidal flats must overcome the extremely harsh environmental conditions that beset their habitat such as the tides and waves, torrential rains, and strong salty winds. They have to adapt to the influx of fresh water from the land and, the sudden flush of salt water from the sea that intermittently shake and change the habitat. Their habitat is practically exposed to the contact between marine and terrestrial format.

Fig. 4.8 Kimpo tidal flat

Several species learned to adapt to the ecological system of the tidal flats like the; large beads of Ark shell, eardrum, baby clam, Cyclina sinensis, Solen strictus, clam, surf clam, oyster, mussel, Note round, pond snail, periwinkle, octopus, etc. These tidal flats became a habitat of these edible species of crabs and clams which in turn provided food and livelihood for local residents who frequently collected them.

Crustaceans are common in the tidal flats and mudflats of the west coast. Other species are reported to invade the area such as; shrimps, prawns, crabs, cutlery, barnacle, and rock crab. The various shellfish and crustaceans attract many local and migrating birds. Among the birds who come to prey on them are; the spoonbill, egret, yellow beak, Spoon-billed Sandpiper, and alrak tail curlew. Large geese from all over the world including endangered birds feed in the tidal flats and mud flats and then breed in the nearby small islands. During winter, spring, and autumn, migratory birds called Snipes and Plovers are reported to fly over here for shelter (Figs. 4.9 and 4.10).

Fig. 4.9 Crab hiding in the reeds of the tidal flats

Fig. 4.10 Gull seeking food on the mudflats

Considerations for habitat conservation

During the 1970 s, tidal flats were considered as useless land, so a massive reclamation project erased many tidal flats from the site. Since the waves and tides continue to develop tidal flats along the long stretch of the coastline, tidal flats reappear in time. Finally, recognizing their uniqueness and significance to biodiversity, so much has been done currently to preserve them.

Facilities had been built along the coastline but the DMZ tidal flats remained as protected areas. Development of new agricultural lands is now prohibited but catching fish and crustaceans are still ongoing. Tidal flats can be maintained continuously through appropriate management, active participation of the people and the community, strict implementation of land use, and environmental policies will surely work.

Rocky Island

Figure 4.11.

Distribution and habitat characterization

Yudo is a small island in Gyeonggi province. Rocky mountains are windy and rocks typify its vicinity. Communities of vegetation are also seen, but only few types of wood are able to grow there. Most of the uninhabited islands have mountains of rocks and because external interference is extremely low, they became a habitat for nesting birds (Fig. 4.12).

Biota in rocky island

It has a small rocky area where spoonbills, cormorants and gulls, among other rare migratory birds, are commonly seen. The site amazed many people, which is the most talked-about scenery in the DMZ. The Black-faced Spoonbill is recognized internationally as endangered species (Natural Monument # 205), and has chosen the rocky mountain as its breeding ground because food is abundant in nearby farmlands and tidal flats, and there is little interference from humans.

Figures 4.13 and 4.14 show the induction of Black-faced Spoonbill and Cormorants.

Fig. 4.11 Rocky island distribution located in DMZ and CCZ

Fig. 4.12 Udo Island, Kimpo

Fig. 4.13 Black-faced spoonbill

Fig. 4.14 Cormorant on Yudo island

Considerations for habitat conservation

This is a habitat that is difficult to recover because it is survived by many other aggregate elements namely; the tidal flats and the mud flats that sustain the food of the birds; the agricultural habitat in the coastal area that also supplies food, the foreshore and the terrestrial water system that neutralize salinity and bring nutrients from the land, the sea habitat that sustains marine life, and the tides and waves that brought marine life and sand to the coast. It is also important to remember that it is the tides that continuously develop and restore the structure of the tidal flats and the mud flats. This natural connectivity and interdependence is like a chain. Its strength lies on its single missing link.

If one of these aggregate links is cut from natural flow of systems or worse, totally destroyed and the eliminated, the dependence chain is also cut. Then it is almost impossible to apprehend the very sad consequent and devastating events. The rest of the habitats will eventually die a natural death. If migratory birds can find another suitable habitat, then there is hope. But if they cannot, they too will be extinct.

The birds came to settle in the rocky mountain because food is abundant and they are not disturbed. The truth is that conservation is actually easy. Just leave the habitats alone because nature creates and finds its own way. If the inevitable catastrophes occur, then let us help restore it by putting the chain back to its original form. The problem only occurs when humans think that modern developments are more important than these habitats. They need to understand the connectivity and dependence of these habitats to produce such a wonder. Such a wonder must be experienced by generations. This should be the most significant tribute of mankind to human existence. Nature is the reason why we survive because we are on top of the food chain. We make and break the chain. All efforts, talents, time, and treasure must be consolidated to preserve these habitats.

Beach and Estuary

Figure 4.15.

Distribution and habitat characterization

The coast harbors a typically sandy beach with some gravel in it. Mud is also observed in some portions of the beach's sidelines. The coastline is slowly inclined

Fig. 4.15 Beach and estuary distribution located in DMZ and CCZ

toward the land forming a crescent shape. In many spots coastal sand dunes are created naturally by the tides and the strong sea wind. Several types of algae inhabit on rocks near the lower portions of the coastal dunes moving toward the pebbles down the sea.

The rivers coming out from the different parts of the land always find its way to the sea. In the area where the sea and the rivers meet, a different kind of water environment is created with a mixture of fresh and salt water. Hence, freshwater and marine life also meets. This is called the brackish water, where various species from the sea and rivers inhabit. This is a special kind of habitat for shrimps, sea shells, and crabs including marine fishes and their fingerlings. In fact, this is a typical area where man-made fish ponds are developed beside it. These are seen in Figs. 4.16, 4.17, 4.18, and 4.19.

Fig. 4.16 Estuary of the Han river in Paju

Fig. 4.17 Estuary linked to Goseong sea

Fig. 4.18 Beach and sand dune in Goseong DMZ

Fig. 4.19 East coast beach which is occupied by barbed wire

Fig. 4.20 Fish in the different parts of estuary in the East coast

Biota in coastal habitat and estuary
The aquatic ecology of these waterways and estuaries made the environment suited for marine and fresh-water invertebrates like the shrimps.

The lamprey and Chinese ninespine stickleback in particular are considered as endangered species by the Ministry of Environment Wildlife (Fig. 4.20).

Considerations for habitat conservation
The biodiversity of the estuaries and coastal habitats are numerous and rarely seen inhabiting together in a single habitat. However, in areas where development is under way, the environmental status is obviously seen as poor. In particular, the Donghae and other interconnecting roads aggravated the situation of the coastal areas in the DMZ. Many coastal ecosystems had been damaged and disturbed because of too much exposure to human activities and to its waste. Added to this miserable condition, contaminants increasingly drains and accumulate in the coastal habitats which threatens irreversible and damaging impact to these ecosystems.

If it is inevitable for the present developments to be relocated, there should never be future plan for developments in these areas. To prevent this clear and eminent danger in the current development areas, the pollutants should be stored properly, transported regularly to other designated places outside DMZ, and recycled or purified before they drain into the ecological environment. Implementation of laws and regulations must be strictly and rigidly followed with a strong political will to impose legal fines, sanctions, and punishments. Laws should ensure that developments are within 5 km or more, distant from the coastal and estuarine habitat conservation areas.

The environment can be preserved and protected further with the active participation of the community, and the cooperation of the business proprietors, and

Fig. 4.21 Distribution of lagoons located in DMZ and CCZ

guests. Regulations on the protection and preservation of the environment should be embraced as an ethical-legal responsibility of the people. This duty should be prudently and constantly oriented and reminded to all.

Lagoons
Figure 4.21.

Distribution and habitat characterization
Lagoons are found in the East Coast. There are only few lagoons found in the DMZ. It is usually found near large bodies of water in the northern parts of the DMZ and CCZ. Reeds surround the wide and open body of water that has been accumulated naturally by the sea and by the rivers nearby. This is now a tourist destination where Hwajinpo is also in full view. A lagoon in Gamho is shown in Fig. 4.22.

Fresh water currents from rivers and waterways move toward the sea carrying silt that accumulates in certain foregrounds of the beach, while the winds from the sea bring sand and salt water toward the foregrounds in opposing direction. Eventually, a lagoon is developed with a suitable salinity for important species. Algae flourished from the minerals of the silt and sands. A planktonic freshwater biological environment is found where small aquatic plants, plankton, and eutrophic lake evolved (Fig. 4.22).

Biota in lagoons
Direct ecological surveys have not comprehensively identified all the life species that inhabit the lagoon. Nevertheless, in conjunction with its surrounding areas, migratory birds had been identified inhabiting these areas such as the mandarin, billed ducks, and geese.

Fig. 4.22 Gamho brackish lake in DMZ (North side)

The lagoon is naturally wide open water and reeds are also observed widely, so a number of migratory birds flock, swim and are hang about in the water surface. Commonly seen are the Eouleojyeo Geese. Sport Konishi and mallard are winter migratory birds that are found frequenting the lagoon. Sport Konishi is recognized as a Natural Monument (No. 201). Particularly, one of the four kinds of rare turtles is found breeding in sandy sections of a lagoon in the East Coast (Hankyoreh Daily 23/11/2007).

Considerations for habitat conservation
According to the 2007 survey results, ragweeds slowly colonize the East Coast lagoon ecosystem disturbing natural flora and fauna of the creatures living there. The lagoon is a critical habitat for rare migratory birds and turtles hence, these weeds have to be removed. At the pace these weeds are growing, soon the lagoon ecosystem will be jeopardized.

Introduction of invasive species must be closely monitored and quickly removed. Human activities like religious rituals, science field trips, and other disturbances should also be strictly prevented. Conservation management should not be compromised in order to preserve the hydrological, topographical, and native habitat.

Coastal Sand Dunes
Figure 4.23.

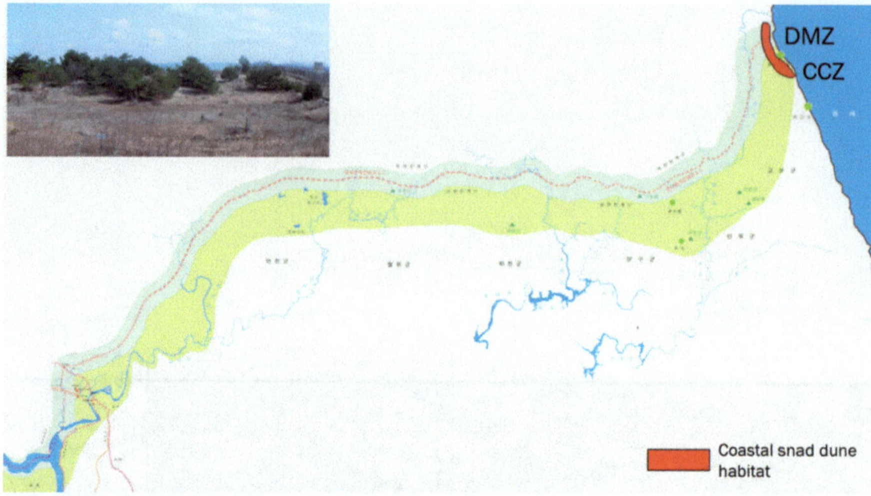

Fig. 4.23 Coastal dunes distribution located in DMZ and CCZ

Distribution and habitat characterization

The coastal areas consist of low sand hills and sand beach accumulated through the strong flow of tides forced by the wind on to the background of the beaches. Black pines are covered widely behind the coastal dunes and beaches, while typical plants creep over the sands under them. The trees in the landscape usually appear along rows of coastal dunes in the shoreline.

The coastal sand dunes specifically in the Donghae portion of DMZ are noted mostly in the coastal areas where there is a wider beach formation. Water is trapped in these dunes that consequently develop its own special vegetation that is fit for specific aquatic plants. This is made possible because of the negligible human interference.

Widely developed vegetation is pine, oak, and most especially the Rosa rugosa which made the dunes' landscape very attractive (Figs. 4.24 and 4.25).

Biota in coastal sand dunes

The trees help to prevent the strong winds from interfering with the natural development of the other vegetating species of plants while the freshwater entry to the ponds formed by the dunes controlled the salinity of the water thereby contributing further to biodiversity. Rosa rugosa community is noticeably developing in these coastal dunes covering widest in the South and scattered therewith are barley, sedges, Elymus geuryeong black pine, and oak. Egret, black-tailed gull, buzzards, and other birds cover a relatively wide radius in these coastal regions (Figs. 4.26, 4.27 and 4.28).

Considerations for habitat conservation

The Coastal dune habitat is a very fragile habitat. Therefore, if these coastal dunes will be destroyed or some elements halted, it would be very difficult to recover.

Fig. 4.24 Sand dune distribution in the Goseong DMZ (1)

Fig. 4.25 Sand dune distribution in the Goseong DMZ (2)

Plans for development in the vicinity of the coastal dunes should be prevented. The farmlands and buildings near the coastal dunes continue to break down the coastal dunes' habitat. The construction of coastal roads to connect North and South, and the restoration of East Coast railroad prevented the sand formation that serves as the only lifeline of the dunes (Fig. 4.29).

Salt Marsh
Figure 4.30.

Distribution and habitat characterization
The coastal wetlands are formed by the frequent flash of tidal salt water from the sea that accumulates in the lower and flatter areas at the offing of the land near the beach. It is inhabited by plant species such as lush grasses and reeds that can withstand the ever changing salinity of the water (brackish water). Salt ingredients are too much in these areas due to the constant overflowing and the sediments of salt are saturated on the floors of the wetland especially in the delta zone.

Fig. 4.26 Community of Sweetbrier in sand dunes of the East Coast

Fig. 4.27 Sea bells

Fig. 4.28 A white-tailed sea eagle flying in the sand dune

동해안 비무장지대 해안선 따라 1.2km

국내최대 해당화 군락 발견

50여년 접근차단 지역

'생태계 寶庫' 재확인

Fig. 4.29 Article on *Rosa rugosa* THUNB in DMZ

Fig. 4.30 Salt marsh distribution map in the DMZ and CCZ

Coastal wetlands often extend inland and the salt undergoes various changes thus forming the estuarine wetland and inland brackish wetland of DMZ. Salt wetlands are also formed in the tidal flats and mostly developed within the vicinity of the estuary of the Han river area. Halophyte Salicornia herbacea and Suaeda japonica are dominant in these areas despite salt saturation. The salts of these tidal flats made a big impact on river reeds (Figs. 4.31 and 4.32).

Biota in salt marsh

Beside Halophyte Salicornia herbacea, and Suaeda japonica other common reeds found in the country are also widely dominant in the salt wetlands. Tidal flats and salt wetlands of DMZ show similar forms of vegetation and animal. Hydropotes inermis egret ducks, and geese appear in these salt wetlands (Figs. 4.33 and 4.34).

Fig. 4.31 Salt marsh in the estuary of the Han river

Fig. 4.32 Salt marsh in Janghang wetland

Considerations for habitat conservation

The habitat developed in tidal flats is of the similar nature and characteristics with those found in the salt wetlands of the DMZ. Phragmites communis community is also noted in the DMZ salt wetland. This played a significant role in sustaining migratory bird habitats. Many areas of the salt wetlands in the vicinity of DMZ are transformed into farmlands infusing pollutants to this precious habitat. This greatly reduced the function of the salt wetland for the vegetation and for the creatures that depended so much on it.

Clearing activities and farming should be prevented in the DMZ areas because it nurtures many types of habitat for very rare species. Salt wetland farming should be absolutely prevented including other plans of development. Salt wetland should be allowed to function properly in order to maintain a continuous hydrologic process. Structures like the sea wall hinder the unique natural phenomena occurring in the salt wetlands.

Fig. 4.33 Community of reeds in the salt marsh in the estuary of Han river

Fig. 4.34 *Suaeda japonica* and reeds

River and Stream Habitat

Figure 4.35.

Each watershed and its associated subwatersheds across the DMZ contain different rivers and streams in its basin. These watersheds offer unique hydrological systems with a wide variety of tributaries of the rivers and streams shown in Fig. 4.36.

The Imjin river watershed contains Sacheon stream, Sami stream, and Myeolgong stream.

Fig. 4.35 The upstream of North Han river

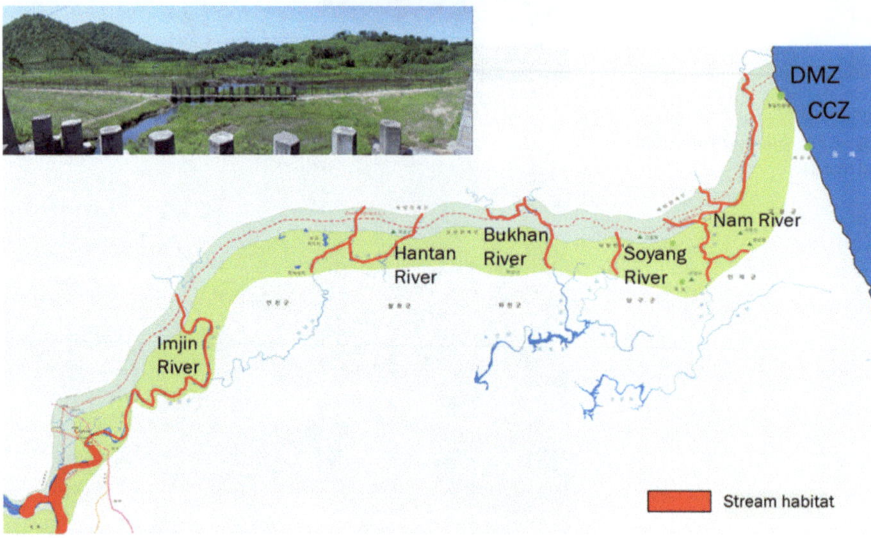

Fig. 4.36 Distribution map of stream habitats in the DMZ and CCZ

Stream

Distribution and habitat characterization

Paju bioregion is greatly affected by the rivers in this region. In Imjin river watershed, the Imjin river changes direction almost at right angles near Cheongok of the CCZ. It is joined by Hantan river moving toward the Yellow Sea.

A riparian wetland is a wetland formed by a river behind it. Watershed riparian wetland is a typical example of a wetland formed by river waters overflowing and settling in the plains of the watershed.

In Hantan river watershed, the Hantan river develops the Keumhwa reservoir, while Mt. Baekam river formed the Daegyocheon stream. They provided a big impact in the biological developments of Cheorwon. The orthogonal flow down the DMZ that is between North Korea's Pyeonggang river and Mt. Ohseong, penetrates and passes through the Hak reservoir and joined the Namdaechon flow that also originated from the tributaries of the eastern parts of the south CCZ. These tributaries form the mountain valley wetlands, the riparian wetland, and sandbar wetland.

Fig. 4.37 Sacheon stream in the western bioregion

Fig. 4.38 Yeokgok stream in the Central Bioregion

In the North Han river watershed Mt. Baekam is divided forming Hwacheon Paro lake and a brook which flows into the mountain forming valley wetlands and sandbars. Tributaries from the coast are joined other tributaries namely; Chiljeolbong, Hyangrobong, Maebongsan and seohwacheon. In the Nam river watershed, Nam river originates from the valley between Guksabong summit and Mt. Kumgang. The Nam river flowing to Donghae forms the alluvial valley wetland (Figs. 4.37, 4.38, 4.39 and 4.40).

Fig. 4.39 The Han river in the eastern bioregion

Fig. 4.40 Hwacheon stream in the Eastern Bioregion

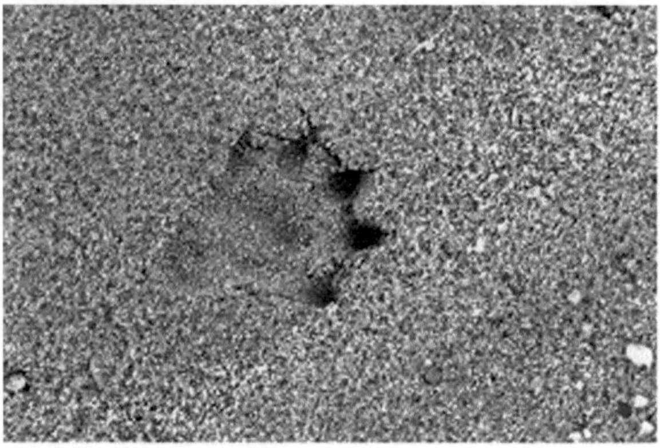

Fig. 4.41 Footprints of *Lutra lutra*

Biota in stream

Communities of reeds widely form a line in the Imjin river watershed. In this watershed, a wide variety of species live, such as swallows, herons, deer, moles, dragon flies, and mandarin ducks.

In the central area of the Hantan river watershed basin of the Namdaechon and Yeongokcheon, bird species such as grebe, gray heron, little egret, white-billed ducks, huinmok plover and dishwasher alrak, black and yellow wagtail, kingfishers, and swallows appear in these places.

In eastern bioregion, bird species, such as cuckoo, blue bird, heron, medium-large egret, and mandarin duck inhabit here.

The water is very clean and the habitat harbors a variety of fish such as carp. The wild boar and wild rabbit, Lutra lutra, ladybug, sliver butterfly, and black locust are found in this basin (Figs. 4.41, 4.42, 4.43 and 4.44).

Considerations for habitat conservation

In the Imjin river basin, several business establishments exist and the influx of tourists is increasing. There are also farmlands aggravating to the pollution of the environment and the water that sustains the habitats due to the rampant use of chemical fertilizers and agricultural pesticides. Pollution then is a serious concern. Species like Bass found in the Sacheon reservoir are of foreign origins. This aggregate collection of unacceptable and pollutant species were possibly introduced and brought about by farming and the businesses' activities. Since they are already there, a facility for the purification of the water must be installed before

Fig. 4.42 Korean buzzard

Fig. 4.43 *Lobocla bifasciata*

Fig. 4.44 *Microphysogobio tungtingensis*

these pollutant compositions add more damage to the river habitats and the wetlands. Destructive weeds of foreign origin must be closely monitored and removed. Urgent action is needed.

The upstream region of the Hantan river basin is still in its natural state. The status of the habitats here are good including further down the midstream and downstream. The farmland irrigation system is enclosed in concrete revetment but the rivers supplying it are diverted toward the farms and exits to the wetlands and rivers. The rivers receiving the outflow from the farm irrigation system are not in its best quality and composition for the sustenance of the habitats; instead it damaged the nearby habitats. Urgent action is needed in preventing water influx to farmland that diverts and drains the streams and preventing outflows that pollute the wetlands due to chemical fertilizers and agricultural pesticides, and other various organic materials. Environment-friendly farming through organic farming must be legally supported and strictly implemented. Contaminants must be purified within the range that habitats can tolerate.

The habitat conditions are poor in forests near military facilities. Throughout the damaged area, growth conditions are far worse where there are frequent human activities going on. Accordingly, important areas such as these habitats must be declared as protected areas. Education campaign must be extensively organized for residents and visitors. Orientations must include the damaging effects of landslides to humans and their properties if forests are gone hence, landslide prevention measures must be taught seriously. People must learn from the miserable experiences of countries that encountered landslides and must not wait until they experience it themselves.

Lake and Reservoir Habitat

Lake and reservoir ecosystem is stagnant water. Its size is more than 8 hectares. On the other hand, the size of Palustrine wetland is less than 8 hectares. A relatively wide area of the body of water is not covered with woody vegetation, mosses, and aquatic plants. Because of the depth of the water, local water channels blocked by dams create vast areas of wetlands.

Most lakes and reservoirs in the DMZ and CCZ, which were created for the farmland irrigation, became the major habitats for fish, reptiles, and amphibians. Naturalization of those habitats here was made possible because these areas have the least human interference. Several lagoons created by the sand dunes can be found in the east coast.

Lakes and Reservoirs

Figure 4.45.

Distribution and habitat characterization

Currently, reservoirs formed in the DMZ are used for agricultural purposes. Reservoirs in watershed were built considering topographical features. Most of these reservoirs are remnants of the war that have been around for many years. The wetlands around the reservoir are well developed unlike those found in the concrete revetment with sandbar formation. Natural reservoir wetlands were formed

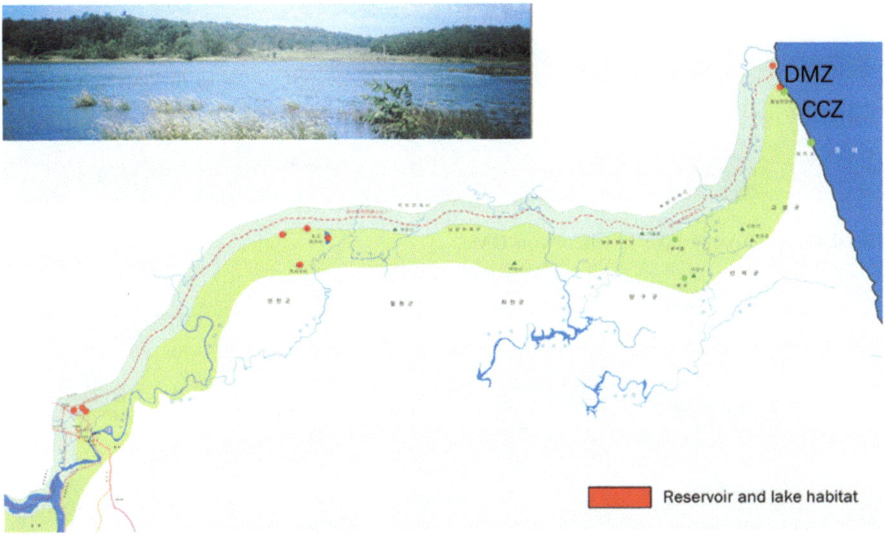

Fig. 4.45 Distribution map of lakes and reservoir within the DMZ and CCZ

Fig. 4.46 Daeseong reservoir in Paju DMZ

by these reservoirs and are being used as critical habitat for fish, reptiles, and amphibians (Figs. 4.46, 4.47 and 4.48).

Biota in lakes and reservoir

The reservoirs that are formed both in the DMZ and CCZ are described as; wide open water with Trapa japonica Filow, zizania latifolia, irises, dalppul ripple, reed, and a variety of creatures were also identified to inhabit the area. Mammals such as leopard cats, wild boar and wild cat, and flydropotes inermis appear along with fish including wicker match, medaka, and the crucian carp. Reptiles and amphibians co-inhabit the area such as several frog species. Snakes appear together with Micorphysogobio yaluensis and Sualidus gracilis majimae. Little egret, mandarin ducks, and cormorants were also observed. A special and rare ecological network has been formed with the rich biodiversity in and around the forests and grasslands (Fig. 4.49).

Fig. 4.47 Sanmyung lake in Cheorwon DMZ

Fig. 4.48 Anho reservoir in Goseong CCZ

Considerations for habitat conservation

Kim (2007) conducted the ecological survey in the DMZ where he found cemented banks for the efficient irrigation of farmland. Disturbance in aquatic ecosystems had been consistently observed due to the invasion of nonnative species that are increasingly developing in Sacheon reservoir. The DMZ survey team actually saw bass in Sacehon reservoir across the north to south of the DMZ affecting badly the aquatic ecosystems. More serious encroachment of exotic species is happening in the Cheorwon reservoir where nonnative fish overpopulate compared to the native fish.

Reservoir and lake ecosystems' health can be improved through the restoration of the wetlands and discontinuation of concrete revetments. To manage and prevent the influx of exotic species, periodic monitoring and immediate action is needed at once.

Wetland Habitat

Palustrine type of wetland ecosystems is a relatively small area. In certain periods of the year, they are becoming dry. In these wetlands, wide variety of vegetation develops with low water depth and less frequent waves. Bog, swamp,

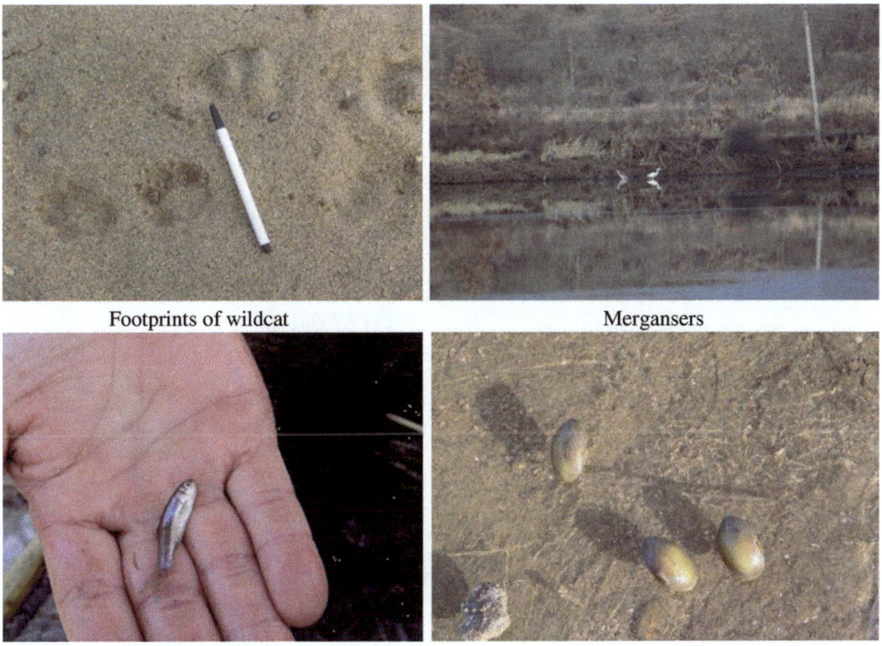

Fig. 4.49 Bio-species living in the lakes and reservoirs within the DMZ and CCZ

marsh, and fen are regularly seen in these wetlands throughout the year. This specific wetland discussed here are the wetlands formed naturally in the abandoned rice paddy fields during and after the war. Because these areas had been secluded and very seldom visited by humans for a long time, the ecosystem changed and developed habitats for many species. Trees here are predominantly willows and alder wood, and many creatures are found in the reeds as their suitable habitats. A large variety of acidic wetlands such as Yongneup (also called Dragon Swamp) is also another contributing element of the habitation.

Strangely, even in the mountainous region of the east DMZ and CCZ, wetlands are also formed throughout the alluvial valleys.

Forested Wetland

Figure 4.50.

Distribution and habitat characterization

Forested wetlands formed in the mountains are reported as generally wooded swamps accounting to 30 % of the entire localized vegetation. These forested wetlands prevent flooding and secure the integrity of the wetlands. They are also believed to have contributed to the purification of the waters in the wetlands and rivers. The "forest-wetland-river-estuary ecological network" sustains highly diverse ecosystems wherever the water flows, that in turn sustain an equally rich biodiversity of habitats.

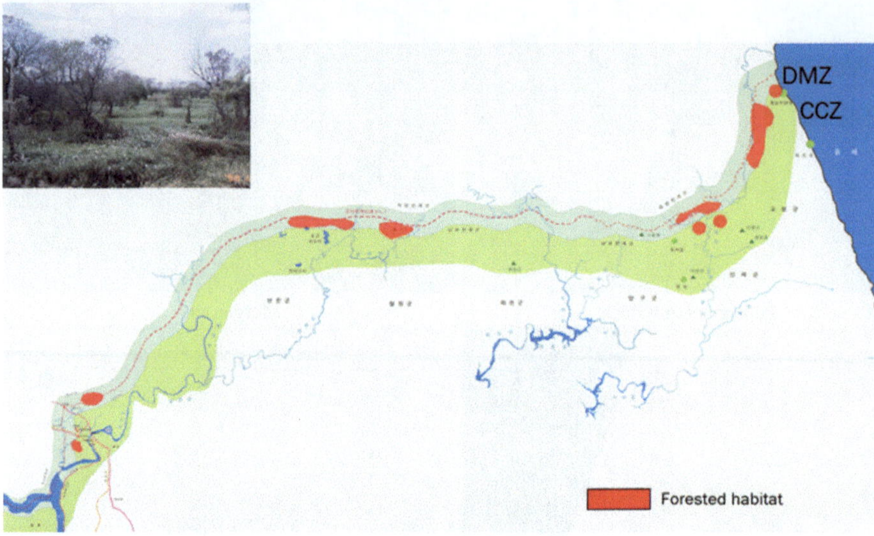

Fig. 4.50 Forested habitats distribution map in DMZ and CCZ

The bioregion of Paju in the DMZ is mainly forested wetlands from its hills down to the plains. In these regions, the forests are dominated by willow and alder trees which play as an important medium to large mammals and birds at its ecological corridors. In the plains, the forested wetlands are dominated by willow trees, considered as the central biological community. Other plants are mixed with reeds that made up the herbaceous landscape with high ecological value.

Most of the wildlife found in the forest is large mammals, but they are also seen in the meadows and plains, even at the edge of agricultural lands feeding together with the birds, reptiles, and amphibians (Figs. 4.51 and 4.52).

Biota in forested wetland

Forested wetlands support medium and large mammals and birds because this area is connected with surrounding mountains and well sheltered for large animals. In particular, reed and rush communities growing under trees feed herbivores. Figure 4.53 and 4.54 show these interactions with water deer and goral in the forested wetland.

Considerations for habitat conservation

Along the interconnecting roads in Gyeonggi and Donghae, the forested wetland is a fragmented in terms of the composition of the endemic trees and plants. In fact, exotic vegetation has been introduced and is gradually dominating the forested wetland known to have a high ecological and biological value before the construction began. In particular, false indigo and false acacia increasingly dominate the area and have become a problem.

Fig. 4.51 Forested wetland in Paju DMZ

Fig. 4.52 Forested wetland in Goseong DMZ

Fig. 4.53 Water deer

Fig. 4.54 Goral

The ecological network, that is the connectivity of the forests, wetlands, rivers, and estuaries must be restored. Unacceptable invasive plants, weeds, reeds, and trees must also be removed immediately as soon as they are seen. Significant consideration and importance should be on the survival of the species and the environment. Human being must realize that they need these ecosystems to maintain human life.

It is not only the birds and the bees, the flowers and the beasts that needed quality water and air. It is not only the wetlands and rivers that require the integrity of the land, or who does not want landslides and floods, but also humans most importantly.

Then, the important role of humans must be emphasized in sustaining the ecosystems that truly sustain them too. They must find the best possible means and strategies to revive the interconnectivity of the ecosystems in the DMZ and CCZ. People must accept the responsibility and must actively participate in the conservation of the sustaining ecosystem and the restoration of those that had been damaged, and the protection of these environments must be handed down to the next generations to come.

Paddy Field Wetland
Figure 4.55.

Distribution and habitat characterization The wetlands that formed in the abandoned rice paddies are called idle rice paddy field wetlands. Local farmers returned to these sites to reclaim their lands and to find economic opportunities that the land could possibly give. According to the local economic development authorities, farming contributed to the improvement of the local socioeconomic status.

The abandoned paddy field wetland ecosystem was virtually forced by the political and military situation. Unlike other areas, for 60 years, the paddy field wetlands were free from human interference that in turn made significant ecological developments and natural transitions possible. So many people have died here and its memories are so meaningful, in fact divided its peoples. History should have taught people lessons from a very sad past. Historical aspects should

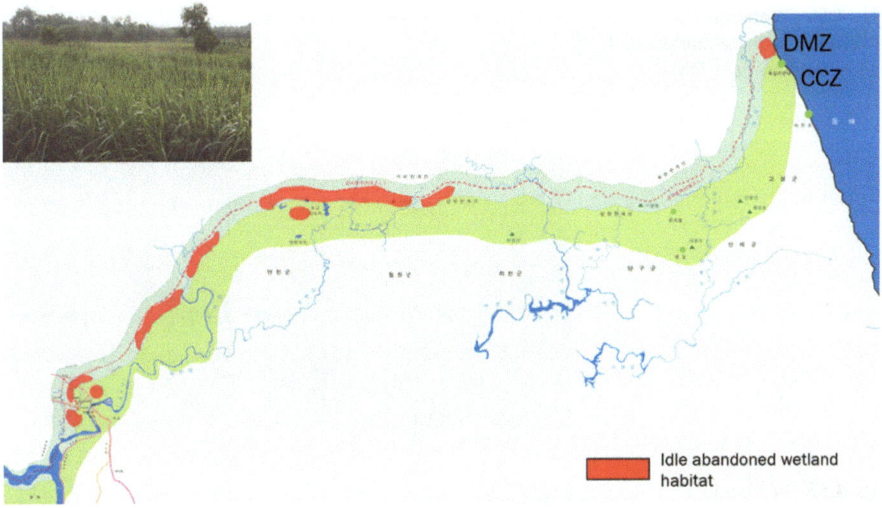

Fig. 4.55 Rice paddy field wetland distribution map in DMZ and CCZ

be emphasized. The DMZ is a memorial that deserves to be preserved to constantly remind the public as well as governments of the effects of war. Hate and misunderstanding, pride and prejudice must be buried here, and a new beginning that is happening at the DMZ brought about the natural miracle that is to be a symbol of hope. Then it should be left this way forever.

These wetlands harbor a variety of vegetation such as willow, alder wood, and reeds. They became a favorite habitat for migratory birds. Most of the abandoned paddy field wetlands believed to be minefields are the only ones left at the DMZ and CCZ, while all the rest of the cleared areas are now farmlands. Clearing the fields from mines continues, which means, in the near future, these wetland habitats will expectedly be gone to oblivion (Figs. 4.56 and 4.57).

Agricultural lands have been largely converted to wetlands (Figs. 4.58, 4.59, 4.60, and 4.61).

Water spider in Eundaeri, Yeoncheon, is designated as Natural Monument No. 412 by the Culture Property Authority. The abandoned paddy field wetland is the favorite habitat for these spiders. They are also found around the military training base. In addition to the spiders, a variety of fish, aquatic organisms, and insects also co-exist here (Figs. 4.62 and 4.63).

Biota in paddy field wetland

During the time when less human interference was imposed, a collective habitat of wildlife animals are found in the abandoned rice paddy wetland well equipped for shelter and food, while the hydrological development acts as corridors.

The paddy field wetland plate is interconnected with the willow communities that are dominated by local birds among other rare birds formatting collectively

Fig. 4.56 Wetland developed in the abandoned paddy field in Paju DMZ

Fig. 4.57 Wetland developed in the abandoned paddy field in NNSC footbridge of Paju DMZ

Fig. 4.58 Wetland developed in the abandoned paddy field in Cheorwon DMZ (1) during the wet season

such as Little Egret, Medium-Large Egret, and Anas Poeilorhyncha at approximately 200 birds. The organisms and vegetation of the wetlands are the lifeline of these birds.

The shallow depressions at Cheorwon CCZ have similar habitat and groups of herons are also found in these areas. Reeds also dominate the wetland areas so that a group of waterfowls are also observed living here such as; warblers, starlings, and spot bills. The collective habitats may be favored by these birds and

Fig. 4.59 Wetland developed in the abandoned paddy field in Cheorwon DMZ (2) during the wet season

Fig. 4.60 Wetland developed in the abandoned paddy field in Goseong DMZ (1)

Fig. 4.61 Wetland developed in the abandoned paddy field in Jeojin area of Goseong DMZ (2)

waterfowls, but biodiversity does not stop there, other terrestrial animals are found to be co-existing here such as water deer, wild boars, and other medium and large size mammals (Fig. 4.64 and 4.65).

Fig. 4.62 Egret habitat in the wetland of abandoned paddy field in Panmun field, Paju DMZ

Fig. 4.63 Water spider habitat

Considerations for habitat conservation (Fig. 4.66).

The idle paddy field wetland areas have rich biodiversity because they are relatively wide and highly interconnected to the surrounding natural ecological environments and interdependent to the surrounding natural ecological environments. But these areas are threatened by reclaimed farming practices (Fig. 4.67) and road constructions. A variety of fish, reptiles, and amphibians, including padlocks that used to inhabit the paddy field wetland before they are diminishing at its critical levels. In the food chain, these species are food for many birds and waterfowls. That is why they flock to these areas. If the food chain will be totally cut, then these birds will not come back.

Very important feature of the wetland area as a habitat should be designated as a conservation area. Reclamation of wetlands and construction of roads must be halted and prevented. A serious restoration plan should be made and strictly implemented. Laws should support the move to prevent constructions and farming in these wetlands.

Peatland

Figure 4.68.

Fig. 4.64 Egret habitat in the wetland of abandoned paddy field in Cheorwon CCZ

Fig. 4.65 Water spider

Fig. 4.66 Depression wetland developed in the abandoned rice paddy field

Distribution and habitat characterization

Peatlands are inland wetlands containing thick deposits of slowly decaying plant material called peat. "Bogs" and "Fens", major types of peatlands, occur in the DMZ area. Fens are richer in nutrients and less acidic than bogs, and are more typically covered by sedges, willows, grasses, and reeds.

Fig. 4.67 Reclamated farmland

Fig. 4.68 Peatland in the DMZ and CCZ

Peatlands are found in Yongneup, Mt. Daeam, in Yanggu, and some other portions of the DMZ and CCZ. Yongneup is located in the mountainous terrain which is the country's first Ramsar site. Its peat layer is an average of 1 meter in depth. The first pollen was extracted from the peat layer at Yongneup wetland. The results of the peat analysis were found to be formed in about 4,200 years ago. The survey conducted by the Ministry of Environment of the Republic of Korea revealed a variety of microorganisms that inhabit the 112 species including 22 species of wetland plants, and declared the peat wetlands are very ecologically important (Fig. 4.69).

Fig. 4.69 Peatland around Gyeonggi Railroad, Paju city (to be scientifically confirmed)

Fig. 4.70 Korean sedge community in the Dragon swamp of Mt. Daeam

Biota in peatland

Carex dispalata Boott is found in the peatlands because of very unique soil environment of the Youngeup. Hanabusaya asiatica f. alba T.B. Lee, willow, and maple trees are also appearing. There are also water toads, different types of frogs, and salamander. The natural ecosystem characteristic only to peat wetlands are the primary basis in designating Mt. Daeam Yongneup as a Natural Environment Conservation Area in 1989. In 1997, it was registered as a Ramsar Site, for the first time in Korea. This area also has been designated as Wetland Protection Area by the Ministry of Environment (Figs. 4.70 and 4.71).

Considerations for habitat conservation

Despite the fact that the Yongnuep has been declared as a conservation area, it dries up. Therefore, *Ambrosia trifida* L., azaleas, maple and willow trees are increasingly invading the peat in the past few years. The peatland is further severed by the large amount of sediment outflow coming from the military facilities, road constructions, and the hydrological disturbance.

Therefore, the inflow of sediments and organic matters into the peat wetlands should be strictly prohibited. Restoration measures must be quickly installed in damaged area. Floodgates must again be reconnected and blocks should not be allowed to hinder normal waterways. Woody plants must be uprooted meticulously and eliminated, including all other exotic plants introduced in the areas (Fig. 4.72).

Fig. 4.71 Sphagna in dragon swamp of Mt. Daeam

Fig. 4.72 Life on the edge. Landification of dragon marsh in Mt. Daeam due to the influx of ragweed and azalea

Grassland Habitat

Grassland can be divided into several classifications. The first one is natural grassland which is composed of pampas grass and bamboo grass species (*Miscanthus sinensis* Andersson). The second one is pasture grassland which is known as the second nature since it is cultivated in order to produce grasses for livestock pasture and grazing. The grasses can be collected and stored as feed for the livestock like the hay or silage.

Grasslands found in the DMZ and CCZ are the secondary grasslands which developed by firing for military operation during summer time. These grasslands

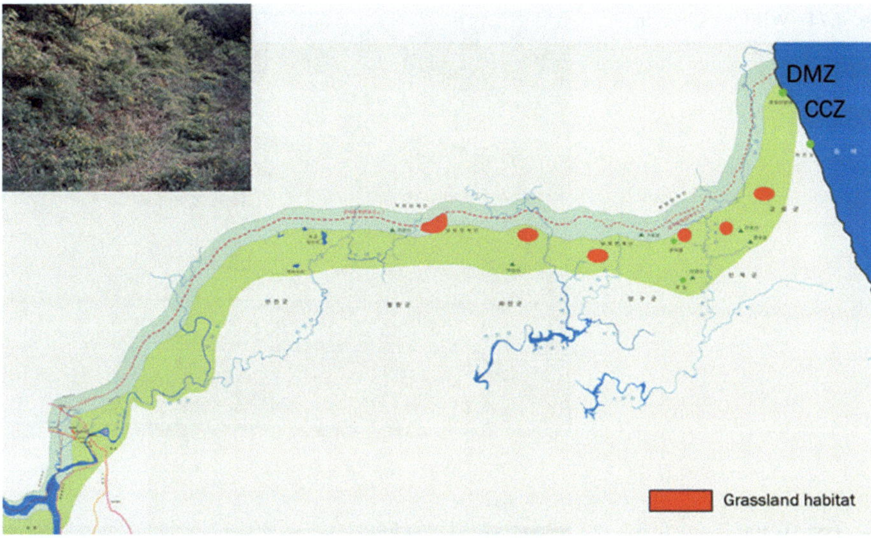

Fig. 4.73 Grassland distribution map in DMZ and CCZ

are now becoming semi-natural grasslands with no human disturbance. Reeds and *Miscanthus sacchari-florus* Benth are dominant in the wet grasslands. Wild meadows are also noted living on the mountain slopes where there are open spaces of land . The wild herbaceous plants that can be seen in these slopes are Chelidonium majus var. asiaticum (Hara) Ohwi, and violets.

Grassland

Figure 4.73.

Distribution and habitat characterization

The Hyangrobong forest road cut the slopes and created the grasslands on a large scale. Recently, the frequent activities of people in these areas induced the surging invasion of exotic species, such as evening primrose and fleabane (Figs. 4.74 and 4.75).

Biota in grassland

Grassland needs a lot of space and sunlight. Many flowering plants are seen in areas that are directly exposed to the sun. Because of this attraction, many insects live in the grasslands, including butterflies, Coleoptera, Hemiptera, and ant aneurysm. Since these grasslands are connected to the forests and rivers, many more kinds of insects can be observed. This ecological connectivity brought the silver butterfly and the Nannophya pygmaea Ramber into the flowering vicinity, which are rare species. These species inhabit the grassland and turned out to be of high natural value (Fig. 4.76).

Fig. 4.74 Wild meadow in Cherowon CCZ

Fig. 4.75 Wild meadow on the slope of Mt. Geonbong CCZ

Fig. 4.76 Insects living in the grasslands of DMZ and CCZ

Fig. 4.77 Evening primrose in the grassland

Fig. 4.78 Colt's tail community in the grassland

Considerations for habitat conservation

Developments in slopes and grasslands disturbed the ecology in the DMZ and CCZ. Human activities should be prohibited and the constructions must be stopped and erected elsewhere distant from the DMZ and CCZ.

Aggravating the human-induced disturbance, the grassland is not free from exotic species like the Evening primrose, *Ambrosa trifida* L.,*Conyza canadensis* (L.) Cronquist, and *Bidens frondosa* L. These alien species invading the natural habitat of native plants and disrupting natural ecosystems and their pollens are blown by the wind to human population and causing allergic reactions like asthma and rhinitis. Restoration's focus is mainly to eradicate exotics plants, that is, to remove and prevent their spread and, allowing the indigenous natural plants to spread instead (Figs. 4.77 and 4.78).

Farmland Habitat

Farmland refers to areas cultivated for agricultural purposes such as; rice paddies, crop fields, and all arable lands. According to the Forest Service, the area occupied by the agricultural farms within the DMZ is 2,495 ha,; 588 ha to the south and

Fig. 4.79 Farmland in Paju CCZ

1,907 ha to the north, respectively. Considering that total area is 90,703 ha, the farmlands occupy a considerably large area, enough to interfere with the natural connectivity of ecological networks.

Paju and Cheorwon plains are formed by river flood plain and lava, in which many organic matters and wetlands are found. Most of the rice paddies in the mountains and valleys, and even upland areas are farmed. Migratory birds feed on these farmlands. These birds include cranes, geese, and ducks.

Farmland

Figure 4.79.

Distribution and habitat characterization

From the ancestors of Koreans, farming has been a way of life and their farms were normally around critical habitats of wildlife animals including feeding grounds of the migratory birds.

TM image data revealed that 2.8 % of the total area of the DMZ is occupied by the farmlands. More than three quarters of the farmlands are in the north. Farmlands are distributed near the local villages of Panmunjom and Daeseongdong, south of DMZ. Currently, many areas of CCZ are used as farmlands except for some unidentified minefields that haveyet to be cleared. Cheorwon plain is very famous for stopover of red-crowned and white-naped cranes (Figs. 4.80, 4.81, 4.82 and 4.83).

Fig. 4.80 Farmland in Paju CCZ

Fig. 4.81 Farmland in Cheorwon CCZ

Fig. 4.82 Red-crowned cranes in Paju CCZ

Fig. 4.83 Community of red-crowned and white-naped cranes in Cheorwon CCZ: They may act as flagship species in this landscape

Biota in farmland

Cranes and geese feed left-over rice in the farmlands which are located near reservoirs and waters.

The cranes, on the other hand, are naturally wetland birds, so they feed on wetland plants or small animals found there. The farmlands have plenty of roots that birds also eat. However, at the advent of the reduction of wetlands while agricultural lands increasing, the birds will gradually lose their food sources. Geese flocks are moving toward the wetlands and farmlands occupied by the cranes. It means that the birds are now invading each others' feeding grounds and compete on the food available.

Considerations for habitat conservation

Most of the farmlands are in CCZ. The farmlands use agricultural pesticides and chemical fertilizers. Some migratory birds die during winter, mostly attributed to drinking of contaminated water. This has become a problem. Croplands are irrigated and the canals are made of concrete. Concreting has greatly reduced the habitat for cranes and geese. The habitat of the damper reotte is a reducing trend and pattern.

The farmlands being used as feeding grounds require consultation with the local residents and the members of the conservation management group to resolve their difficulties in maintaining the balance between the economic viability of the farm and the preservation of the birds.

A biodiversity management agreement may be considered, specifically designating the wetland as an ecological park. According to a biodiversity management agreement that is based on the Natural Environment Conservation Act, the following actions can be recommended:

A. Protection of endangered plant and animal species in the farm lands
B. Proper way of cultivation
C. Less use of chemicals
D. Implementation of proper land use

The creation of the wetland ecological park that provides a habitat of birds, reptiles and, amphibians, and ecological programs, such as eco-tourism and eco-education to run can be activated by the surrounding wetland and farm lands.

Forest, Shrub, and Scrub Habitat

Forest, shrub, and scrub habitat occupies a wide area actually accounted for about 75 % of the ecosystem of the DMZ.

From the East coast to Cheorwon beyond the Taeback mountain range is the fairly rugged mountains. Beyond Cheorwon plain, rolling hills are formed toward the West coast. There are many mountains whose height is more than 1,000 M above sea level. Yanggu in a coastal basin is shaped like a punch bowl, so it is named as punch bowl. The hills from Cheorwon to the estuary of the Han river are low relatively less than 400 M.

Forest, Shrub, and Scrubs

Distribution and habitat characterization

Approximately 60 % of the DMZ forest is mixed forest of the coniferous and broad-leaved forest.

At a regional point of view, the Paju bioregion is made of low hills where human interference is frequent, and is dominated by bush types of forest and broad-leaved forest species. Cheorwon area where military facilities are operational showed signs of active human disturbance, so mixed forest and scrub forest are observed. Most of the hilly parts have simple vegetation communities of Quercus mongolica—Azalea. In the highlands, coniferous and deciduous forest is abundant and mixed forest is also present. Maple-Ash community and Fraxinus rhynchophylla Hance-Philadelphus schrenckii Rupr. Community occupy the East coast and eastern mountain areas (Figs. 4.84 and 4.85).

Biota in forest, shrub, and scrubs

Large, medium, and small mammals and mountain birds compose the wildlife that inhabits most of different species. They occupy a huge area and their activities are

Fig. 4.84 Fragmented landscape: Hyangro mountain summit in Inje

Fig. 4.85 Mt. Geonbong shows altitudinal distribution of plants

Fig. 4.86 Goral

observed at a wider radius covering practically the entire topography including cliffs and steep slopes.

DMZ and CCZ's forests used to be a part of the Baekdudaegan mountain range in the past, and tigers were reported to roam in these forests. Currently, the movement of large mammals is very limited compared to other areas. Yet, compared to other regions, the conservation potential of the DMZ and CCZ is still high because of the presence of some large-, medium-, and small-sized mammals that inhabit here. The endangered animals designated by the Ministry of Environment which are found in the eastern mountainous region, include mountain goats and musk deers (Figs. 4.86 and 4.87).

Considerations for habitat conservation
The conservation of ecosystems and biota of the CCZ forests are actually being undermined by the presence of military facilities, military roads, farmlands, and railroad projects. They cut off the connectivity of the ecosystems and marginalizing the very important biota that normally inhabits here. Forest fires frequently happen, killing the natural flora of the DMZ and CCZ, not to mention the

Fig. 4.87 Musk deer

unwanted exotic species of plants introduced here. The ecological bridges built to promote ecological connections may have helped but is yet to be investigated through a full monitoring. The habitat impact assessment should be conducted to ensure the integrity of the habitats. Important forest and bush ecosystem should be proclaimed legally as conservation areas and designated as protected areas because proper conservation management is required.

Village Ecosystem
Village ecosystem is also happening near man-made structures. Natural ecosystem elements had developed favorably for specific creatures like the lark, sparrow, and snakes that are not often seen in the vicinity. In the agricultural society of the ancient towns, animals, and humans were living together and co-exist. But these creatures are now disappearing because of a wide variety of developments.

Many areas of the DMZ and CCZ are originally village areas. At the advent of the Korean War, villages were abandoned but the remains of the villages and trains remain. Habitats develop in these remains and creatures eventually inhabit this artificial ecology of old buildings and walls. It is now a very valuable historical and cultural remembrance. Gyeonggi train (Fig. 4.89), Unification Observatory, Jangdan-myeon office (Fig. 4.91), Unification village, Haeoreum village in Paju county, the villages in Cheorwon, old castle ruins of Gungyeo, and the Koguryo remains are now ecological marvels.

Artificial Structures
Figure 4.88.

Distribution and habitat characterization
These **artificial** village habitats of the DMZ and CCZ are taken advantage by wildlife animals. The old abandoned home sites became wetlands with no residents mainly due to landmines. These areas are currently being used as a habitat of wildlife animals (Figs. 4.89, 4.90, 4.91).

Fig. 4.88 Artificial structure distribution in DMZ and CCZ

Fig. 4.89 Old locomotive in Paju DMZ

Biota in artificial structures

Artificial village habitats have a variety of creatures. For example, the Jangdan-myeon office forms the habitat for pheasants, ants, bees, raccoons, and water deers. A lot of artificial structures provide animals with shelters and the dietary needs of specific wildlife. Jujube trees, peach trees, and persimmon trees are found in the old home sites and feed animals. This area were used as old home site and vegetations like the painter, jujube trees, tree spade, peach tree, and arboreal habitats provide the dietary needs of specific wildlife. Unlike in the wetlands, a lot of leverage has been enjoyed by the feeding ground compared to the other wildlife feeding there.

Fig. 4.90 Ice Cream Hill in Cheorwon with an observatory

Fig. 4.91 Species-rich former district office of Jangdan-myeon in Paju city

Considerations for habitat conservation

The artificial structures are currently being used as habitats of wildlife animals, but it is the history and culture of the Korean peninsula that remain in ruins. Therefore, priority is cultural and ecological evaluation. A dilemma is to determine which is important, restoration of cultural relics or restoration of important habitats of animal species.

There is a need to make sure that the current habitats of wildlife animals that flourish in the artificial structures must not be interrupted. In order to prevent interruptions in corridors, stepping stones can be made, so it should establish a habitat network throughout the DMZ and CCZ.

4.2.3 *Type, Function, and Value of Wetlands in Photos of the DMZ*

Much of the DMZ consists of a patchwork of rice paddy fields, grasslands wetlands and forests. It is a convenient place to test the succession model of wetlands.

Type of Wetlands in the DMZ

A wide variety of wetlands have formed across the DMZ due to regional and local differences in topography, hydrology, soils, vegetation, water chemistry, and other factors. Although there are many different wetland types, they can be divided into two broad categories: NON-TIDAL WETLANDS and TIDAL WETLANDS.

NON-TIDAL WETLANDS

NON-TIDAL WETLANDS account for most of the wetlands of the DMZ and are found throughout the DMZ's interior beyond tidal effects:

Peatlands is inland wetlands containing thick deposits of slowly decaying plant material called peat. "Bogs" and "Fens" the major types of peatlands occur in the DMZ area. Fens are richer in nutrients and less acidic than bogs, and are more typically covered by sedges, willows, grasses, and reeds (Figs. 4.92, 4.93 and 4.94).ount for most of the wetlands of the DMZ and are found throughout the DMZ's interior beyond tidal effects.

Deepwater Swamps are wooded wetlands of the DMZ that have standing water for meet, if not all, of the growing season. Characteristic trees of these swamps are bald cypress, tupelo gum, and water oak (Figs. 4.95 and 4.96).

Inland Freshwater Marshes include a variety of wetlands that are full of soft-stemmed plants like grasses, rushes, cattails, and water lilies which are found in the DMZ area. Wet "mountain meadows" are high elevation freshwater wetlands found in forested and non-forested mountain regions (Figs. 4.97, 4.98, 4.99, 4.100, 4.101, 4.102).

Riparian Wetlands form on the floodplains of rivers and streams, and are often dry for portions of the growing season in the DMZ area (Figs. 4.103, 4.104, 4.105, 4.106, 4.107, and 4.108).

TIDAL WETLANDS

TIDAL WETLANDS are found along the DMZ's coasts within reach of the ocean's tides:

Tidal Salt Marshes develop near river mouths, behind barrier islands, in bays, on coastal plains, or in other areas affected by tides and protected from the full force of the surf. Typically vegetated by grasses and other emergent plants adapted to salt water, they can range from narrow fringes on steep shorelines to nearly flat expanses several kilometers wide (Figs. 4.109, 4.110 and 4.111).

Tidal Freshwater Marshes form inland from salt marshes and mangrove swamps, but are still affected by ocean tides. Grasses and floating-leaved aquatic plants typically dominate these wetlands, which are found in bays, inlets, and along tidal rivers (Figs. 4.112 and 4.113).

Fig. 4.92 Yongneup (dragon marsh) in Mt. Daeam, Yanggu county

Fig. 4.93 Peatlands in Paju Gyeonggi railroad site (in-depth survey required)

Fig. 4.94 Peatlands in Paju Gyeonggi railroad site (in-depth survey required)

Fig. 4.95 Willow tree, alder tree, Amur Maple community in the water

Fig. 4.96 Alder tree community in Sewol-cheon watershed

Fig. 4.97 Replacement wetlands in front of Geumgang gate

Fig. 4.98 Anho reservoir in Goseong county

Fig. 4.99 Daeseong reservoir in Paju city

Fig. 4.100 Eoryong reservoir in Paju city

Fig. 4.101 Paju story shooting range in Paju city is a source of life

Fig. 4.102 Wetland in Anho stream

Fig. 4.103 Riparian Wetland in Sacheon stream

Fig. 4.104 Riparian Wetland in Imjin river

Fig. 4.105 Riparian Wetland in Sami stream (1)

Fig. 4.106 Riparian Wetland in Sami stream (2) : Original meandering channel

Fig. 4.107 Riparian wetland in Namdae stream

Fig. 4.108 Riparian wetland in Yeokgok-cheon

Fig. 4.109 Ganghwa reed wetland

Fig. 4.110 Reed wetland in Ganghwa county

Fig. 4.111 Tidal Wetland in Han river of Paju city

Fig. 4.112 Donghae railroad area in Goseong county

Fig. 4.113 Wetland in Anho stream area, Goseong county, inflowing to the eastern coast of Korea

Function and Value of Wetlands in the DMZ

Environmental Quality

DMZ Wetland Which Regulates Flood and Erosion

Intertidal habitats such as salt marshes and mudflats provide buffer function from the tidal energy of the ocean to the coast. Also, a geological process such as a shingle beach and creation of beach plays an important function of protecting the coast.

Water Quality

Wetlands act as natural water purifiers, filtering sediment, and absorbing many pollutants in surface waters. In some wetland systems, this cleansing function enhances the quality of groundwater supplies as well.

Wildlife Habitat

Wetlands provide habitat for many species of amphibians, reptiles, birds, and mammals that are uniquely adapted to wet environments. Upland wildlife such as Chinese water deer commonly use wetlands for food and shelter. Wetlands are particularly vital to many migratory bird species.

Specialized Plant Habitat

Habitat for Threatened and Endangered Species

There are only 1700 ~ 2000 number of cranes are left in the world, and it is included in the IUCN Red List as an endangered species. Cranes of 200 ~ 300 are found as living in the DMZ, and 300–350 cranes are living in North Korea.

Many idle abandoned rice paddy fields have been converted into the wetlands of ecological importance

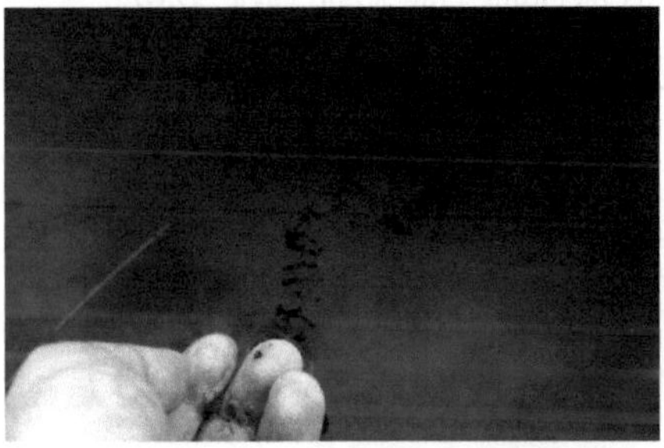

Socioeconomic Function and Value

Historical Value

Historical relics found in the DMZ area provide myth, meaning of place, regional characteristics, historical background, etc.

Ceasefire line from the east to the west represents the sadness of the separation, and the remaining of the Gyeonggi train shows the traces of time passing. Unlike the fast-changing cities in ROK, DMZ, that contains the appearances from the past, is the history and culture, and symbolizes the spirit of Korean people.

Recreational Opportunities

DMZ and CCZ areas have become more popular as people began to recognize its values. Paju city and Cheorwon county areas, in particular, have attracted many people with birds including cranes and eagle. DMZ experience tour has become one of the most well-liked tours in ROK.

Education

DMZ provides an important resource for general education to everyone. For instance, the area can be used to such education program as improving recognition of environmental issue. There are currently various visit programs in the CCZ areas.

Chapter 5
Species in the DMZ Area

Survey and analysis on species are to be conducted by category. It is even more important that surveys on specific species are carried out if the purpose is to manage and restore the habitat of such species. This chapter presents the results of the survey conducted on the animals and plants that thrive in the DMZ and the analysis of their respective habitats.

5.1 Overview of Survey and Analysis Methods of Species

5.1.1 The Flora and Vegetation Survey and Analysis

The Flora and vegetation are very important elements in the management and restoration of the environment. They served as an important determinant of the success or failure of the management and restoration initiatives implemented. Survey on flora and vegetation is the foundation of a management and restoration plan and thus, must be detailed as possible.

5.1.1.1 Flora Survey

In general, survey on flora covers all plant species distributed in targeted areas. According to the Guidelines on National Basic Survey of Natural Environment (The Korean Association for Conservation of Nature 1977), the procedure of the survey has to be divided into the following specific indicators; (a) field investigation, (b) specimen production process, (c) identification process, (d) to input the contents of survey sheet into plant species DB program, (e) specimen label printing process and, (f) completion of proof specimen.

In field investigation, selection of survey routes, actual field investigation, and collection are important. Survey routes are to be selected based upon where many

K.-G. Kim, *The Demilitarized Zone (DMZ) of Korea*,

DOI: 10.1007/978-3-642-38463-9_5,

species as possible can be collected and studied. In analyzing the existing data, the local geography was considered as well as its respective ecological habitats and landscapes such as; ridge of mountains, slopes and valleys, wetland, among other topographies peculiar to the research locale.

The actual survey on plant species carried out in the field is a comprehensive observation noting on everything that is present on site. These observations are written objectively and those that need to be scrutinized like the quality of soil and water, etc., are sent to the laboratory for further study. The name of plant species, its identification, and listing sequence are based on the species identified and illustrated in the atlas of plants and habitats.

5.1.1.2 Vegetation Survey

To prepare for a vegetation survey, the following materials and gadgets are to be made available namely; vegetation survey sheet, topographic map, plant book, camera, forest type map altimeter, altimeter, goniometer, clinometer, tape measure, altitude gauge, plant collection tools, writing supplies, and other things predetermined to be necessary. A commonly used vegetation survey method is the plant sociological method, the Braun-Blanquet Z-M school method. For more details, refer to *Method of Vegetation* (Kim 1987).

It is advisable that the existing vegetation map be drawn on the forest type map first and placed onto the topographic map. The existing vegetation map shall be written only within the scope of the investigation based on the forest type map and appearance survey. On the other hand, specific methods used in the vegetation survey shall be dependent on the characteristics of the main habitat in targeted areas. For example, it is advisable that the wetland habitat survey follows the Guidelines on National Inland Wetland Survey (Ministry of Environment 2001).

5.1.1.3 Analysis of the Vegetation Survey Results

Several important points shall be kept in mind when analyzing and summarizing the results gained from the on-site survey and from the laboratory findings. It is desired to ensure that the following procedures are carefully observed.

Mapping the survey results

Preparation of a vegetation map enables a reiteration analysis with other data (especially wildlife animals, etc.) by mapping the distribution of major communities. Therefore, it is desired that the map is prepared based upon the on-site survey results. Also, it is preferable to use the positioning system when protected species or special species are investigated in order to enter into correct locations. It is very helpful for the researcher to prepare the existing vegetation map as well as the potential vegetation map drawn in the environmental management and restoration plan.

Table 5.1 Various types of vegetation structure

Classification	Characteristics
Forest	Forests on the average of 4.6 m (15 ft) form at least 60 % of canopy
Woodland	Forests on the average of 4.6 m (15 ft) form 20–60 % of canopy
Plantation	Similar form with forests or woods but formed at regular intervals
Brush	Formed with trees and shrubs below 4.6 m (15 ft), high density of stems but various forms of crowns built
Fence rows	Trees and shrubs mixed along with boundaries such as roads, paddy fields, playgrounds, etc.
Wetland	High density plants dominant in overall low and wet grounds
Grassland	Dominant herbs
Field	Being cultivated or recently cultivated farmland

Arrangement of photographs

Photographs taken on-site are helpful in understanding the geographic information of the flora. Although it is necessary to write important plant species in the forest type map or topographic map on the spot, photographs are useful when reviewing missing points and in understanding the characteristics of habitat. Photographs are further needed in the presentation of data where significant features of the flora are to be emphasized. It is important that major plants are categorized by habitat. In order to utilize it effectively in targeted areas, photos must be sorted out by types of habitat, by specie groups, or by individual plant which can be further categorized by vegetation types like; forest, wood, orchard, wetland, or paddy field as shown in Tables 5.1, 5.2, 5.3 and Fig. 5.1.

5.1.2 Fauna Survey

5.1.2.1 Insect Survey

Insects have various habitats for instance, a variety of insects thrive in certain trees. There are specific insects thriving in the tree's underground, another species in its surface ground and other different varieties can be found up to the branches and leaves of the tree itself.

Establishment of Survey Range

In general, survey, which is all areas in survey subjects can be divided by the entire area; by section where field survey is being carried out and by survey point (light trap, trap collection, etc.). It is desired that the scope of field survey is pre-established in consideration of the subjected environment's vegetation and landform in order to be able to observe as many insects as possible. To balance the field survey points, the following should be considered;

Table 5.2 A sheet arranged and analyzed based on survey results—Example 1

Site no.	I-I	Subregion	I	Name of administrative district	Pyeongsan-ri, Hyunkyung-myeon, Muan-gun, Jeonnam
Ecosystem type	Stream (Link canal)				
Water environment	**Water temperature**		**pH**	**Dissolved oxygen**	**Electrical conductivity**
	17.6 ℃		7.2	9.8 mg/L	423 μS/cm
Plant species	Hydrophyte (1)	Submerged plants			
		Floating plants			
		Floating leaved plants			
		Emerged plants		*Zizania latifolia*	
	Hygrophyte (3)	*Persicaria thunbergii*, *Rumex crispus* L., *Stellaria aquatica SCOP*			
	Certain plants	None			
Vegetation cross section			**Picture of current status**		
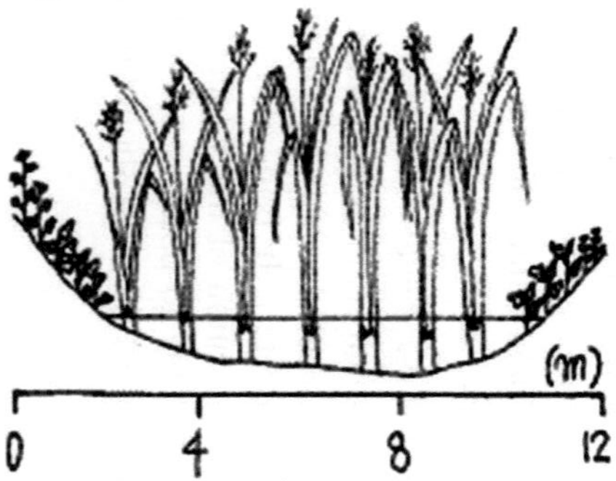					
Habitat evaluation	Zizania latifolia, an emerged plant is flourishing in link canal with slow velocity Habitat for a variety of plants required to be formed due to low diversity of plant species				
Management practices	Introduction to the natural purification technique for water improvement Application of method of construction for development of riffle and pool				

(continued)

Table 5.2 (continued)

Site no.	I-I	Subregion	I	Name of administrative district	Pyeongsan-ri, Hyunkyung-myeon, Muan-gun, Jeonnam
※ Changes of plant cover degree (%) depending on habitat cross section					
Landform	Left bank Slope	Water route			Right bank Slope
Distance (m)	2	4	6	8	10	12
Persicaria thunbergii	60					60
Artemisia princeps var. orientalis (Pampan.) Hara	30					
Rumex crispus L.	20					10
Stellaria aquatica SCOP.	10					40
Zizania latifolia		60	60	70	60	
Height (cm)	30	210	230	200	210	30
Water level (cm)	30	30	40	25		

Table 5.3 Relationships between major communities and infrastructure of ecological pond in Seoul

Name of Group		Location	Water level	Slope	Shore protection material	Relativeness to planting
Emerged plant	*Phragmites australis*	About 5–6 m from the edge	About ~60 cm aboveground part	1:8–1:16	Fertile soil	Distribution proliferation
	Typha angustifolia L.	About 2 m in the water from edge	About 10–50 cm	1:2.6–1:16	Fertile soil	Distribution proliferation
Floating leaved plant	*Nymphaea tetragona* var.	1 m in the water-center from the edge	20–120 cm	1:3.5–1: 6	Fertile soil	Distribution proliferation
Hygrophyte	*Miscanthus sacchariflorus* Benth.	1.8–5 m out of water from the edge	Above ground part	About 1:18	Fertile soil	Distribution proliferation
	Salix gracilistyla Miq.	100 cm out of water from the edge	Above ground part	About 1:4	Fertile soil, natural stone	Located in planting point

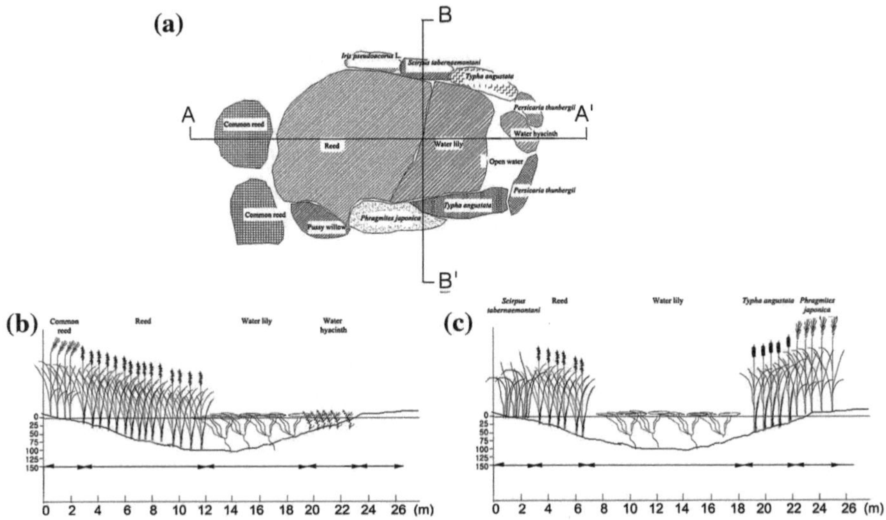

Fig. 5.1 A sheet arranged and analyzed based on survey results—Example 2. **a** Horizontal distribution map of major groups in ecological pond announced by Seoul city. **b** A-A′ lined section **c**. B-B′ lined section

- Areas with big/small artificial variations and areas with changed alluvial environment by river improvement.
- Various areas to be set in consideration of the situation of the stream (vegetation, downstream, river bed materials, width of stream, velocity and form inside the bank), urban area, farmland, mountains, etc.
- Areas with a variety of environments such as grassland, forest, sand, and pebble, to be set as a survey point.
- Balance among river mouth, downstream, midstream, and upstream.

Survey Time and Frequency

In principle, survey time includes spring, summer and autumn, as no insect is observed in their natural environments in winter, as they may be hibernating, dugged themselves into deeper grounds for warmth or are into any forms of adaptations in order to survive. It is desired that the survey will be carried out more than three times per specific time of survey for time series analysis. It is further suitable for a survey to be carried out when they are developing at different stages of growth depending on the region where they thrive. It is important to classify and categorize per survey subject according to its growth and development within the specified survey time. Thus, through this procedure, analyses are made accordingly. Proper reference with a conglomeration of experts will be necessary and the comparison of the present findings with previous findings as well, could give a clear understanding of the adaptive mechanisms and successions of specific species.

Office Investigation

Understanding entomofauna is primarily expected. This is done by using 1:2,500 plane map, 1:25,000–1:50,000 topographic map, aerial photographs, and vegetation maps of targeted areas. Information and guidance from experts who had ample experience in the specific survey point area is very helpful and a review of research abstracts and related literatures on this field of exploration is really a must. In undergoing an on-site survey at targeted areas, analyzing light traps, trap collections, and pictures can be based on the knowledge gained from literatures and abstracts to enhance a general understanding of the overall condition of the survey points.

Survey methods for various insect species

Field surveys shall be conducted in various areas in the census tract by using random collection, sweeping, beating, light trap, pit-fall trap, etc. As many as possible, proper collections shall be made utilizing appropriate survey methods, depending on the type of insect and on other specific local situations. Specific methods of collection are as follows.

Random collection

The most common method used by researchers when targeting various species in many different environments, is the insect net. It has a diameter of about 30–40 cm. The number of researchers should be identified and survey time shall be appropriately assigned to each researcher, depending on the vastness of their assigned survey spot.

Sweeping

This is a method that shakes and snatches plants like that of sweeping floors using an insect net swung ten times. After the tenth swinging, the content of the net is investigated. This technique is used primarily in forests, shrubs, and grasslands. Insects trapped in the net are transferred into a vinyl bag which has cotton inside wet with a fungicide, ethyl acetate, or ammonia. It is advisable that quantitative collection is done in 30 sets and done three times before changing location.

Beating

Branches or grasses are beaten ten times by a stick so that fallen insects are collected in the net. Again, the collected insects are put into a poison bottle. The same method is repeated three times for the various trees.

Light Trap

Certain types of insects are difficult to collect so that their attraction to light is the best way to catch them. This is also used in areas where there are no moonlight and strong winds that penetrate the specific research location. Dark spots in the research area shall be identified where this can be the most effective method to use.

In the light trap method, the researcher uses a transparent mercury lamp or a 'black light' for about 3 h in the specified location. A curtain and a bucket will be used as traps for fallen insects. Appropriate technique shall be used between the curtain and the bucket in such a way that the insects could be trapped. There should be no other light nearby. The brightness of illuminant should be set up.

Table 5.4 Classification scheme in consideration of relativeness between flora and insects

Classification item	Function	Insects
Pollinators	Visits flowers, eats honey and pollen, and is in charge of moving the pollen	Butterflies, honeybees, syrphid
Juice suckers	Inoculates juice from plants	Cicada, soldier bug, aphid, leafhopper
Leaf feeders	Eats leaves of plants	Butterfly, fireworm, leaf beetle
Predators parasitoids	Feeds on other insects or small animal, preys on them	Harmonia axyridis (Pallas), Parasitoids bee
Ants	Does social life and is omnivorous	Camponotus japonicus Mayr, Formica japonica
Hygienic insect	Has harmful effects on human's health or gives an unpleasant feeling	Mosquito, tussock moth, chironomid
Aquatic insect	Spends all or most of its life cycle in the water or on the surface of the water	Semblis phalaenoides (Linne), mayfly, dragonfly
Others	Ants, insects that burrow into the woody parts, soil insects, mycophagous insects, corrosive insects are included	Longicorn beetle, Melanagromyzasp, leaf mite, springtail, corrosive mulberry thrips

Proper directions should be followed so as not to disturb or stress the surrounding habitat and other inhabitants.

Pit-fall Trap

Nocturnal or geophilous insects are the targets of this method. It is a method that buries a cup or an empty case in the ground with its height at the same level as the ground and, put feed in it so that insects are driven into the trap. Spoiled meat (fish or beef) and others are used as feed. About 30 traps are set in various environments and in different locations in the target area. The cups are picked up on the next day to collect the insects that were trapped inside. If this method is used for longer periods of time, a hole shall be made on the sides of a 1.5 L plastic bottle that has antifreeze in it to prevent trapped insects from decomposing. This requires a schedule for routine check up for data collection.

Other Methods

The Malaise trap is for insects that fly with the wind. A flat yellow-pan-vessel with water and a sticky material is placed in strategic areas where wandering insects such as flies fly over. This method is used depending on the purpose of survey and local situations.

Relativeness Between Flora, Habitat, and Insect Species

Flora and habitat shall also be considered for insect surveys. Table 5.4 is an example of the relativeness between flora and insects.

5.1.2.2 Fish Survey

Fish survey shall be divided into; (a) on-site understanding and, (b) identification after collection. In this section, methods of collection are briefly introduced.

When collecting fish, a triangular net (length 50 cm, network 10 × 10 cm) and two gill nets (50 m each, network 10 × 10 cm) are installed in different sites and collected for 1–10 days per season. Tour fishing survey shall also be conducted once a year to check into the condition of bass-introduced fish species that cannot be collected by nets.

Where direct collection is possible, cast net (network 6 × 6 cm) and skimming nets (network 5 × 5 cm), fishing net (network 5 × 5 cm) shall be used for collection. Identification shall be done directly on-site with reference to the classification key by Kim (1997), Choi (1990), etc.

5.1.2.3 Amphibians and Reptiles Survey

Amphibians and reptiles survey can be either direct check or indirect check.

Direct Check Method

Anuran Amphibians

Batrachia can be collected by using an insect net, including entities that are moving at 10 m intervals from side to side, in the rocks or in streams, waterways, and near wetlands along accessible areas near the targeted location.

Urodele Amphibians

Small rocks in slow streams can be raised to see larvae of urodele (*Hynobius leechii*) and *Onychodactylus fischeri*. Spawned eggs can be found in small depressions where water is collected. Adults can be observed in fallen old trees in the shade where broad-leaved forests or dents in the rocks.

Reptile

Lacertid (Lizard)

Lacertid can be found in abandoned fields, near grassland, rivers, and sunshiny places by raising rocks, and collected moving individuals on the roadside using an insect net.

Serpents

Snake tongs and insect nets near a hole in lowland areas of forest edge and abandoned fields are used to collect serpents, and they can also be seen in stone piles in farmlands, specifically under rocks and slates.

Chelonian

A pair of binoculars (Nikon 10 × 40 5.9°, 8 × 32 10°) are used to see chelonian living in lakes, ponds, flumes, or streams.

Indirect Check Method

Call of anuran amphibians

Amphibian (Batrachia) may be identified by their call as they cry in groups at night near rice paddies and crop fields on the waterside and in waterways and depressions.

Reptile vestigium (Skin)

A reptile, especially serpents, frequently slips out of its skin while growing. If it is well-nourished, molted skins can be collected to check species.

Audience Survey

For species that were not collected or observed during the survey period, existence of species can be checked with nearby residents or audiences.

5.1.2.4 Birds Survey

The Birds survey relies on point and line census. Observation can be made within a horizontal range of 25 m along with the survey route by using binoculars (Nikon 10 × 40 5.9°, 8 × 32 10°) or by following their type of call.

To understand the distribution of birds in the area, the survey is based on line census. However, point census is considered more effective than line census, but still depending on the situation. Though both point and line census can be used at the same time.

Point record method

As a method of checking birds nearby, one or more strategic survey points of observation are set in order to be able to observe a specific range. Using binoculars or an outdoor telescope from a certain point of observation for about half an hour is considered as an "observation point" in this method. One must stay in a certain point and record observations, before transferring to another point of observation.

Line census

This is a method of observing the frequency of birds appearing in a specific skyline or area, or the frequency of bird calls heard along the survey line. It is desired that various environments become the targets of the survey considering the landform, vegetation, among other pertinent considerations in the line census method. An ocular survey can be conducted by just walking along the survey line at about 1.5–2.5 km/h.

5.1.2.5 Mammals Survey

Direct and indirect observations are carried out for mammals. Details are as follows.

Direct observation method

Observation

Species are directly observed on the spot while moving along the selected survey route. Information collected regarding the surroundings, altitude and, number of individuals is critically written down.

Capture

Capture is the most accurate method in checking mammals. A trap is used to capture them alive. Sorex, moles, and rodents that are not observed through observation and field-sign are targeted. Various forms of traps exist depending on targeted animals. The trap can be left all night long. The species can be collected, identified, and written down every morning. The following are traps that can be used in capturing mammals.

Live trap and Snap trap

Sherman live trap or Sherman snap trap are used to capture rodents and small mammals. Dead fish and anchovy may be used in the seaside area, while peanuts and sausages may be used as a decoy at the riversides in forestlands. That is because feed is dependent on the condition of inhabiting environment. Twenty traps per survey area are adequate. The places can be selected where rodents are active such as; thickets near hole of houses, coppice bush, and fallen trees but avoid open area. Also, the spots can be included where there is soft soil, a bit dark silva, and wet grassland in order to see as many species as possible.

Pitfall Trap

It is mainly targeted to sorex (insectivora) which can be captured by a small plastic cup or paper cup as it is not so good at jumping. A pitfall trap can be set up by burying the cup where fallen leaves are piled and the soil is soft and a lid is put on it to prevent rainwater from sinking the cup.

Mole Trap

Moles are seen abundantly in spring when they reproduce and from late autumn to early winter they expand since they do not hibernate. Traps should be set up where moles are active in dense and new mole tunnels are situated.

Indirect observation method

Field sign method

This is to identify species through footprint, excretion along alleys, and roads. The habitat of mid-sized and large-sized mammals are also identified using field-sign method as mammals are mostly nocturnal, and the researcher can conduct an audience survey with the residents in the area to verify unidentified species during the survey period. The following indicators are noted in field-sign method:

- Fecal material or sight (foot track, feather etc.).
- Tree nests (squirrels, vole).
- Ground nests (mole, vole).
- Rodent mounds and tunnels.

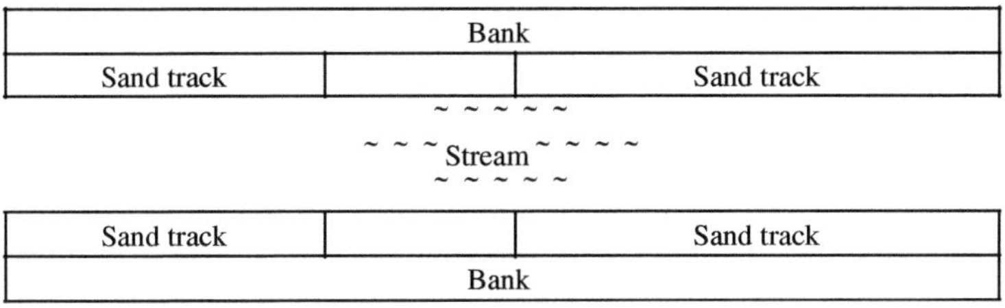

Fig. 5.2 Where to and how to set up sand track on the riverside (Jae-Han Shim 2003)

- Squirrel middens (caches).
- Ground dens (rabbit, bear, weasel, raccoon dog, badger, small-eared cat).
- Snag (mammals).
- Deer rubs.

Sand track

Although field-sign method is mainly used to identify species, by writing down the number of individuals observed and, by checking their footprint, there is however, one drawback. The field-sign method cannot quantify the species (Korea National Park Authority 2002). Therefore, sand track is used. It is an artificial method used to estimate the relative density of targeted mammals identified in targeted areas.

Setup and its scale

Sand track is set up as a rectangular form with 10 $\times$ 2 m on riverside flatland where two forests are connected to each other, and where an alluvial island is developed as in Fig. 5.2.

The vegetation should be removed on the surface layer with a sickle once decided where to set it up and make sure no plant is left over. All pebbles should be removed and the layer should be ploughed over by the depth of 5 cm, the surface is puddle and it should be softened to ensure that footprint is left after wildlife animals pass by (Korea National Park Authority 2002).

The footprints left on the sand track should be checked the next morning and the direction, type, and size should be written down. Connected footprints are regarded as identical to one individual mammal regardless of its number. All sand tracks should be removed after investigation so that nothing is duplicated on the next day's observation.

5.1.2.6 Analysis of the Fauna Survey Results

Biota and habitat type

The type of species and the number of amphibians, reptiles, birds, and mammals per collection should be written down, and the biological, physical, and chemical elements of habitat should be evaluated. The data should be compared to the typical types of species (to already known biological behaviors) to find adaptive behaviors and investigate any possible environmental disturbance.

Diversity index of species

Diversity of species is also referred as species heterogeneity. High diversity means that there are plenty of identical or similar species. Diversity of species is a criterion of group stability and degree of group normality. The reason why varied species exist is because of the healthy interaction between species even if the species vary. Such interactions appear intricately including energy movement, food web, relationship of predator, competition, ecological niche, etc.

Therefore, it is essential that the interrelation between species and between habitats be established, quantitatively analyzed and ecologically measured.

Ecological measurement through biological standard extraction

There are important measurements in describing population and groups of species such as; density, dominance index (DI), relative density, and biodiversity.

Dominance index

The deeper the change of environment becomes, the more certain species become superior. It is an index calculated from the perspective that relative ratios of certain dominant species in groups can be used as a clear index toward the change of environment. Dominance value is calculated by the standing stock of the number of individuals per each survey point (McNaughton 1967). The formula is:

$$\mathrm{DI} = \mathrm{ni}/N$$

DI: Dominance value, N: Total number of individuals, ni: The number of individuals i

Biodiversity index

It is calculated through Shannon–Weaver function (Pielou 1966) introduced by Margalef (1968) information theory. It states that the relative balance of species' abundance in groups and the number of individuals represent the complexity of groups. The formula is:

$$D' = \_\mathrm{Pi}\ :\ \mathrm{Ratio}\ (\mathrm{ni}/N)\ \text{of individuals that belong to}\ I$$

(N: Total number of individuals in group, ni: The number of each species).

Evenness index

Evenness is expressed as the actual ratio of the maximum value of each index, and the diversity index becomes maximum when the number of individuals of all species in group is identical. EI is then the degree of uniformity in terms of species in groups which is calculated by the formula of Pielou (1975) as follows:

$$E' = D'/\mathrm{Ln}(S)$$

E´: Evenness, D´: Diversity, *S*: Total number of species.

Richness index

It is an index which indicates only the status of groups with total number of individuals and species . The richness index is based on an assumption that the environment is good and as higher the value is, the more plentiful the composition of species becomes. It is calculated by typical index introduced by Margalef (1958) as follows:

$$R' = (S - 1)/\text{Ln}(N)$$

R´: Richness, *S*: Total number of species, *N*: Total number of individuals.

Mapping habitat distribution and spawning grounds

Natural ecological map is a map which classifies the value about each ecology (forest, stream, farmland, town, lake, and ocean) in consideration of natural view, rarity, endemism, abundance, and representativeness. Produced in the Netherlands in 1968, Natural ecological map is an ecological and environmental map in which vegetation and birds are marked on the 25 km^2 of national grid system. Ecological map or Biotope mapping has been generally accepted in some European countries. Ecological zoning map is developed to be utilized as a useful spatial information bank to combine, comprehend, restore, and manage the information and knowledge about natural ecosystem.

When a modeling should be done at landscape ecology level for amphibians and reptiles, however, relevant areas can be evaluated by forecasted distribution map developed by Row and Yuill (1997) of Virginia University.

Altitude and location are observed and marked on the distribution map about the amphibians, reptiles, and mammals' spawning ground, in accordance with the criteria of Shim (1997a, b).

Territory Analysis

Basic concept

Territory can be defined as all or part of the radius of action of wildlife animals, which functions as a protector from invasion of other animals.

Territorial access

The reason why access is needed is as follows:

First, wildlife animals have appropriate activities and physiology against environment. By understanding all this, you can reinforce the management of wild animals by providing territorial access as the main requirement.

Second, territorial access helps you to look after young individuals, let adult individuals live and prepare reasons to protect and manage their environments.

Comparative analysis on method of measurement and requirement habitat area

The author has compared and analyzed in detail the required habitat areas for Amur tiger in Sikhote Alin in Russia and *Grus japonensis* in Cheorwon in Korea. Figure 5.3 illustrates an individual's radius of action and the required habitat areas of minimum 50 individuals. The radius of action of an Amur tiger in Sikhote Alin

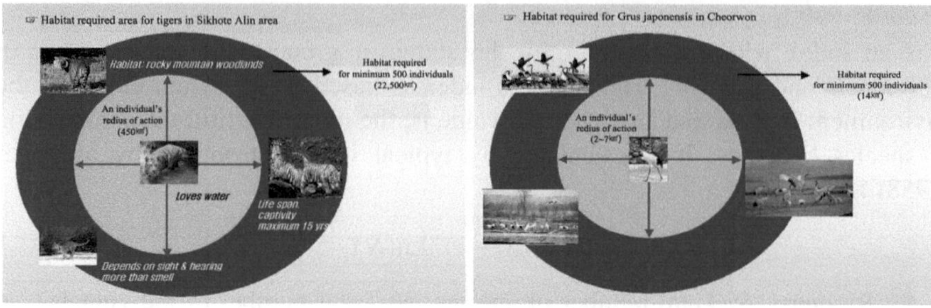

Fig. 5.3 Comparison of required habitat area for Amur tigers in Sikhote Alin area to that for *Grus japonensis* in Cheorwon (Kwi-Gon Kim et al. 2003)

is 450 km^2. These tigers like water, live in a dense forest, rely on hearing and seeing rather than smelling, and can live up to 15 years at the maximum. The habitat area that 50 individuals require is 500 km^2, which is quite large.

On the other hand, the feeding area radius of action of a *G. japonensis* in Cheorwon is 2–7 km^2. A lot of other birds visit this area and they have established a family of wildlife that harmoniously lives together. The area of inhabitation takes only less than 1 % of the radius of action. The feeding area of 50 individuals required is only about 5–8 km^2. If we look at the ecological characteristics, they are different from each other and territorial access shall also be different in accordance with the requirements of the habitat. Therefore, at least 22,500 km^2 of forest, where territorial water and rock are, shall be planned for Amur tigers of minimum 500 individuals in Sikhote Alin area. In terms of *G. japonensis*, at least 14 km^2 of the habitat area near wetland where people do not interfere and where farmland is nearby is recommended to be secured for minimum 500 individuals.

5.2 The Application

5.2.1 Methods Used in Analyzing Data

In this book, the items below were selected based on the Guidelines on The 3rd National Natural Environment Survey (National Institute of Environmental Research, Ministry of Environment 2006).

1. Landscape.
2. Habitat.
3. Vegetation and flora.
4. Land insects.
5. Fish and aquatic organism.
6. Amphibian and reptile.
7. Birds.

8. Mammals.
9. Ecological connectivity.
10. Cultural assets.

5.2.2 Tool Development

A Survey form was developed and tested in the DMZ and CCZ. It is composed of two parts namely; the on-site form and the audience form.

An on-site survey form has been developed based upon the habitat survey in Leicester, UK.

On the other hands, an audience survey form has been developed for soldiers in command at the site.

5.2.3 Survey Results

The ecological study in the DMZ and CCZ was conducted by the Environmental Ecology Research Group that the author led and the results are summarized in Table 5.5. Although biota living in the entire DMZ area could not be fully identified in such a research timeframe, the author found that the area had more biota and more number of individuals than the CCZ had. In the future, the biota data in DMZ and CCZ may need to be combined. It is important to note that the characteristics of the study on characterization of the wetlands'aquatic plants which are truly representative of DMZ and CCZ.

5.2.3.1 Composite Table

Tables 5.6, 5.7, 5.8, 5.9.
Figures 5.4, 5.5, 5.6, 5.7, 5.8, 5.9, 5.10, 5.11 show distribution of plant and animal species by major natural area of the DMZ and CCZ respectively, which is the same information in Table 5.9, in graphic forms.

5.2.4 Total Species by Western-Central Area of the DMZ

This section addresses total species in the western-central area of the DMZ, number of plant, mammals, birds, reptiles and amphibians, insects, fish, and fresh invertebrates by major natural area of the west-central area of the DMZ.

Table 5.5 Summary of status of biota in DMZ and CCZ by survey agency

Agent	Survey area	Period	Method	Results	
				DMZ	CCZ
Forestry Researcher	▫ Major forest, valley, stream, island of CCZ	1995–2000	Literature search and on-site survey		▫ Total : 1,866 species Plants : 1,194 Animals : 672 (Mammals : 52, Birds : 201, Reptiles and amphibians : 28, Freshwater fish : 67, Insects : 324)
UNDP, Seoul National University	▫ DMZ : Eoryong reservoir in DMZ, Near Panmunjeomin DMZ, Upstream of Sacheon river in DMZ and wetland nearby ▫ CCZ : Paju city Jangdan-myeon Jangdan peninsula, Paju city Jangdan-myeon Nosang-ri Dukhyeon-dong Paju city Jindong-myeon Hapo-ri Jung-dong, Paju city Jindong-myeon Yongsan-ri Bangchuk-dong, Paju city Gunnae-myeon Jeomwon-ri Dohwa-dong, etc.	1996–1999	Focused on on-site survey	▫ Total : 87 species Plants : 31 Animals : 56 (Mammals : 2, Birds : 14, Reptiles and amphibians : 5, Freshwater fish : 15, Insects : Uninvestigated)	▫ Total : 699 species Plants : 470 Animals : 229 (Mammals : 11, Birds : 58, Reptiles and amphibians : 22, Freshwater fish : 24, Insects : 114)

(continued)

Table 5.5 (continued)

Agent	Survey area	Period	Method	Results	
				DMZ	CCZ
Gyeonggi Line North–South Road Environment Joint Research Group	▫ DMZ : Paju city Jangdan-myeon the Dora Mountain ri, Paju city Jangdan-myeon Ha-dong, Paju city Jangdan-myeon Gubong-chon, etc. ▫ CCZ : Paju city Gunnae-myeon Baekyeon-ri, Paju city Gunnae-myeon Gunryang-dong, etc.	September 25, 1999–December 1	Literature search and on-site survey	▫ Total : 259 species Plants : 183 Animals : 76 (Mammals : 4, Birds : 32, Reptiles and amphibians : 10, Freshwater fish : Uninvestigated, Insects : 30)	▫ Total : 312 Plants : 225 Animals : 87 (Mammals : 5, Birds : 39, Reptiles and amphibians : 16, Freshwater fish : 7, Insects : 20)
Gyeonggi Line North–South Road Environmental Impact Survey	▫ DMZ : Paju city Jangdan-myeon the Dora Mountain ri, Paju city Jangdan-myeon Ha-dong, Paju city Jangdan-myeon Gubong-chon, etc. ▫ CCZ : Paju city Gunnae-myeon Baekyeon-ri, Paju city Gunnae-myeon Gunryang-dong, etc.	June, 2001–2003	Literature search and on-site survey	▫ Total : 434 species Plants : 183 (Surveyed community only) Animals : 251 (Mammals : 4, Birds : 32, Reptiles and amphibians : 4, Freshwater fish : Uninvestigated, Insects : 211)	▫ Total : 497 Plants : 225 (Surveyed community only) Animals : 272 (Mammals : 2, Birds : 35, Reptiles and amphibians : 2, Freshwater fish : 22, Insects : 211)
Gyeonggi Line Post-Environmental Effect Survey	▫ DMZ : Paju city Jangdan-myeon Dora Mountain ri, Paju city Jangdan-myeon Ha-dong, Paju city Jangdan-myeon Gubong-chon, etc. ▫ CCZ : Paju city Gunnae-myeon Baekyeon-ri, Paju city Gunnae-myeon Gunryang-dong, etc.	July, 2004–December, 2005	On-site survey	▫ Total : 463 species Plants : 196 Animals : 267 (Mammals : 14, Birds : 32, Reptiles and amphibians : 4, Freshwater fish : 6, Insects : 211)	

(continued)

Table 5.5 (continued)

Agent	Survey area	Period	Method	Results	
				DMZ	CCZ
Donghae railroad line Post-Environmental Impact Survey	▫ DMZ : Goseong-gun Hyunnae-myeon, Songdojin-ri (GP Line ~ MDL) ▫ CCZ : Goseong-gun Hyunnae-myeon Sacheon-ri	January, 2004–December, 2005	On-site survey	▫ Total : 463 species Plants : 163 Animals : 300 (Mammals : 12, Birds : 92, Reptiles and amphibians : 8, Freshwater fish : 10, Insects : 178)	
DMZ Neighboring Area (Natural Ecosystem Survey in Goseong-gun)	▫ CCZ and some DMZ : Restricted area for civilians from Gangwon province Goseong-gun Hyunnae-myeon Songhyeon-ri to SLL railings, South-North 4.7 km, width 4.4 km area	October 15, 2001–Present	On-site survey	▫ Total : 357 species Plants : 288 Animals : 69 (Mammals : 10, Birds : 31, Reptiles and amphibians : 18, Freshwater fish : 10, Insects : Uninvestigated)	
DMZ Panmunbul Ecosystem Survey	▫ DMZ Panmunbul area	November, 2005–March, 2006	On-site survey	▫ Total : 75 species Plants : 6 Animals : 69 (Mammals : 4, Birds : 27, Reptiles and amphibians : Uninvestigated, Freshwater fish : 20, Insects : 2)	
The 1st DMZ Internal Ecosystem Survey (Ministry of Environment, DMZ Ecosystem Research Group)	▫ DMZ : Gokreung-cheon in Paju estuary of the Han river, Manwoo-ri area in downstream of Paju Imjin river, Paju Daeseong reservoir, Paju Jangdan-myeon Geogok-ri area and near the Dora Mountain Station, Paju Sacheon-cheon and Sewol-cheon, Yeoncheon Sami stream river basin and Yeoncheon plain, Pilseung bridge near the Imjin river and Gowang mountain Choryang valley	November, 2008	On-site survey	▫ Total : 348 species Plants : 208 Animals : 140 (Mammals : 16, Birds : 47, Reptiles and amphibians : 4, Fish : 31, Insects : 25, Freshwater invertebrate : 17)	

(continued)

Table 5.5 (continued)

Agent	Survey area	Period	Method	Results	
				DMZ	CCZ
The 2nd DMZ Internal Ecosystem Survey (Ministry of Environment, DMZ Ecosystem Research Group)	▫ DMZ : Cheorwon-gun Cheorwon-eup Deoksan-ri area, Cheorwon-eup Jungse-ri area, Cheorwon-eup Woljeong-ri area, Cheorwon-gun Geunbuk-myeon dandelion field area, Geunbuk-myeon Yugok valley field area, Geunbuk-myeon Mando field area, Cheorwon-gun Geundong-myeon Gimhwa Namdae area, Geundong-myeon Gimhwa Namdae Gwangsam plain area, Geundong-myeon Taehyun-ri area, Cheorwon-gun Wonnam-myeon Ssangyong dam area, Cheorwon-gun Wondong-myeon Jupa-ri area	September, 2009	On-site survey	▫ Total : 450 species Plants : 334 Animals : 116 (Mammals : 11, Birds : 45, Reptiles and amphibians : 5, Fish : 20, Insects : 87, Freshwater invertebrate : 30)	

Table 5.6 Total Biota species discovered in the first DMZ internal ecosystem survey

Category	Name of species	Number of species
Vegetation and flora	*Phacelurus latifolius* (Steud.) Ohwi, *Phragmites australis*, *Lespedeza bicolor* Turcz., *Artemisia princeps* var. orientalis (Pampan.) Hara) (*Artemisia scoparia* Waldst. & Kit., *Artemisia princeps* var. orientalis (Pampan.) Hara, American *Aster pilosus* Willd.), *Carex scabrifolia* Steud., *Cardamine amaraeformis* Nakai)(*Wasabia japonica* (Miq.) Matsum., *Conyza canadensis* (L.) Cronquist, *Cnidium japonicum* Miq., *Polypogon fugax* Nees ex Steud., *Erigeron annuus* (L.) Pers., *Lespedeza bicolor* Turcz.)(*Lycopus europaeus*, *Aster tripolium* L., *Miscanthus sacchariflorus* var. gracilis Y.N.Lee, *Quercus acutissima* CARRUTH, *Miscanthus sacchariflorus* var. gracilis Y.N.Lee)(*Miscanthus sacchariflorus* Benth., *Miscanthus sacchariflorus* var. gracilis Y.N.Lee, wild millet, *Calamagrostis pseudophragmites* (Haller f.) Koeler, *Artemisia princeps* var. orientalis (Pampan.) Hara, *Robinia pseudoacacia* L., *Sium suave* Walter, *Phellodendron amurense* Rupr., *Quercus mongolica* Fisch. ex Ledeb., *Cyperus difformis* L., *Artemisia keiskeana* Miq., *Symplocos chinensis* for. pilosa (Nakai) Ohwi, *Quercus aliena* Blume, *Artemisia feddei* H.Lev. & Vaniot, *Morus bombycis* Koidzumi, *Prunus padus* L. for. Padus, four seasons *Artemisia capillaris* Thunb., hazel, Pinus rigida Mill., *Cardamine amaraeformis* Nakai)(Wasabia japonica (Miq.) Matsum., *Ailanthus altissima*, *Acer tataricum* subsp. ginnala (Maxim.) Wesm., *Persicaria thunbergii* (Siebold & Zucc.) H. Gross ex Nakai, *Zanthoxylum schinifolium* Siebold & Zucc., Betula davurica Pall. *Ambrosia trifida* L., Elaeagnus typical, a wild rosebush, pine, *Calamagrostis epigeios* (L.) Roth, cocklebur, *Lepidium apetalum* Willd.)(*Wasabia japonica* (Miq.) Matsum., *Setaria viridis* (L.) P.Beauv., Aster subulatus MICHX, *Dendranthema boreale* (Makino) Ling ex Kitam., *Rorippa palustris* (Leyss.) Basser, Swinhoe´s willow-warbler, *Rubus oldhamii* var.), *Imperata cylindrica* var. koenigii (Retz.) Pilg., *Trifolium pratense* L., *Mosla dianthera* (Buch.-Ham. ex Roxb.) ex Maxim., *Medicago lupulina* L., *Quercus dentata* Thunb. ex Murray, *Rhododendron mucronulatum* Turcz. var. mucronulatum, *Humulus japonicus* Sieboid & Zucc., *Quercus variabilis* Blume, *Plantago asiatica*, *Chloris virgata* Sw., *Spiraea thunbergii* Siebold ex Blume, *Viburnum erosum* var. taquetii (H.Lev.) Rehder, *Viola rossii* Hemsl., *Rhamnus davurica* Pall., *Artemisia gmelini* Weber, *Cyperus microiria* Steud., *Lonicera maackii* (Rupr.) Maxim., *Prunus sargentii* Rehder, *Trifolium pratense* L., *Styrax japonicus* Siebold & Zucc., *Oxalis corniculata* L., *Oenothera biennis* L., *Rhus tricocarpa* Miq., *Scirpus wichurae* var. asiaticus (Beetle) T.Koyama, *Forsythia koreana* (Rehder) Nakai, *Clematis apiifolia* A. P. DC., *Scirpus karuizawensis* Makino, *Zizania latifolia*, *Phragmites japonica* Steud, *Paris verticillata* M.Bieb., *Tofieldia nuda* Maxim., *Rubus idaeus* var. microphyllus Turcz., *Spodiopogon cotulifer* (Thunb.) Hack., *Salix koreensis* Andersson, *Rubus crataegifolius* Bunge, *Spodiopogon cotulifer* (Thunb.) Hack., *Alnus japonica* Steud., *Calamagrostis arundinacea* (L.) Roth), *Melica onoei* Franch. & Sav., *Alnus japonica* Steud.)(Alnus sibirica Fisch. ex Turcz., *Scirpus maritimus* L., *Typha angustifolia* L., *Juncus effusus* var. decipiens Buchenau, Pueraria lobata (Willd.) Ohwi, *Veratrum maackii* var. japonicum (Baker) T.Schmizu), *Juncus effusus* var.	208

(continued)

Table 5.6 (continued)

Category	Name of species	Number of species
	decipiens Buchenau, *Cirsium japonicum* var. maackii (Maxim.) Matsum., *Carex lanceolata* Boott, *Trapa japonica* Flerow, *Ceratophyllum demersum* L., Carex glabrescens Ohwi, *Nymphoides peltata* (J.G.Gmelin) Kuntze, *Elsholtzia splendens* Nakai, *Carex tuminensis* Kom., *Hydrilla verticillata* (L.f.) Royle, *Celastrus orbiculatus* Thunb., *Carex tuminensis* Kom., *Potamogeton crispus* L., *Lonicera japonica* Thunb., *Carex siderosticta* Hance, *Vallisneria natans* (Lour.) H.Hara, *Persicaria sagittata* (L.) H.Gross ex Nakai, *Vicia unijuga* A.Braun, *Sagina japonica* (Sw.) Ohwi, *Potentilla anemonefolia* Lehm., PhytolaCCZ americana L., *Potamogeton malaianus* Miq., *Stellaria aquatica* (L.) Scop, *Viola kapsanensis* Nakai, *Smilax nipponica* Miq., *Potentilla fragarioides* var. major, *Betula costata* Trautv., *Clerodendrum trichotomum* Thunb. ex Murray, *Aster koraiensis* Nakai, *Leonurus japonicus* Houtt., *Aristolochia contorta* Bunge, *Ligustrum salicinum* Nakai, *Fraxinus rhynchophylla* Hance, *Isodon inflexus* (Thunb.) Kudo, *Rubus hirsutus* Thunb., *Salix hallaisanensis* H.Lev. for. Hallaisanensis, *Lindera glauca* (Siebold & Zucc.) Blume, *Saussurea pulchella* (Fisch.) Fisch., *Arachniodes aristata* (G. Forst.) Tindale, *Solidago virgaurea* var. coreana Nakai,. *Arachniodes aristata* (G.Forst.) Tindale, *Chelidonium majus* var. asiaticum (Hara) Ohwi, *Solidago serotina* Aiton, *Stephanandra incisa* (Thunb.) Zabel, *Atractylodes ovata* (Thunb.) DC., *Isachne globosa* (Thunb.) Kuntze, *Lindera obtusiloba* Blume, *Rhus javanica* L., Aster scaber Thunb., *Eupatorium japonicum* Thunb. ex Murray, *Populus davidiana* Dode, *Salix integra* Thunb., *Meliosma myriantha* Siebold & Zucc., *Salix gracilistyla* Miq., *Salix caprea* L., *Hypericum ascyron* L., *Potentilla fragarioides* var. major, *Smilax sieboldii* Miq. for. Sieboldii, *Agrimonia pilosa* Ledeb., *Pueraria lobata* (Willd.) Ohwi, *Myriophyllum verticillatum* L., *Peucedanum terebinthaceum* (Fisch.) Fisch. ex DC., *Thalictrum filamentosum* var. tenerum (Huth) Ohwi. *Equisetum palustre* L., *Serratula coronata* var. insularis (Iljin) Kitam. for. Insularis. *Silene jenisseensis* Willd., *Carex humilis* var. nana (H.Lev. & Vaniot) Ohwi. *Pica pica*. *Staphylea bumalda* DC., *Youngia japonica* (L.) DC. Wasabia japonica (Miq.) Matsum., Lespedeza maximowiczii C.K.Schneid., Persicaria perfoliata (L.) H.Gross. Lespedeza cyrtobotrya Miq., Clematis terniflora var. mandshurica (Rupr.) Ohwi. Rubia akane Nakai. *Lespedeza bicolor* Turcz., *Celtis edulis* Nakai. *Actinidia polygama* (Siebold & Zucc.) Planch. ex Maxim., *Arachniodes aristata* (G.Forst.) Tindale. *Adenophora verticillata* var. angustifolia Regel. *Oplismenus undulatifolius* (Ard.) P.Beauv., *Aralia cordata* continentalis. *Melampyrum setaceum* (Maxim.) Nakai. *Woodsia ilvensis* (L.) R.Br., *Crepidiastrum sonchifolium*. *Dioscorea tenuipes* Franch. & Sav., *Sigesbeckia glabrescens* (Makino) Makino, *Panicum bisulcatum* Thunb., *Quercus xurticaefolia* Blume. *Heracleum moellendorffii* Hance. *Aralia elata* (Miq.) Seem., *Hemerocallis dumortieri* Morren. *Adenophora remotiflora* (Siebold & Zucc.) Miq., *Phasianus colchicus* tree. *Aconitum jaluense* Kom. subsp. Jaluense. *Smilax china* L., *Euonymus alatus* for. ciliatodentatus (Franch. & Sav.) Hiyama. *Clematis trichotoma* Nakai. *Petasites japonicus* (Siebold & Zucc.) Maxim., *Syringa reticulata* var. mandshurica (Maxim.) H.Hara. *Ulmus davidiana* var. japonica (Rehder) Nakai	

(continued)

Table 5.6 (continued)

Category	Name of species	Number of species
Mammals	*Hoplopleura affinis* (Burmeister), *Crocidura suaveolens* shantungensis, *Rattus norvegicus* Berkenhout, *Hydropotes inermis*, *Lutra lutra*, *Felis catus*, *Sus scrofa*, *Mustela nivalis*, *Nyctereutes procyonoides*, *Apodemus peninsulae* peninsulae, *Micromys minutus*, *Felis bengalensiseuptilura*, *Microtus fortis* pelliceus, *Mustela sibirica*, *Mogera woguracoreana*, *Meles meles*	16
Birds	*Artemisia princeps* var. orientalis (Pampan.) Hara bird, *Hypsipetes amaurotis*, *Paradoxornis webbianus*, *Parus palustris*, *Circus aeruginosus*, *Mergus merganser*, *Aquila chrysaetos*, *Larus argentatus*, *Aegypius monachus*, *Dryocopus martius*, *Streptopelia orientalis*, *Accipiter nisus*, *Alauda arvensis*, *Coccothraustes coccothraustes*, *Accipiter gentilis*, *Buteo buteo*, *Turdus naumannieunomus*, *Pica pica*, *Pica pica*, *Troglodytes troglodytes*, *Aix galericulata*, white-naped crane, *Grus japonensis*, *Ardea cinerea*, *Urosphena squameiceps*, *Uragus sibiricus*, *Tachybaptus ruficollis*, *Egretta albamodesta*, *Egretta albamodesta*, *Anas poecilorhyncha*, *Anser albifrons*, *Veratrum oxysepalum* Turcz., *Phoenicurus auroreus*, *Passer montanus*, *Pica pica*, *Anser fabalisserrirostris*, *Anser cygnoides*, *Phalacrocorax capillatus*, *Phasianus colchicus*, *Corvus corone* Carrion Crow, *Emberiza cioides*, *Corvus corone* Carrion Crow, *Parus ater*	47
Reptiles and Amphibians	*Glandirana rugosa*, *Rana coreana*, *Pelodiscus sinensis*, *Hyla japonica*	4
Fish	*Acanthogobius hasta* (Temminck et Schlegel), *Zacco koreanus*, *Rhinogobius brunneus*, *Chelon hematocheilus*, *Hemibarbus mylodon* (Berg), *Acheilognathus lanceolatus*, *Tridentiger bifasciatus*, *Acheilognathus lanceolatus*, *Zacco platypus*, *Carassius auratus*, *Macropodus chinensis*, *Pseudorasbora parva*, *Oryzias sinensis*, *Odontobutis platycephala*, *Hemibarbus labeo*, *Aphyocypris chinensis*, *Rhynchocypris oxycephalus*, *Silurus asotus*, *Misgurnus anguillicaudatus*, *Siniperca scherzeri* (Albino type), *Squalidus gracilis* majimae, *Microphysogobio yaluensis*, *Acheilognathus lanceolatus*, *Hemibarbus longirostris*, *Pseudogobio esocinus*, *Gobiobotia macrocephala*, *Iksookimia koreensis*, *Pungtungia herzi*, *Coreoleuciscus splendidus*, *Psudopungtungia tenuiooupus*	31
Insects	*Colias erate* (Esper). *Artogeia rapae* (Linnaeus). Ants. *Parapolybia varia* (Fabricius). *Sympetrum infuscatum* (Selys). *Sympetrum eroticum* eroticum (Selys). *Oxya japonica* japonica (Thunberg). *Harmonia axyridis* (Pallas). *Polygonia c-aureum* (Linnaeus). *Sympetrum parvulum* Bartenef. *Sympetrum depressiusculum* (Selys). *Nannophya pygmaea* Rambur. *Ognevia longipennis* Shiraki. *Vespa mandarinia* Cameron. *Vespula koreensis* koreensis Radoszkowski. *Callipogon relictus* Semenov-Tian-Shansky. *Aquarius paludum* (Fabricius). *Ducetia japonica* (Thunberg). *Acrida cinerea* cinerea (Thunberg). *Geoclemys* (Chinemys) *reevesii Aiolocaria hexaspilota* (Hope). *Meloe proscarabaeus* Linne. *Vespula flaviceps* lewisii (Cameron).	25
Freshwater Invertebrate	*Notonecta triguttata* Motschulsky, *Muljarus japonicus* (Vuillefroy), *Neocaridina denticulate*, *Semisulcospira libertine*, *Semblis phalaenoides* (Linne), *Daphnia pulex*, *Ranatra chinensis* Mayr, *Laccotrephes japonensis*, *Lethocerus deyrollei*, *Ranatra chinensis* Mayr, *Nepa hoffmanni* Esakii, *Ranatra chinensis* Mayr, *Freshwater shrimp*, *Sesarma dehaani*	17
Total		348

Table 5.7 Total Biota Species Discovered in the 2nd DMZ Internal Ecosystem Survey

Category	Name of species	Number of species
Vegetation and Flora	*Zanthoxylum schinifolium* Siebold & Zucc., *Perilla frutescens* var. acuta Kudo, Lactucascariola L., *Ambrosia trifida* L., *Bidens tripartita* L., *Ailanthus altissima* (Mill.) Swingle for. altissima., *Oenothera biennis* L., American *Aster pilosus* Willd., *Erigeron annuus* (L.) Pers., *Fallopia dumetorum* (L.) Holub, *Vicia villosa* Roth, *Senecio vulgaris* L., *Ambrosia artemisiifolia* L., *Galinsoga parviflora* Cav., *Chloris virgata* Sw., *Conyza canadensis* (L.) Cronquist, *Erechtites hieracifolia* Raf., *Wasabia japonica* (Miq.) Matsum., American *Panicum bisulcatum* Thunb., *Trifolium pratense* L., *Taraxacum officinale* Weber, *Robinia pseudoacacia* L., *Euphorbia supina* Raf., *Solidago altissima* L., *Medicago lupulina* L., *Lespedeza bicolor* Turcz., *Crassocephalum crepidioides* (Benth.) S.Moore, *Plantago lanceolata* L., *Persicaria orientalis* (L.) Spach, *Trifolium pratense* L., *Nepeta cataria* L., *Cirsium nipponicum* (Maxim.) Makino, *Epimedium koreanum* Nakai, *Artemisia princeps* var. orientalis (Pampan.) Hara *Aster fastigiatus* Fisch., *Polygonatum robustum* (Korsh.) Nakai., *Campanula glomerata* var. dahurica Fisch. ex KerGawl., *Aristolochia contorta* Bunge, *Sparganium erectum* L., *Philadelphus schrenckii* Rupr., *Cirsium setidens* (Dunn) Nakai, *Thalictrum rochebrunianum* var. grandisepalum (H.Lev.) Nakai, *Berberis koreana* Palibin, *Weigela subsessilis* L.H.Bailey, *Paulownia coreana* Uyeki, *Populus davidiana* Dode, *Salix koriyanagi* Kimura for. Koriyanagi, *Aruncus dioicus* var. kamtschaticus (Maxim.) H.Hara, *Cirsium vlassovianum* Fisch., *Quercus aliena* Blume, *Quercus dentata* Thunb. ex Murray, *Quercus variabilis* Blume, *Quercus acutissima* CARRUTH, *Rhus tricocarpa* Miq., *Viburnum erosum* var. taquetii (H.Lev.) Rehder, *Fraxinus rhynchophylla* Hance, *Aster scaber* Thunb., *Synurus deltoides* (Ait.) Nakai, *Solidago virgaurea* subsp. asiatica Kitam. ex Hara var. asiatica, *Quercus mongolica* Fisch. ex Ledeb., *Lespedeza bicolor* Turcz., *Lespedeza cyrtobotrya* Miq., *Prunus japonica* var. nakaii (H. Lev.) Rehder, *Bidens tripartita* L., *Artemisia princeps* var. orientalis (Pampan.) Hara, *Aster meyendorfii* (Regel & Maack) Voss), *Artemisia princeps* var. orientalis (Pampan.) Hara, *Aster incisus* Fisch.), *Eupatorium japonicum* Thunb. ex Murray, *Persicaria thunbergii* (Siebold & Zucc.) H.Gross ex Nakai, *Phragmites japonica* Steud, *Miscanthus sacchariflorus* Benth., *Miscanthus sacchariflorus* var. gracilis Y.N.Lee, *Miscanthus sacchariflorus* var. gracilis Y.N.Lee, *Alnus japonica* Steud., *Salix koreensis* Andersson, *Acer tataricum* subsp. ginnala (Maxim.) Wesm., *Peucedanum terebinthaceum* (Fisch.) Fisch. ex DC., *Picris hieracioides* var. koreana Kitam., *Pimpinella brachycarpa* (Kom.) Nakai, *Zizania latifolia*, *Pueraria lobata* (Willd.) Ohwi, *Adenophora verticillata* var. angustifolia Regel, *Persicaria sagittata* (L.) H.Gross ex Nakai, *Silene jenisseensis* Willd., Cyperus iria L., *Phragmites australis*, *Salvinia natans* (L.) All., *Ranunculus tachiroei* Franch. & Sav., *Typha orientalis* C.Presl,	334

(continued)

Table 5.7 (continued)

Category	Name of species	Number of species
	Actinidia polygama (Siebold & Zucc.) Planch. ex Maxim. Tree, *Scirpus karuizawensis* Makino, *Dendranthema boreale* (Makino) Ling ex Kitam., *Arachniodes aristata* (G.Forst.) Tindale, *Arthraxon hispidus*, Chloris virgata Sw, *Oenothera odorata*, *Ixeridium dentatum*, *Lactuca raddeana* Maxim., *Prunus sargentii* Rehder, *Celastrus orbiculatus* Thunb., *Allium sacculiferum* Maxim., *Juncus effusus* var. decipiens Buchenau, *Meliosma myriantha* Siebold & Zucc., *Betula davurica* Pall., *Prunus serrulata* var. spontanea (Maxim.) E.H.Wilson, *Thalictrum rochebrunianum* var. grandisepalum (H.Lev.) Nakai, *Isodon japonicus* (Burm.) Hara, *Inula britannica* var. japonica (Thunb.) Franch. & Sav., *Vitis amurensis* Rupr., *Aconitum jaluense* Kom. subsp. Jaluense, *Persicaria perfoliata* (L.) H.Gross, *Pedicularis verticillata* L., *Geranium sibiricum* L., *Sanguisorba tenuifolia* Fisch. ex Link var. tenuifolia	
Mammals	*Felis bengalensiseuptilura*, *Hydropotes inermis*, *Nyctereutes procyonoides*, *Meles meles*, *Mustela sibirica*, *Sus scrofa*, *Lepus coreanus*, *Mogera woguracoreana*, *Crocidura suaveolens shantungensis*, *Felis catus*, *Sciurus vulgaris*	11
Birds	*Egretta albamodesta*, *Ardea cinerea*, *Actits hypoleucos*, *Anas poecilorhyncha*, *Aix galericulata*, *Corvus corone* Carrion Crow, *Corvus corone* Carrion Crow, *Pica pica*, *Garrulus glandarius*, *Accipiter gentilis*, *Accipiter nisus*, *Streptopelia orientalis*, *Veratrum oxysepalum* Turcz., *Parus palustris*, *Emberiza sulphurata*, *Artemisia princeps* var. orientalis (Pampan.) Hara bird, *Dendrocopos leucotos*, *Dendrocopos kizuki*, *Paradoxornis webbianus*, *Pica pica*, *Hypsipetes amaurotis*, *Passer montanus*, *Hirundo rustica*, *Hirundo daurica*, *Motacilla alba*, *Cyanistes varius*, *Pica pica*, *Phasianus colchicus*, *Picus canus*, *Motacilla cinerea*, *Emberiza elegans*, *Emberiza sulphurata*, *Aegithalos caudatus*, *Eurystomus orientalis*, *Acrocephalus bistrigiceps*, *Phoenicurus auroreus*, *Phylloscopus coronatus*, *Arachniodes aristata* (G.Forst.) Tindale	45

(continued)

Table 5.7 (continued)

Category	Name of species	Number of species
Reptiles and Amphibians	*Hyla japonica*, *Rana nigromaculata*, *Glandirana rugosa*, *Rana catesbeiana*, *Takydromus amurensis*, *Rhabdophis tigrinus*, *Elaphe schrenckii*, *Elaphe schrenckii*	10
Fish	*Rhynchocypris oxycephalus*, *Zacco koreanus*, *Zacco platypus*, *Carassius auratus*, *Acheilognathus rhombeus*, *Acheilognathus lanceolatus*, *Squalidus gracilis majimae*, *Gnathopogon strigatus* (REGAN), *Hemibarbus labeo*, *Pungtungia herzi*, *Microphysogobio* springerI Banarescu et Nalbant, *Pseudogobio esocinus*, *Coreoleuciscus splendidus*, *Iksookimia koreensis*, *Coreoperca herri* (Herzenstein), *Silurus asotus*, *Odontobutis intereupta*, *Odontobutis platycephala*	20
Insects	*Leptidea amurensis*	86
Benthic Micro-invertebrate (freshwater invertebrate)	*Ephemera strigata* Eaton, *Ephemera orientalis* McLachlan, *Molanna moesta*, Notopsyche KUa) KUa, *Anax parthenope* Julius, *Cercion calamorum* calamorum, *Epeorus pellucidus*, *Palaemon paucidens* De Haan, *Laccotrephes japonensis*, *Nepa hoffmanni* Esakii, *Aquarius paludum* (Fabricius), *Gerris latiabdominis* Miyamoto, Tipula KUb KUb, *Semblis phalaenoides* (Linne) KUa, *Semblis phalaenoides* (Linne) KUa, *Micronecta scholtzi*, *Caridina denticulata* denticulata De Haan, *Calopteryx atrata* Selys, *Copera annulata*, *Cybister brevis*	30
Total		537

Table 5.8 Stauts of protected species in DMZ and CCZ

Category	DMZ West Bioregional approach (Gimha, Paju, Gimpo)	DMZ Central Bioregional approach (Yeoncheon, Cheorwon, Hwacheon)	DMZ East Bioregional approach (Yanggu-, Inje, Goseong)	Total number of species
Vegetation and Flora	*Aristolochia contorta* (Korea Forest Service)	*Nepeta cataria* L. (Korea Forest Service), *Cirsium japonicum* var. maackii (Maxim.) Matsum., (Korea Forest Service), *Epimedium koreanum* (Korea Forest Service), *Artemisia princeps* var. orientalis (Pampan.), *Aster fastigiatus* Fisch. (Korea Forest Service), *Polygonatum robustum* (Korsh.) Nakai. (Korea Forest Service), *Campanula glomerata* (Korea Forest Service), *Aristolochia contorta* (Korea Forest Service). *Sparganium erectum* L, (Korea Forest Service)	*Thuja koraiensis* Nakai (Korea Forest Service), *Anemone koraiensis* Nakai (Korea Forest Service), *Epimedium koreanum* (Korea Forest Service), *Pueraria lobata* (Willd.) Ohwi (Korea Forest Service), *Rodgersia podophylla* A.Gray (Korea Forest Service), *Acanthopanax senticosus* (Korea Forest Service), *Glehnia littoralis* Fr. Schm. (Korea Forest Service), *Syringa velutina var. kamibayashii* (Korea Forest Service), *Syringa reticulata* var. mandshurica (Maxim.) H.Hara (Korea Forest Service), *Scopolia japonica* Maxim. (Ministry of Environment,Korea Forest Service), *Scrophularia koraiensis* Nakai (Korea Forest Service), *Patrinia saniculaefolia* Hemsl.(Korea Forest Service), *Hanabusaya asiatica* (Nakai) Nakai (Korea Forest Service), *Hanabusaya asiatica* (Nakai) Nakai Flower (Korea Forest Service), *Sinosenecio koreanus* (Kom.) B.Nord. (Korea Forest Service), *Lilium distichum* Nakai ex Kamib. (Korea Forest Service), *Lilium leichtlinii* var. *tigrinum* (Korea Forest Service), *Trillium kamtschaticum* Pallas.ex Pursh.(Korea Forest Service)	27 (except duplication)
	1	9	19	

(continued)

Table 5.8 (continued)

Category	DMZ West Bioregional approach (Gimha, Paju, Gimpo)	DMZ Central Bioregional approach (Yeoncheon, Cheorwon, Hwacheon)	DMZ East Bioregional approach (Yanggu-, Inje, Goseong)	Total number of species
Mammal	*Felis bengalensis euptilura* (Ministry of Environment), *Lutra lutra* (Ministry of Environment,National Heritage)	*Naemorhedus goral* (Ministry of Environment, National Heritage), *Felis bengalensis euptilura* (Ministry of Environment)	*Lutra lutra* (Ministry of Environment, National Heritage), *Naemorhedus goral* (Ministry of Environment, National Heritage), *Moschus moschiferus parvipes* (National Heritage), *Felis bengalensis euptilura* (Ministry of Environment)	4 (except duplication)
	2	2	4	
Birds	*Egretta eulophotes* (Ministry of Environment,National Heritage), *Grus japonensis* (Ministry of Environment, National Heritage), *Grus vipio* (Ministry of Environment), *Platalea minor* (Ministry of Environment, National Heritage), *Anser fabalis* (Ministry of Environment), *Aegypius monachus* (Ministry of Environment, National Heritage), *Aquila chrysaetos* (Ministry of Environment, National Heritage), *Anser cygnoides* (Ministry of Environment, National Heritage), *Accipiter soloensis* (National Heritage), *Accipiter nisus* (National Heritage),	*Grus vipio* (Ministry of Environment, National Heritage), *Grus japonensis* (Ministry of Environment,National Heritage), *Cygnus cygnus* (National Heritage, Ministry of Environment), *Aix galericulata* (National Heritage), *Accipiter soloensis* (National Heritage), *Falco tinnunculus* (National Heritage), *Otus scops* (National Heritage), *Ninox scutulata* (National Heritage), *Dryocopus martius* (National Heritage, Ministry of Environment), *Ixobrychus eurhythmus* (Ministry of Environment), *Buteo buteo* (Ministry of Environment), *Falco subbuteo* (Ministry of Environment), *Gallicrex cinerea*	*Haliaeetus pelagicus* (Ministry of Environment, National Heritage), *Accipiter gentilis* (National Heritage), Falco tinnunculus (National Heritage), *Otus scops* (National Heritage), *Cygnus olor* (Ministry of Environment, National Heritage), *Accipiter soloensis* (National Heritage), *Aix galericulata* (National Heritage), *Cygnus cygnus* (National Heritage), *Haliaeetus albicilla* (National Heritage,), *Aegypius monachus* (Ministry of Environment, National Heritage)	31 (except duplication)

(continued)

Table 5.8 Stauts of protected species in DMZ and CCZ

Category	DMZ West Bioregional approach (Gimha, Paju, Gimpo)	DMZ Central Bioregional approach (Yeoncheon, Cheorwon, Hwacheon)	DMZ East Bioregional approach (Yanggu-, Inje, Goseong)	Total number of species
	Circus aeruginosu (National Heritage), *Aix galericulata* (National Heritage), *Falco tinnunculus* (National Heritage), *Bubo bubo* (National Heritage), *Falco subbuteo* (Ministry of Environment), *Accipiter gentilis* (National Heritage, Ministry of Environment), *Buteo buteo* (Ministry of Environment), *Circus aeruginosus* (National Heritage, Ministry of Environment)	(Ministry of Environment), *Charadrius placidus* (Ministry of Environment), *Terpsiphone atrocaudata* (Ministry of Environment)		
	18	9	10	
Reptiles and Amphibians	*Elaphe schrenckii* (Ministry of Environment). *Pelophylax chosenicus* (Ministry of Environment). Geoclemys (Chinemys) *reevesii* (Ministry of Environment)	*Elaphe schrenckii* (Ministry of Environment)	*Rana amurensis* (Ministry of Environment). *Hynobius leechi* (Ministry of Environment). Dinodon rufozonatum) (*Elaphe schrenckii* (Ministry of Environment)	6 (except duplication)
	3	1	3	

(continued)

Table 5.8 (continued)

Category	DMZ West Bioregional approach (Gimha, Paju, Gimpo)	DMZ Central Bioregional approach (Yeoncheon, Cheorwon, Hwacheon)	DMZ East Bioregional approach (Yanggu-, Inje, Goseong)	Total number of species
Fish	*Acheilognathus lanceolatus* (Ministry of Environment). *Gobiobotia macrocephala* (Ministry of Environment). *Psudopungtungia tenuiooupus* (Ministry of Environment). *Hemibarbus mylodon* (Berg) (National Heritage). *Siniperca scherzeri* (Albino type) (National Heritage). *Rhodeus uyekii*. (Endemic species) *Squalidus gracilis* majimae. (Endemic species) Microphysogobio springerI Banarescu et Nalbant (Endemic species). Odontobutis intereupta (Endemic species) 9	*Acheilognathus lanceolatus* (Ministry of Environment) 1	*Pungitius sinensis* (Ministry of Environment). *Cottus hangiongensis* Mori (Ministry of Environment) 2	11 (except duplication)
Insects	*Parnassius bremeri Bremer* (Ministry of Environment). *Nannophya pygmaea* Rambur (Ministry of Environment) 2	Undefined	Undefined	2
Total number of species	35	22	33	81 (except duplication)

Table 5.9 Distribution of plant and animal species by major natural area of the DMZ and CCZ

	Total	Vegetation	Mammals	Birds	Amphibians reptiles	Insects	Fish	Invertebrate
Panmunjeom area (1996–1999)	699	470	11	58	22	114	24	–
Gyeonggi line area (September 1999)	312	225	5	39	16	20	7	–
Gyeonggi line area (2001–2003)	434	183	4	32	4	211	–	–
Gyeonggi line area (2004)	463	196	14	32	4	211	–	–
Goseong area (October 2001)	357	288	10	31	18	–	10	–
Donghae railroad line (2004–2005)	463	163	12	92	8	178	10	–
Panmunbul area (2005–2006)	75	6	4	27	–	20	2	–
Western area (November 2008)	348	208	16	47	4	25	31	17
Central area (September 2009)	502	334	11	15	5	87	20	30
Donghae railroad line (September 2010)	269	223	8	11	8	15	4	–

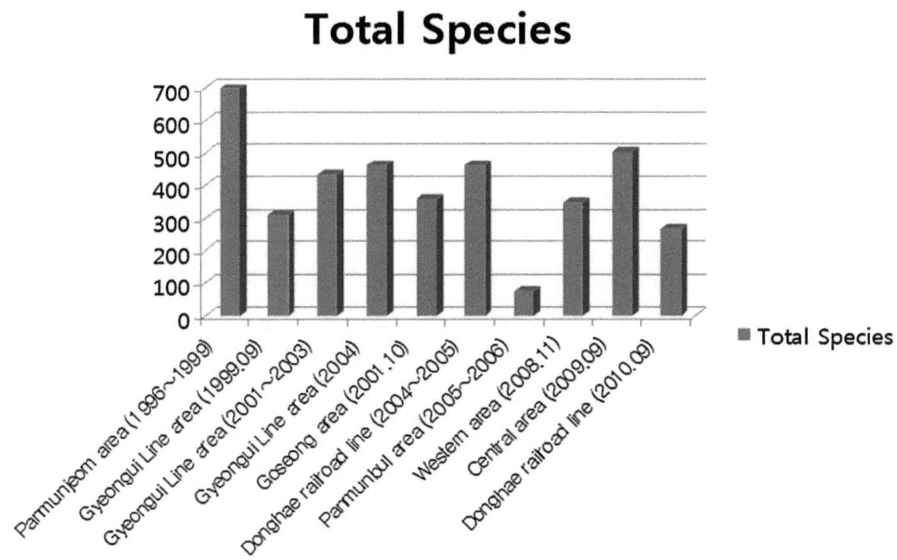

Fig. 5.4 Total species by major natural area

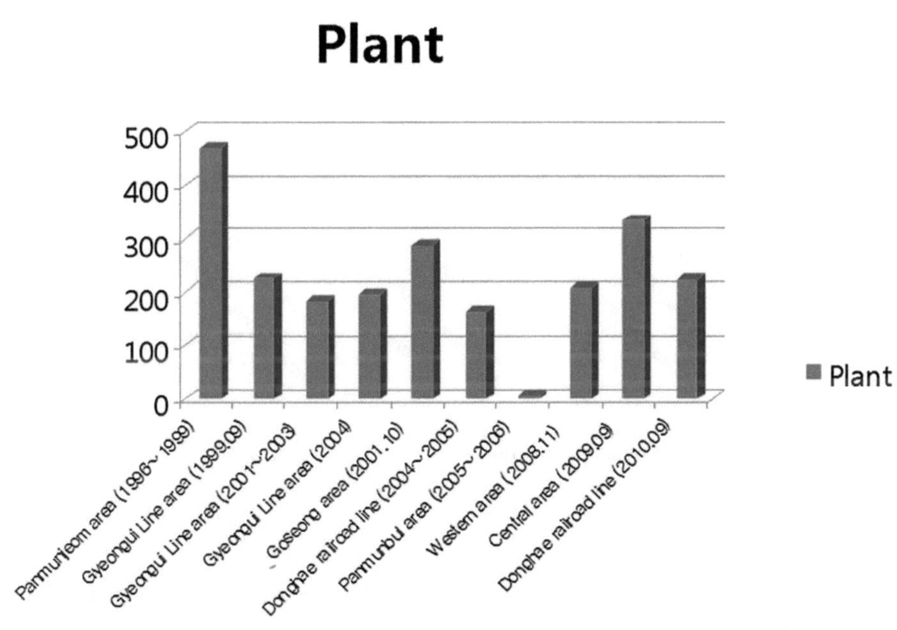

Fig. 5.5 Plant distribution by major natural area

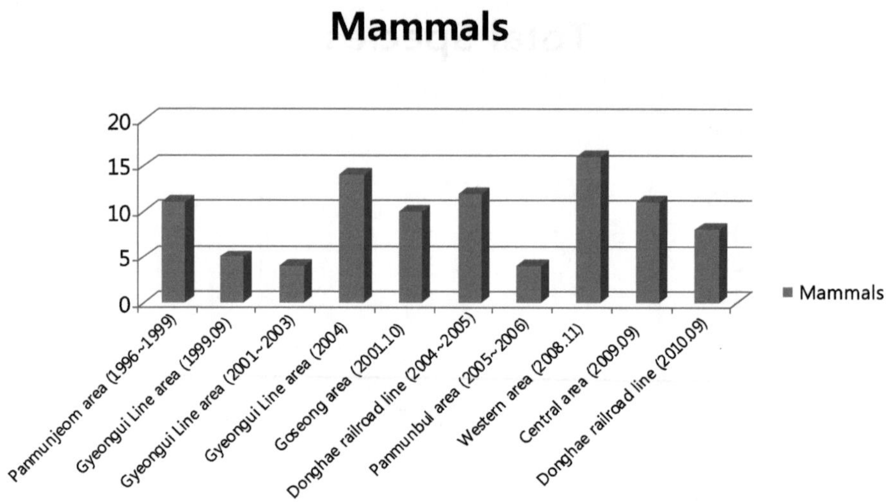

Fig. 5.6 Mammal distribution by major natural area

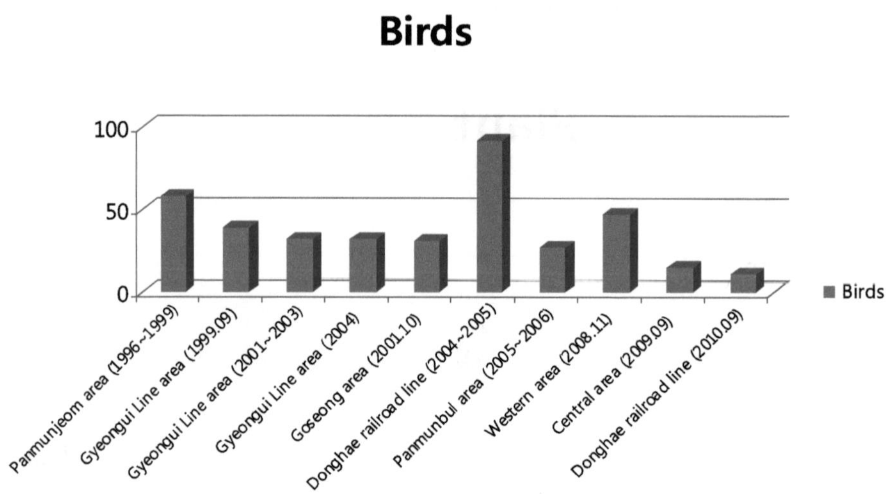

Fig. 5.7 Bird distribution by major natural area

5.2.4.1 Total Species

Total species by western-central area of the DMZ is shown in Table 5.10, while the same information is depicted in graph forms of Figs. 5.12, 5.13, 5.14, 5.15, 5.16, 5.17, 5.18 and 5.19.

Table 5.10, Fig. 5.12.

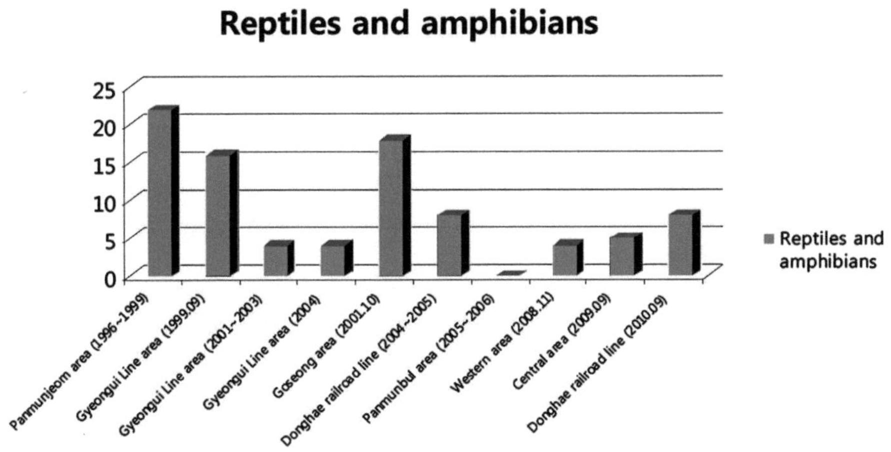

Fig. 5.8 Reptiles and amphibians distribution by major natural area

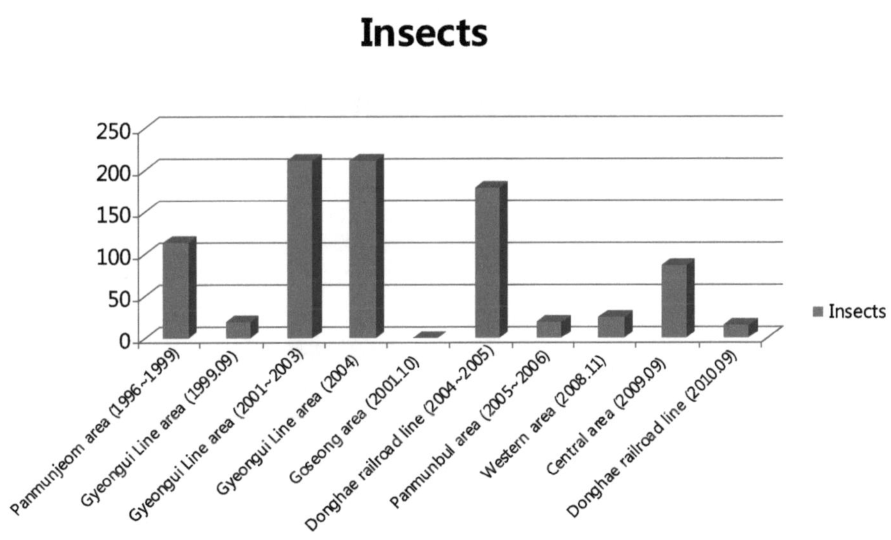

Fig. 5.9 Insects distribution by major natural area

Plants
Figure 5.13.
Mammals
Figure 5.14.
Birds
Figure 5.15.

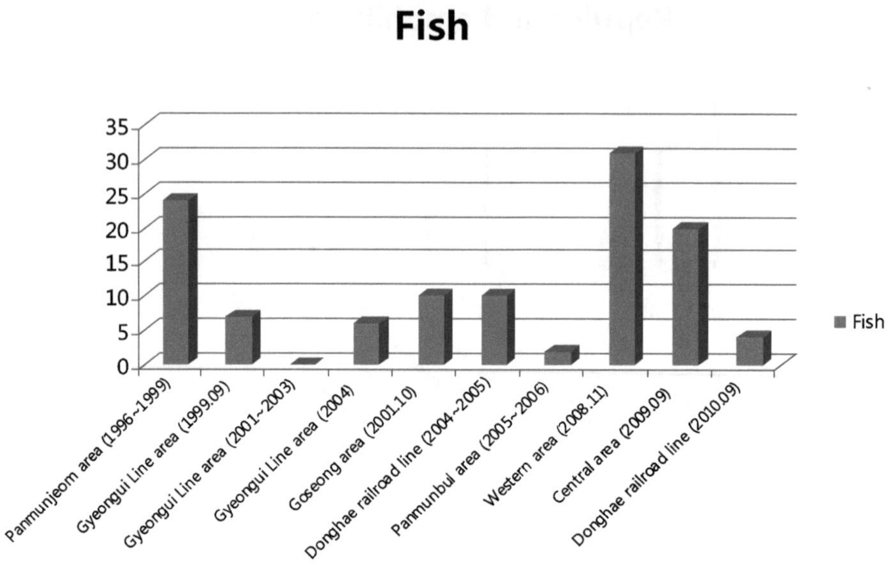

Fig. 5.10 Fish distribution by major natural area

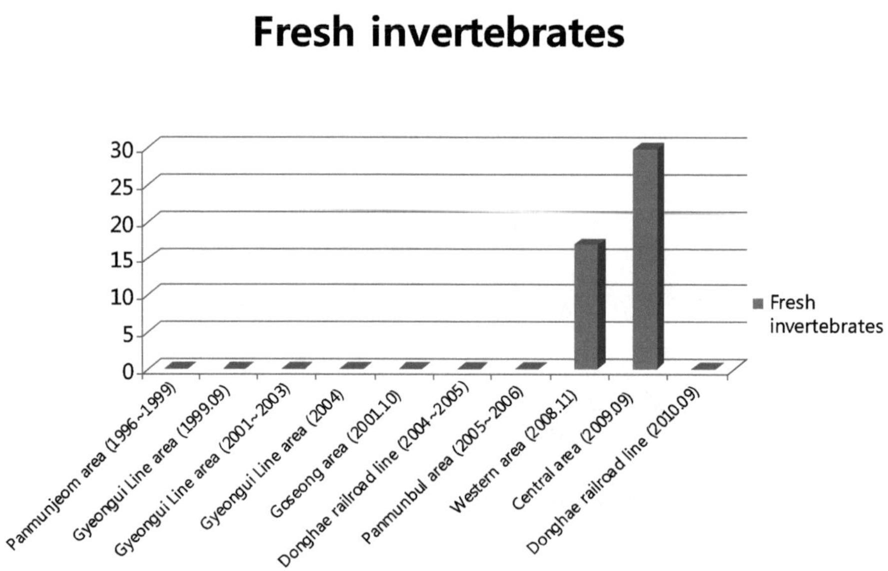

Fig. 5.11 Fresh invertebrate's distribution by major natural area

Table 5.10 Total species by western-central area of theDMZ

Category	Total	Vegetation	Mammals	Birds	Amphibians reptiles	Insects	Fish	Invertebrate
Gokreung-cheon natural region	35	15	2	7		4	4	3
Manwoo-ri natural region	21	9	4	5		2		1
Geogok-ri natural region	43	20	4	8	1	2	4	4
Dora mountain natural region	35	16	6	6	1	1	2	3
Daeseong reservoir natural region	47	16	4	19		2	3	3
Sewol stream natural region	53	14	5	13	5	1	13	2
Sami stream Yeoncheon plain (West) natural region	67	20	3	17		9	14	4
Sami stream Yeoncheon plain (East) natural region	21	8	2	6		1	1	3
Gowang mountain Choryang valley natural region	51	12	5	12		7	9	6
Imjin river (West) natural region	57	18	6	18		2	9	4
Yeokgok stream natural region	187	160	4	8	3	7	5	
Deokeun valley natural region	163	150	1	4		3	1	4
Naepogangsan natural region	77	19	9	17	8	6	12	6
Mindeulle plain natural region	22	9	1	12				
Yugok valley plain natural region	39	15	1	6	6	2	8	1
Mando plain natural region	40	15	5	10	2	3	3	2
Gimhwa Namdae natural region	30	13	3	10	1		1	2
Hajinhyeon valley natural region	35	15	8	6	3		3	
Northern Han river upstream natural region	19	7	9	2			1	

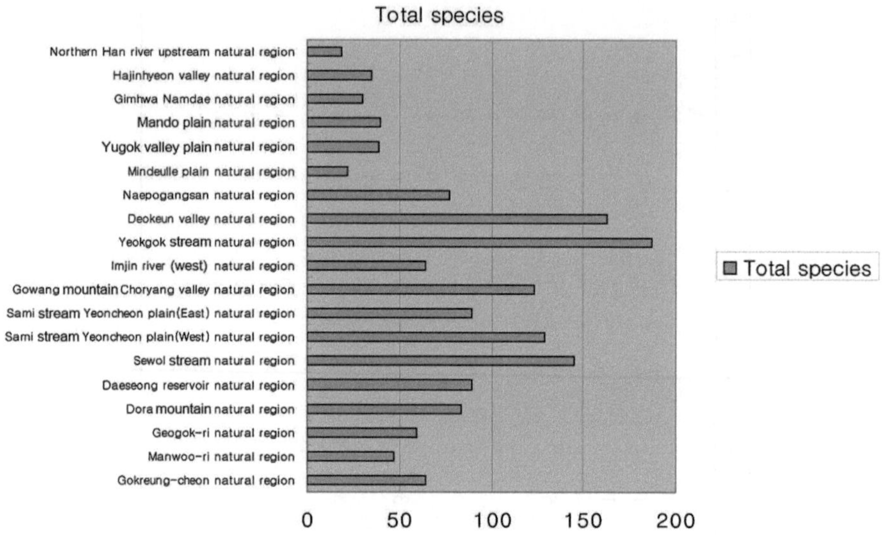

Fig. 5.12 Total species in the western-central area of the DMZ

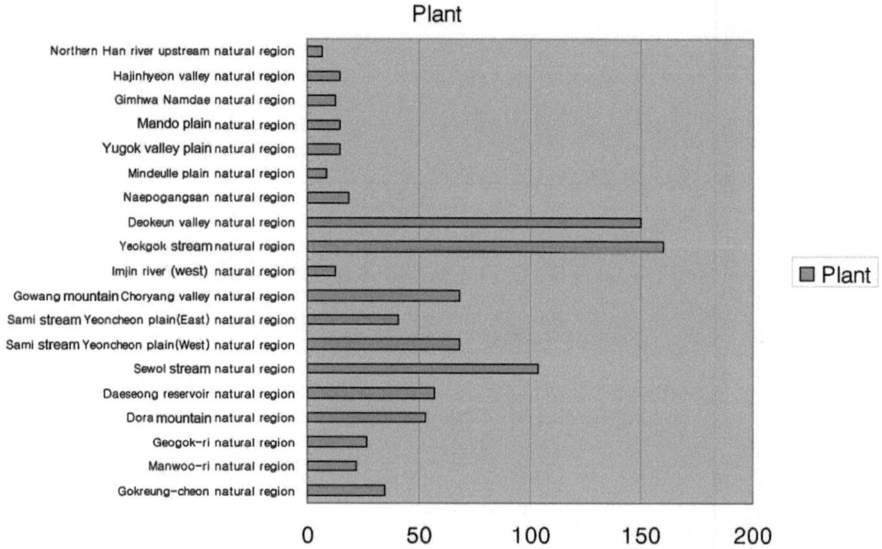

Fig. 5.13 Number of plant species by major natural area of the west-central area of the DMZ

Reptiles and amphibians
Figure 5.16.

Insects
Figure 5.17.

Fish

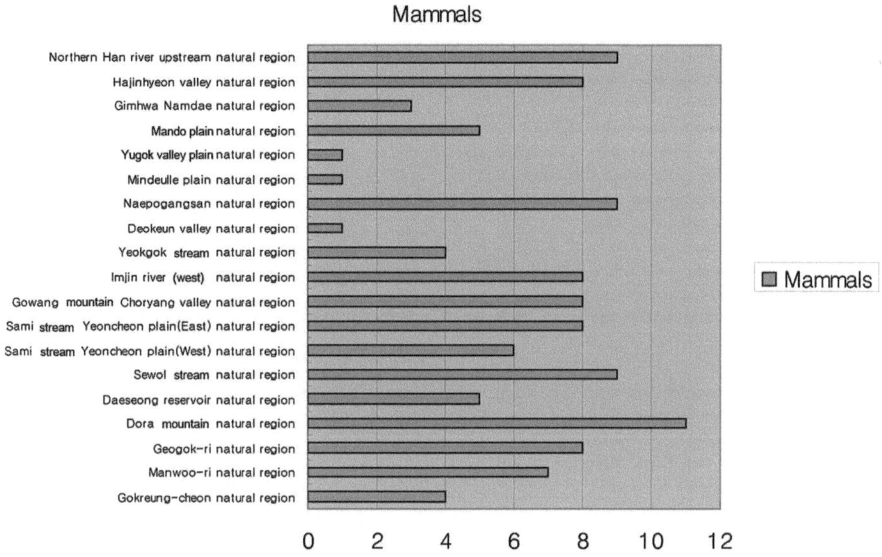

Fig. 5.14 Number of mammals by major natural area of the west-central area of the DMZ

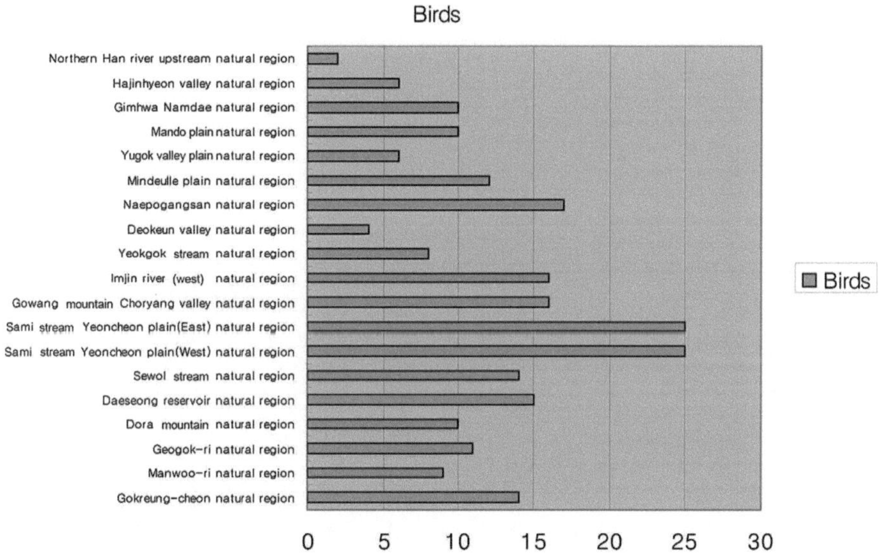

Fig. 5.15 Number of birds by major natural area of the west-central area of the DMZ

Figure 5.18.

Fresh invertebrate

Figure 5.19.

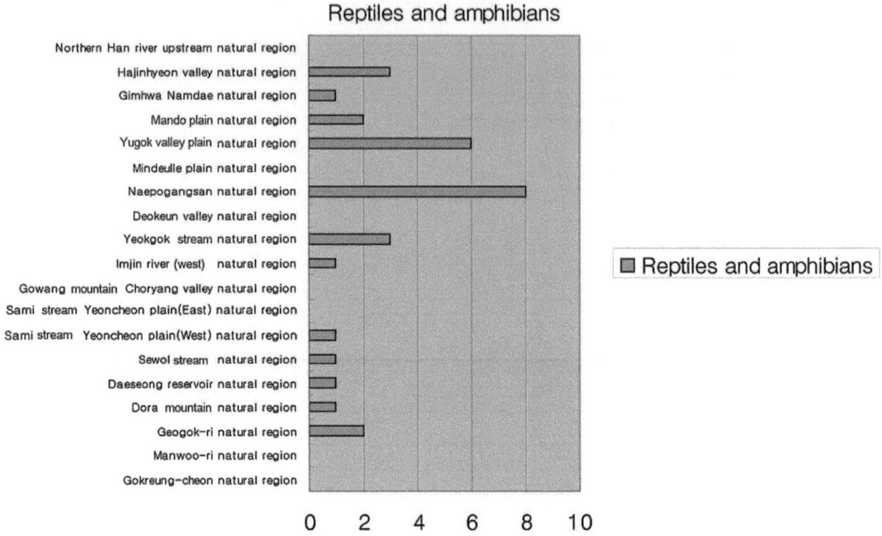

Fig. 5.16 Number of reptiles and amphibians by major natural area of the west-central area of the DMZ

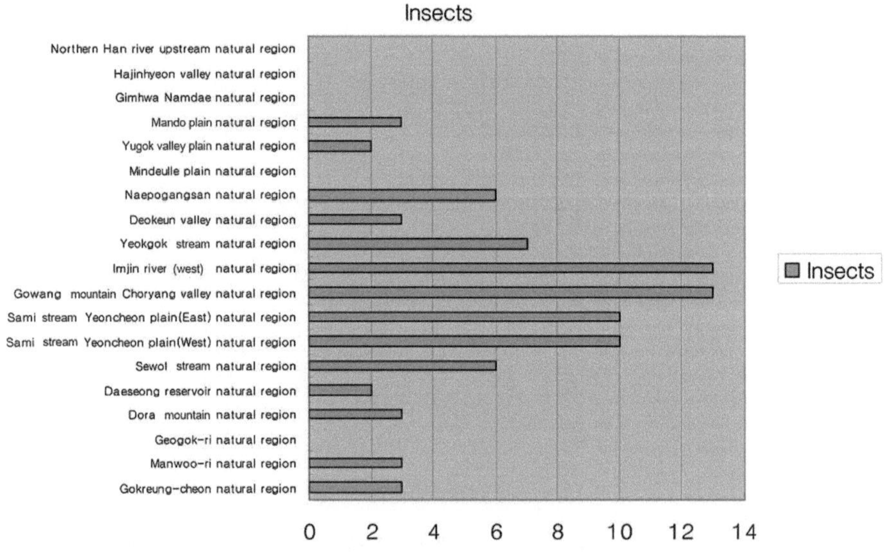

Fig. 5.17 Number of insects by major natural area of the west-central area of the DMZ

Legally protected species

Number and names of legally protected species by major natural area of the western-central area of the DMZ are shown in Table 5.11, while the same information is depicted in a graphic form of Fig. 5.20.

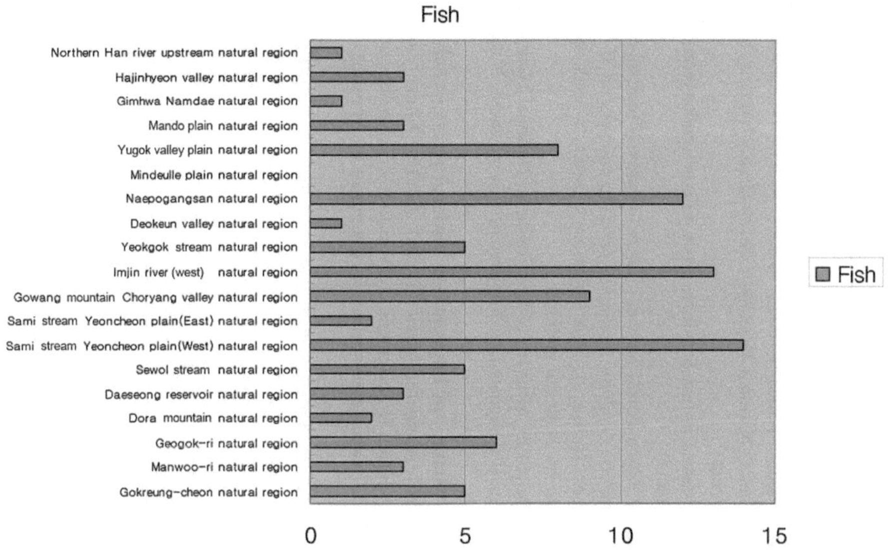

Fig. 5.18 Number of fish by major natural area of the west-central area of the DMZ

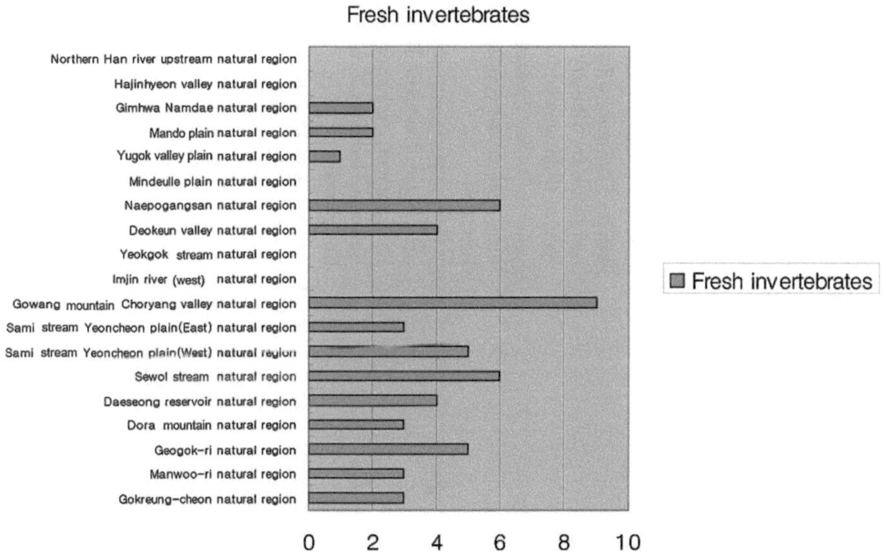

Fig. 5.19 Number of fresh invertebrates by major natural area of the west-central area of the DMZ

Table 5.11 Number and names of legally protected species by major natural area of the western-central area of the DMZ

Survey area	Number of species	List of legally protected species
Gokreung stream natural region	3	*Phacelurus lalifolius* (rare species designated by Korea Forest Service), *Anser cygnoides* (endangered species II designated by Ministry of Environment, National Heritage), *Anser fabalis* (endangered species II designated by Ministry of Environment)
Manwoo-ri natural region	2	*Phacelurus lalifolius* (rare species designated by Korea Forest Service), *Anser fabalis* (endangered species II designated by Ministry of Environment)
Geogok-ri natural region	4	*Aristolochia contorta* (rare species designated by Korea Forest Service), *Grus japonensis* (endangered species I designated by Ministry of Environment), *Aegypius monachus* (endangered species II designated by Ministry of Environment, National Heritage), *Felis bengalensis* euptilura (endangered species II designated by Ministry of Environment, National Heritage)
Dora mountain natural region	2	*Aegypius monachus* (endangered species II designated by Ministry of Environment, National Heritage), *Felis bengalensis* euptilura (endangered species II designated by Ministry of Environment, National Heritage)
Daeseong reservoir natural region	5	*Grus vipio* (endangered species II designated by Ministry of Environment, National Heritage), *Aix galericulata* (National Heritage), *Anser fabalis* (endangered species II designated by Ministry of Environment), *Accipiter gentilis* (endangered species II designated by Ministry of Environment, National Heritage), *Felis bengalensis* euptilura (endangered species II designated by Ministry of Environment)
Sewol stream natural region	7	*Aristolochia contorta* (rare species designated by Korea Forest Service), *Hemibarbus mylodon* (Berg) (National Heritage), *Grus japonensis* (endangered species I designated by Ministry of Environmen, National Heritage), *Buteo buteo* (endangered species II designated by Ministry of Environment), *Accipiter gentilis* (endangered species II designated by Ministry of Environment), *Felis bengalensis* euptilura (endangered species II designated by Ministry of Environment), *Lutra lutra* (endangered species I designated by Ministry of Environment, National Heritage)

(continued)

Table 5.11 (continued)

Survey area	Number of species	List of legally protected species
Sami stream Yeoncheon plain (West) natural region	16	*Aristolochia contorta* (rare species designated by Korea Forest Service), *Callipogon relictus* Semenov-Tian-Shansky (endangered speciesII designated by Ministry of Environment, National Heritage), *Nannophya pygmaea* Rambur (endangered species II designated by Ministry of Environment), *Lethocerus deyrollei* (Vuillefroy) (endangered species II designated by Ministry of Environment), *Gobiobotia macrocephala* (endangered species II designated by Ministry of Environment), *Hemibarbus mylodon* (Berg)(National Heritage), *Psudopungtungia tenuiooupus* (endangered species II designated by Ministry of Environment), (*Siniperca scherzeri* (Albino type)(National Heritage), *Grus japonensis* (endangered species I designated by Ministry of Environment, National Heritage), *Grus vipio* (endangered species II designated by Ministry of Environment, National Heritage), *Aegypius monachus* (endangered species II designated by Ministry of Environment, National Heritage), *Accipiter gentilis* (endangered species II designated by Ministry of Environment, National Heritage), *Anser fabalis* (endangered species II designated by Ministry of Environment), *Buteo buteo* (endangered species II designated by Ministry of Environment), *Accipiter nisus* (National Heritage), *Dryocopus martius* (endangered species II designated by Ministry of Environment, National Heritage), *Felis bengalensis* euptilura (endangered species II designated by Ministry of Environment)
Sami stream Yeoncheon plain (East) natural region	13	*Aristolochia contorta* (rare species designated by Korea Forest Service), *Callipogon relictus* Semenov-Tian-Shansky (endangered species II, National Heritage), *Nannophya pygmaea* Rambur (endangered species II), *Lethocerus deyrollei* (Vuillefroy) (endangered species II), *Grus japonensis* (endangered species I, National Heritage), *Grus vipio* (endangered species II, National Heritage), *Aegypius monachus* (endangered species II, National Heritage), (endangered species II, National Heritage), *Anser fabalis* (endangered species II), *Buteo buteo* (endangered species II), National Heritage, *Dryocopus martius* (endangered species II, National Heritage), *Felis bengalensis* euptilura (endangered species II)

(continued)

Table 5.11 (continued)

Survey area	Number of species	List of legally protected species
Gowang mountain Choryang valley natural region	9	*Aristolochia contorta* (rare species designated by Korea Forest Service), *Acheilognathus lanceolatus* (endangered species II designated by Ministry of Environment), *Aquila chrysaetos* (endangered species II designated by Ministry of Environment), *Grus japonensis* (endangered species I designated by Ministry of Environment, National Heritage), *Aegypius monachus* (endangered species II designated by Ministry of Environment, National Heritage), *Circus aeruginosus* (endangered species II designated by Ministry of Environment, National Heritage), *Accipiter gentilis* (endangered species II designated by Ministry of Environment, National Heritage), (endangered species II designated by Ministry of Environment), *Felis bengalensis* euptilura (endangered species II designated by Ministry of Environment)
Imjin river (West) natural region	10	*Aristolochia contorta* (rare species designated by Korea Forest Service), *Gobiobotia macrocephala* (endangered species II designated by Ministry of Environment), *Acheilognathus lanceolatus* (endangered species II designated by Ministry of Environment), *Hemibarbus mylodon* (Berg) (National Heritage), *Aquila chrysaetos* (endangered species II designated by Ministry of Environment), *Grus japonensis* (endangered species I designated by Ministry of Environment, National Heritage), *Aegypius monachus* (endangered species II designated by Ministry of Environment, National Heritage), *Circus aeruginosus* (endangered species II designated by Ministry of Environment, National Heritage), *Accipiter gentilis* (endangered species II designated by Ministry of Environment, National Heritage), *Buteo buteo* (endangered species II designated by Ministry of Environment)
Yeokgok stream natural region	2	*Accipiter gentilis* (endangered species II designated by Ministry of Environment, National Heritage), *Felis bengalensis* euptilura (endangered species II designated by Ministry of Environment)
Deokeun valley natural region	1	*Falco subbuteo* (endangered species II designated by Ministry of Environment)

(continued)

Table 5.11 (continued)

Survey area	Number of species	List of legally protected species
Naepogangsan natural region	9	*Lutra lutra* (endangered species I designated by Ministry of Environment, National Heritage), *Moschus moschiferus* parvipes (endangered species I designated by Ministry of Environment), *Vulpes vulpes* peculiosa (endangered species I designated by Ministry of Environment), *Elaphe schrenckii* (endangered species I designated by Ministry of Environment), *Felis bengalensis* euptilura (endangered species II designated by Ministry of Environment), *Martes flavigula* koreana (endangered species II designated by Ministry of Environment), *Acheilognathus lanceolatus* (endangered species II designated by Ministry of Environment), *Buteo buteo* (endangered species II designated by Ministry of Environment), *Aristolochia contorta* (rare species designated by Korea Forest Service)
Mindeulle plain natural region	3	*Aegypius monachus* (endangered species II designated by Ministry of Environment, National Heritage), *Grus vipio* (endangered species II designated by Ministry of Environment, National Heritage), *Haliaeetus albicilla* (National Heritage)
Yugok valley plain natural region	4	*Accipiter gentilis* (endangered species II designated by Ministry of Environment, National Heritage), *Aix galericulata* (National Heritage), *Hemibarbus mylodon* (Berg)(National Heritage), *Sparganium erectum* L.(rare species designated by Korea Forest Service)
Mando plain natural region	5	*Grus japonensis* (endangered species I designated by Ministry of Environment, National Heritage), *Aegypius monachus* (endangered species II designated by Ministry of Environment), *Accipiter gentilis* (endangered species II designated by Ministry of Environment, National Heritage), *Buteo buteo* (endangered species II designated by Ministry of Environment), *Felis bengalensis* euptilura (endangered species II designated by Ministry of Environment)
Gimhwa Namdae natural region	2	*Felis bengalensis* euptilura (endangered species II designated by Ministry of Environment), *Accipiter nisus* (National Heritage)
Hajinhyeon valley natural region	4	*Elaphe schrenckii* (endangered species I designated by Ministry of Environment), *Felis bengalensis* euptilura (endangered species II designated by Ministry of Environment), *Naemorhedus goral* (endangered species II designated by Ministry of Environment), *Martes flavigula* koreana (endangered species II designated by Ministry of Environment)
Northern Han river upstream natural region	4	*Felis bengalensis* euptilura (endangered species II designated by Ministry of Environment), *Naemorhedus goral* (endangered species II designated by Ministry of Environment), *Lutra lutra* (endangered species I designated by Ministry of Environment, National Heritage), *Buteo buteo* (endangered species II designated by Ministry of Environment)

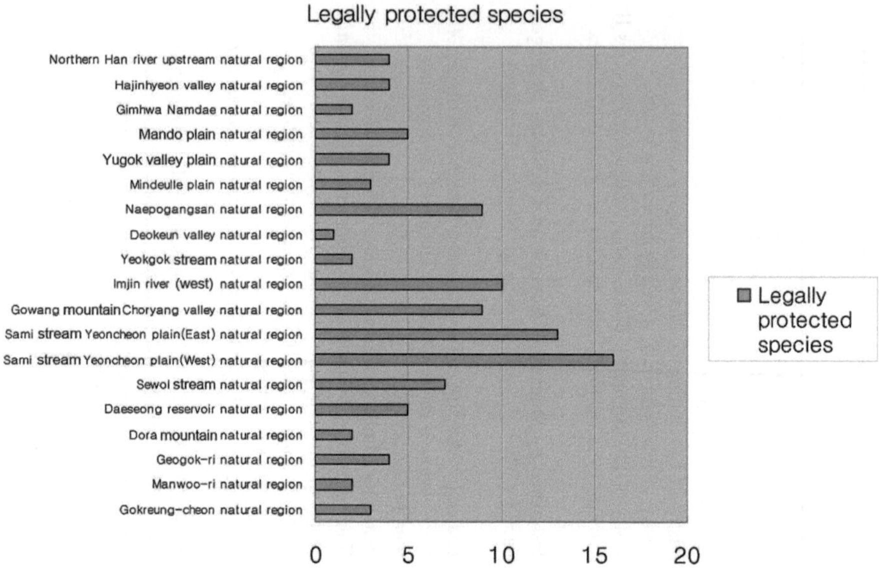

Fig. 5.20 Number of legally protected species by major natural area of the west-central area of the DMZ

5.2.5 Detailed Description by Species

5.2.5.1 Flora

Aquatic plants in the DMZ and CCZ are mainly found in stream, abandoned paddy field wetland, and forest wetland area, and various aquatic plants were able to live in water due to continuous water effect in little disturbance for a lengthy period of time. Although the number of species of aquatic plants living in this area is undefined, yet ,as a detailed investigation has not been carried out, it is assumed that many rare and endangered species designated by the Korean Forest Administration would live in. Major plants classified by life of aquatic plants are shown in Table 5.12.

Emerged Plant

Bur reed (*Sparganium erectum*)

Features of Bur reed Habitat in DMZ and CCZ area

S. erectum L. are found in the wetland formed at the abandoned paddy fields, in the stream wetlands, and in ponds of both DMZ and CCZ. These areas developed a small ecosystem in which various aquatic plants live such as; wild-walnut, Zizania latifolia, *Phragmites australis*, Persicaria thunbergii (Siebold & Zucc.), and H. Gross ex Nakai. The habitat where *S. erectum* L. was found was ecologically in good condition being connected with nearby stream, forest, shrubs, and scrubs. *S. erectum* L. is also found in foot paths along the rice fields from time to time (Figs. 5.21, 5.22).

Table 5.12 Characteristics of DMZ Flora and Vegetation

Classification	Name of species (scientific name)	Characteristics	Distributed area
Emerged plants	*Sparganium erectum* L. (Sparganium stoloniferum)	*Sparganium erectum* L., living in wetland near pond and ditch is a protected species designated by Korea Forest Service even though it is seen in various wetland areas in the country as only seen in clean area where there is organic matter	Asia, Europe, and North America
	Tofieldia nuda Maxim. (*Acorus calamus* var. angustatus)	*Acorus calamus* L., living in wetland near pond and ditch can live in many wetland areas. A long-lived araceous grass with thick midrib, dark green leaves	Korea · Japan · China
	Juncus effusus var. decipiens Buchenau (*Juncus effusus*)	*Juncus effusus* var. decipiens Buchenau, appearing in waterside and wetland of stream is as an emerged plant which has a cylinder-shaped flat green stem. Triangular egg-shaped fruits grow in September–October which has characteristics of being brown and bluntly	Korea (Jeonnam) · Japan · Taiwan · China · Amur River · North America
Floating leaved plants	*Trapa japonica* Flerow (*Trapa japonica* Flerow)	Living in pond and marsh with leaves floated on the water surface. Triangular leaves which has irregular tooth-shaped serra on top. White or strawberry blonde flowers with diameter of 1 cm bloom in July–August	Korea, Japan, China, Europe
	Potamogeton distincuts A.Benn.	Spread roots into the soil in pond and marsh and have leaves exposed to water. The shape of leaves is lanceolate or oval and its stem has a length of 6–10 cm but can be shorter or longer depending on the depth of water	Korea, China, Japan
Floating plants	*Lemna paucicostata*	Floating plant floated on the surface of rice paddy or pond, has leaves with green front and green or purple back. Appears everywhere except polar regions and is distributed throughout Korea	The tropics through temperate all over the world

(continued)

Table 5.12 (continued)

Classification	Name of species (scientific name)	Characteristics	Distributed area
Submerged plants	*Utricularia vulgaris* var. japonica (Makino) Tamura	As a perennial insect-eating plant, it lives in calm water of pond, stream, rice paddy, etc. It floats without roots, leaves grow at the tip of stem in winter which makes round winter bud to sink	Korea, Japan, Manchuria, Sakhalin
Hygrophyte	*Drosera rotundifolia*	As an insectivorous plant with slim and long stem living in sunny wetland. It especially lives in sunny acidic wetland and has long red tentacle on the edges and front to catch insects. Insects are caught by long brown hair under the footstalk of length 3–13 cm	The temperate through warm-temperate on the Northern hemisphere
Salt plants	*Rosa rugosa* var. rugosa (*Rosa rugosa*)	Rosaceous shrub living in a beach's sandy soil, with lengthy of 1–1.5 m downing branches. Dense brown thorns with hair on it. Its fruit belongs to spheroid achene with diameter of 2–3 cm turning red and edible carnose which is feed for wildlife animals	Northeast Asia
	Calystegia soldanella	A convolvulaceae plant living in a beach's sandy soil with thick transcurrent subterranean stem. Stem branches off on the surface or has a characteristics of rolling up others. Dislocated leaves and long footstalk, glossy and kidney-shaped, pink flowers bloom in May	Korea (Jeju · Jeonnam · Jeonbuk · Gyeongnam · Gyeongbuk · Gangwon · Gyeonggi · Hwanghae · Hamnam)

Fig. 5.21 Bur reed in pond habitat connected to Namdae stream in Cheorwon, CCZ

Fig. 5.22 Detailed view of *Sparganium erectum* L

Fig. 5.23 Sweet-flag habitat in pond wetland, Panmunbul field, Paju, DMZ

Fig. 5.24 Flower of sweet-flag

Considerations for securing the habitat
The ponds and marshes where *S. erectum* L. lives have clean waters and are in good ecological connectivity. A lot of wetlands have disappeared due to the inflow of organic matters, chemical fertilizers, and agricultural pesticides which led to water pollution and the cemented waterside. Thus, it is required to block and purify pollutants before it flows into streams and ponds in order to maintain its clean water quality, to alter the concreting of watersides and ultimately restore the wetlands'habitat that ensures the security of its flora.

Sweet-flag (Acorus calamus)

Features of Sweet-flag habitat in DMZ and CCZ
Sweet-flags are also found thriving in wetlands that are sustained by ponds and streams with slow velocity of flow. The color of these plants is purple or yellow which gave it a name "flower sweet-flag or yellow sweet-flag". Living in wetlands throughout Korea, it appears in all the areas in both DMZ and CCZ which formed water, soil and vegetation, a characteristic typical for sweet-flags to live.

Figure 5.23 illustrates a pond wetland in Paju DMZ Panmunbul where sweet-flags live in and Fig. 5.24 illustrates flower sweet-flag.

Considerations for securing the habitat
Sweet-flags are able to live in any wetland and in other places that have structural characteristics of a wetland. At the moment, sweet-flags are transplanted to be used as an ornament in many artificial wetland areas which create beautiful sceneries in the landscape. Thus, the wetland and its clean waters have to be preserved and must be well maintained.

Rush (Juncus effusus)

Features of Rush habitat in DMZ and CCZ
J. effusus variety decipiens Buchenau is easily found throughout wetland areas in DMZ and CCZ, and it especially lives well in areas where sufficient water is provided. The waterway wetland of CCZ in Goseong as shown in Figs. 5.25, 5.26

Fig. 5.25 *Juncus effusus* var. decipiens Buchenau habitat in waterway wetland of Goseong CCZ

Fig. 5.26 *Juncus effusus* var. decipiens Buchenau

is an area where *J. effusus* var. decipiens Buchenau is being found which is located in valleys and mountains, where a stream tributary flows toward downstream. Besides Juntus effuses, there are many other aquatic plants found living here such as the *P. australis*, Japanese bulrush, *Salix koreensis* Andersson. There are also various species of mammals found living co-habiting these areas such as the *Hydropotes inermis*, *Microtus fortis* pelliceus, *Aix galericulata*, mallard, Japanese Bush Warbler, *Cobitis melanoleuca*, *Rhynchocypris steindachneri*, as well as varieties of birds, fish, reptiles, and amphibians.

Considerations for securing the habitat

As *J. effusus* var. decipiens Buchenau lives well in moist wetland, floodgate needs to be connected well and important wetland habitats shall be managed and preserved. Floodgates in damaged or polluted area shall be restored back to wetland structure otherwise *J. effusus* var. decipiens Buchenau will not be seen in these areas.

Floating Leaved Plant

Waternut (*Trapa japonica*)

Features of Waternut in DMZ and CCZ

Waternuts are also found in specific wetlands such as a reservoir or a lake where there is an open surface. Waternuts are floating leaved plants whose leaves cover the water surface in community. In the Anho reservoir of Goseong CCZ (Fig. 5.27), waternuts formed a large-scale community (Fig. 5.28).

The waternut is an important habitat that acts as shelter and provides food for fish and birds. Wildlife animals are also found living in these waternut communities such as chaenogobius urotaenia, hyla japonica, *A. galericulata*, anas platyrhynchos.

Considerations for securing the habitat

Waternut lives in a wetland with relatively open surface exposing its leaves to the sun. Thus, ponds, reservoirs, and lakes are suitable for these plants to live well.

Fig. 5.27 Anho reservoir in Goseong CCZ

Fig. 5.28 Waternut community

Pollutants should be prevented from getting into these habitats and water should not be polluted by eutrophication.

Pondweed (Potamogeton distinctus)

Features of Pondweed habitat in DMZ and CCZ

Pondweed also lives in ponds where its name is taken from. But it also lives in lakes, cultivated lands, and in agricultural waterways of the DMZ and CCZ. As an aquatic plant, it is easily found everywhere. These plants are either submerged in water or live on the water surface and, bloom flowers during the months of April – August.

The Sanmyeong lake in Cheorwon CCZ provides water to nearby farmlands as in Fig. 5.29. *S. koreensis* Andersson, *P. australis,* and Zizania latifolia inhabit the edges while pondweed lives in the waterway connected to the farmlands and even found on the edges.

Considerations for securing the habitat

Water fringe wetland is not developed well as artificial banks have been built near lakes and waterways in DMZ and CCZ. Sometimes pondweed disappears in these artificial banks due to its competition with other exotic species. Thus, artificial concrete banks should not be built on the edges of waters. Water fringe wetland

Fig. 5.29 Sanmyeong lake in Cheorwon CCZ

Fig. 5.30 Duckweek habitat in pond wetland connected to Sanmyeong lake in Cheorwon CCZ

should be preserved in order to preserve pondweed plants. It lives well in calm ponds and it needs to create various habitats.

Floating Plant

Duckweed (*Lemna paucicostata*)

Habitat Feature of DMZ and CCZ area

Other floating plants are found co-habiting the ponds and small streams of DMZ and CCZ. Duckweed is a floating plant easily found everywhere in these areas. Fish, reptiles, and amphibians also live in places where these plants are found.

Pond wetlands are developed by land formation in some parts of the stream tributaries connected to Sanmyeong lake in Cheorwon CCZ. Water is trapped and accumulated in this area as shown in Fig. 5.30. This picture shows a very good wetland area where there are forests nearby and water species like *P. australis*, *Phragmites japonica* Steud, Zizania latifolia living on the edges, while duckweeds and *T. japonica* Flerow thriving on the surface of the water.

Considerations for securing the habitat

Duckweeds live in pond and link canal throughout the country but its habitat is decreasing due to the inflow of pollutants that eventually destroyed the wetlands. The wetlands at the DMZ and CCZ where cultivation of farmlands is rampant consequently threatened the survival of duckweeds together with other aquatic plants. In order to preserve these habitats, it is a salient need to maintain the natural environment of the buffer wetland areas where duckweed lives. Pollutants should not be allowed to flow directly on these wetlands. In addition to that, the open water surface should exceed 50 % of the whole area to provide enough space for these water plants to live.

Submerged Plant

Bladderwort (*Utricularia vulgaris*)

Fig. 5.31 Bladderwort habitat in Paju DMZ

Features of bladderwort habitat in DMZ and CCZ

Bladderwort is another submerged aquatic plant found in ponds, streams, and rice paddies in the DMZ and CCZ. The bladderwort appears in this area because there are lesser pollutants flowing into these areas and because there is less human interference.

In fact, bladderwort was found in community in a stream of Panmunbul area in Paju DMZ as in Figs. 5.31 and 5.32. In this area, it lives in deep water with calm water surface or shallow water with stable water surface, and it eats perennial bugs in water.

Considerations for securing the habitat

DMZ wetlands are very rare semi-natural ecosystems in the country that need to be preserved. The focus of the preservation should be on the DNA sources and ecosystem succession process. This ensures the survival and prevalence of the variety of species living here. The Government designated the DMZ as the Natural Envrionment Reservation Area for two years after re-unification of the Korean peninsula to preserve the species, especially those that are not found in other areas of

Fig. 5.32 Bladderwort

Fig. 5.33 Round-leaved sundew habitat in Young-neup marsh in Mt. Daeam in CCZ

the country like the Bladderworst. More detailed investigation has to be conducted specifically in terms of its distribution and quality of habitat of submerged plant.

Hydrophyte

Round-leaved Sundew (*Drosera rotundifolia*)

Features of Round-leaved Sundew habitat in DMZ and CCZ

The Round-leaved Sundew is a hydrophyte found in Young-neup marsh of Mt. Daeam located in the CCZ (Fig. 5.33). High bog in the mountain forms a special wetland and a unique environment in which peat layer, rainwater, and underground water are provided (Fig. 5.34).

Fig. 5.34 Round-leaved Sundew

Considerations for securing the habitat

As rare species may have adapted to unique environments such as the Young-neup marsh, they can also easily die out if this special ecosystem is destroyed or damaged. Thus, a thorough study on the preservation of the middle-high bog is essential. These studies involve biodiversity maintenance, preservation of the genetic resources, ecological succession, palaeoecology and palaeoclimatology. Restoration of this area is being done.

Halophyte

Turkestan rose (*Rosa rugosa*)

Features of Turkestan rose habitat in DMZ and CCZ

Salt plants are also found in the DMZ and CCZ. Identified Turkestan rose habitat is in the eastern coast of Korea, specifically in Goseong (Fig. 5.35). However, large parts of the sane dunes in the eastern coast had been destroyed and Turkestan rose habitats are also disappearing due to development of beach resorts and farmlands. According to the survey, despite the difficulty of investigation due to the presence of landmines, the largest Turkestan rose community is found in Goseong DMZ area with approximately 1.2 km in length. These communities are observed to live in beaches and sand dunes that lead to the sea. Oak trees and a species of pine are believed to have helped in stabilizing the sand dunes thereby preventing the wind to deform it. A riverine wetland is formed within the perimeter of the sand dunes turning it into an ideal environment for the propagation of the Turkestan rose.

Considerations for securing the habitat

The Turkestan rose community also thrives in hundreds of meters in some of the coastal sand dune in the west coast such as in Anmyeon island, Sindu-ri, Taean, Chungnam province. The community is not easily seen in the east coast as coastal

Fig. 5.35 Turkestan rose bloom in a beach in Goseong DMZ

sand dune decreased and disappeared due to various developments. Thus, it is said that the large-scale Turkestan rose community habitat in Goseong DMZ has a very high value. The aqua-ecosystem and the sand dunes had been preserved in this area because it was isolated for a long time. But recently, the South–North highway and Donghae railroad line do not keep this area safe anymore. Therefore, more detailed investigation must be conducted with an objective to find ways and means to minimize the damage brought about by the railway and roads.

Sea Bindweed (Calystegia soldanella)

Features of Sea Bindweed habitat in DMZ and CCZ area

The Sea Bindweed is one of the plant species that was found abundant in DMZ and CCZ. It was observed in the coastal sand dunes near the beach. These sand dune areas are still in its original form since it has not been infiltrated by humans as well. Contributory to the development of this very suitable habitat is the water which leads to the stream. A Community of Sea Bindweed is easily found in the east coast of Goseong DMZ and CCZ as shown in Fig. 5.36. Other than sea bindweeds, *R. rugosa* was also observed there, as well as species of pine, oak, sea pea, and *Elymus mollis* Trinius (Fig. 5.37).

Considerations for securing the habitat

As beaches are developed, coastal roads and country houses are built in many sand dune areas. Sand has been collected, and sand dunes and hills are damaged. The ecosystem for bindweeds is not found anymore in developed areas. Due to the artificial structures, coastal sand dune changes constantly as the wind and sea waves hit the ground. The coastal sand dune functions as a hill that blocks waves and the wind, covering vegetation and prevents tsunami and typhoon. In order to help sea bindweed live safely in coastal sand dunes, preservation plans need to be developed and implemented soon. Artificial structures should be avoided near beaches and areas where there are coastal sand dunes. The sand dunes have to be designated as conservation area and well protected from human and developmental

Fig. 5.36 Sea Bindweed habitat in beach and coastal sand dune in the east coast of DMZ

Fig. 5.37 Sea bindweed

interferences especially the dunes in DMZ and CCZ where the largest communities of bindweeds are found to be abundantly thriving.

Vegetation

This section aims to provide data for inference necessary for the restoration of the vegetation in damaged forests in DMZ. The vegetation status and diagnosis of this area are revealed through plant community survey and vegetation mapping. The study was conducted by Chang-Hwan Kim and Kwi-Gon Kim (2010) and was focused on the vegetation of designated areas within the DMZ and CCZ, including the forest and wetland vegetation near Northern Donghae railroad line.

Survey and Analysis Method

Survey Location

In order to conduct a more detailed vegetation survey in the Northern Donghae railroad line, the area was divided into DMZ wetland, CCZ coastal sand dune, and CCZ forested wetland (Fig. 5.38).

Survey Method

A Vegetation survey was conducted using 1:25,000 topographic map, 1:5,000 topographic map, and 1:3,000 topographic map made by the National Geography Institute. Plant sociological survey was also conducted (Lim et al. 1989, 1990). Communities were categorized according to dominant and differential species using table control method with data from each quadrate (Kim and Yim 1988; Muller-Dombois and Ellenberg 1974). The existing vegetation map was drawn using the Küchler's (1967) method.

A Sample block was set in the most typical areas. The targeted sample extraction areas are supposed to be homogeneous. The size of the sample plots is categorized as; plat, shrubs and grasslands. The set is depending on the minimal area (Kim et al. 1995). The vascular plant species in the sample plots were the only ones described, and the list of high forests were specified by class such as; tree layer, shrub layer, herbaceous layer in the vegetation survey table. The cover and

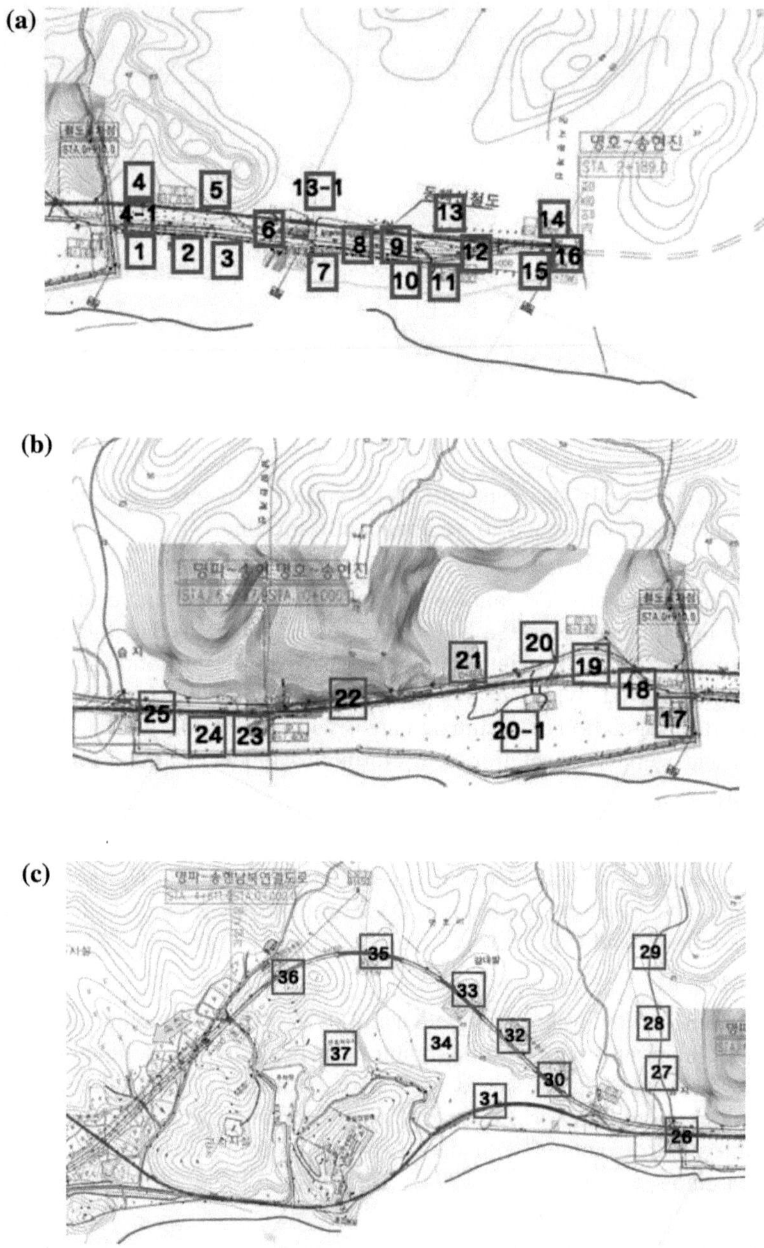

Fig. 5.38 Survey areas in Northern Donghae railroad for a detailed vegetation survey. **a** Wetland vegetation survey area near DMZ Northern Donghae railroad line. **b** CCZ coastal sand dune survey area. **c** CCZ wetland vegetation survey area

sociability of each species were measured and documented in accordance with total estimation method by Braun-Blanquet (1964).

Survey and Analysis Result

Classification of Plant Community

Forest vegetation

In accordance with plant sociological method of Braun-Blanquet (1964), community system of forest vegetation in CCZ near Northern Donghae railroad line and DMZ area were classified as *Quercus mongolica* community, *Q. mongolica-Quercus variabilis* community, and *Quercus acutissima-Q. mongolica* community (Table 5.13).

Table 5.13 Forest vegetation map of Donghae railroad-road construction project area and CCZ (A: *Quercus mongolica* community B: *Quercus variabilis* Blume community C: *Quercus acutissima* CARRUTH community)

Study site no.		1	2	3	4	5	6	7	8
Area (m^2)		25	25	25	25	25	100	100	100
Community classification									
Quercus mongolica Fisch. ex Ledeb.	T1	•	•	•	•	•	2.3	3.3	•
	T2	•	•	•	•	•	+	+	•
	S	3.3	3.4	3.4	2.3	3.4	•	•	3.3
	H	+	+	+	+		+		
Quercus variabilis Blume	T1	•	•	•	•	•	2.2	2.2	•
	S	+	•	+	+	•	•	•	•
Quercus acutissima CARRUTH	T1	•	•	•	•	•	+	•	2.2
	T2	•	•	•	•	•	•	•	+
Companion species									
Quercus dentata Thunb. ex Murray	T1	•	•	•	•	•	1.1	•	•
	T2	•	•	•	•	•	1.2	•	•
	S	+	•	•	1.2	+	+	•	+
	H	•	•	•	+	•	+	+	•
Prunus sargentii Rehder	T1	•	•	•	•	•	+	+	•
Actinidia polygama (Siebold & Zucc.) Planch. ex Maxim.	T2	•	•	•	•	•	•	+	•
Salix koreensis Andersson	T2	•	•	•	•	•	•	+	•
Quercus xurticaefolia Blume	T1	•	•	•	•	•	•	•	1.1
	T2	•	•	•	•	•	•	+	•
	S	+	+	1.2	1.1	+	•	•	+
	H		+	+	+	•	•	•	•
Quercus aliena Blume	T2	•	•	•	•	•	•	•	1.2
	S	•	•	•	•	•	•	•	+
Lindera obtusiloba Blume	T2	•	•	•	•	•	+	•	•
	S	•	•	•	•	•	+	•	•
	H	•	•	•	+	•	•	•	•
Lespedeza bicolor Turcz.	S	+	•	2.2	1.1	•	+	•	1.2
Smilax china L.	S	1.2	+0.2	1.1	1.2	1.1	+	+	2.2
	H	•	•	•	+	•	•	•	•

(continued)

Table 5.13 (continued)

Study site no.		1	2	3	4	5	6	7	8
Lespedeza maximowiczii C.K. Schneid.	S	•	1.2	•	•	1.1	+	+	+
Callicarpa japonica	S	•	•	•	•	•	•	+	•
Lespedeza bicolor Turcz.	S	•	•	•	•	•	•	+	•
Hazel	S	•	•	•	•	•	+	•	+
Staphylea bumalda DC.	S	•	•	•	•	•	+	•	•
Pine	T2	•	•	•	•	•	•	•	+
	S	•	+	•	•	•	•	•	•
	H	•	•	•	+	+	•	•	•
Styrax japonicus Siebold & Zucc.	S	•	•	•	+	+	•	•	•
Phasianus colchicus tree	S	•	•	+	+	•	•	•	•
Kalopanax pictus (Thunb.) Nakai	S	•	•	•	+	•	•	•	•
Rhododendron schlippenbachii	S	•	+	+	•	•	•	•	+
Acer triflorum	S	•	•	+	•	•	•	•	•
Zanthoxylum schinifolium Siebold & Zucc.	S	+	+	•	•	•	•	•	•
Aster scaber Thunb.	H	1.1	+	1.2	+	+	1.1	•	+
Carex humilis var. nana (H.Lev. & Vaniot) Ohwi	H	2.2	+	1.2	1.2	1.2	•	+	+
Potentilla fragarioides var. major	H	+	+	+	+	•	•	•	•
Artemisia princeps var. orientalis (Pampan.) Hara/Artemisia keiskeana Miq.	H	+	+	1.2	1.1	2.2	•	+	+
Arachniodes aristata (G.Forst.) Tindale	H	+	+	1.2	+	+	1.2	•	1.1
Miscanthus sacchariflorus var. gracilisY.N.Lee	H	•	2.2	•	•	•	•	•	+0.2
Bird	H	+	+	+	1.2	+	+	+	1.2
Carex siderosticta Hance	H	•	•	•	•	•	1.2	2.3	+
Chrysanthemum zawadskii var. latilobum	H	2.3	•	1.2	•	1.2	•	•	•
Melampyrum roseum Maxim.	H	+0.2	•	1.2	+	+0.2	•	•	+
Disporum smilacinum	H	+	+	1.1	+	+	1.2	•	•
Clematis terniflora var. mandshurica (Rupr.) Ohwi	H	+	•	•	•	•	+	+0.2	•
Arachniodes aristata (G.Forst.) Tindale	H	•	•	•	•	•	•	2.2	•
Syneilesis palmata	H	•	•	•	•	•	•	2.3	•
Carex lanceolata Boott	H	•	•	•	•	•	1.2	•	•
Convallaria keiskei	H	•	•	•	•	•	1.1	+0.2	•
Hepatica asiatica Nakai	H	•	•	•	•	•	1.2	•	•
Eupatorium japonicum Thunb. ex Murray	H	•	•	•	•	•	+0.2	•	+0.2
Isodon inflexus (Thunb.) Kudo	H	•	•	•	•	•	+	+0.2	1.2
Vitis amurensis Rupr.	H	•	•	•	•	•	+0.2	•	•
Solidago virgaurea subsp. asiatica Kitam. ex Hara var. asiatica	H	+	•	+	+	+	•	•	+
Epimedium koreanum Nakai	H	+	•	+	•	•	•	•	•

(continued)

Table 5.13 (continued)

Study site no.		1	2	3	4	5	6	7	8
Atractylodes ovata (Thunb.) DC.	H	•	+	•	+	+	•	•	•
Synurus deltoides (Ait.) Nakai	H	+	•	•	+	+	•	•	•
Pica pica mustache	H	+	•	+	•	+	+	•	•
Platycodon grandiflorum	H	+	•	+	•	+	•	•	•
Vicia unijuga A.Braun	H	•	•	+	+	•	•	+	•
Adenophora verticillata var. angustifolia Regel	H	•	•	+	•	•	•	•	•
Peucedanum terebinthaceum (Fisch.) Fisch. ex DC.	H	•	•	•	+	•	•	•	•
Veratrum oxysepalum Turcz.	H	•	•	•	+	•	•	•	•
Siphonostegia chinensis	H	•	•	•	•	+	•	•	•
Patrinia scabiosaefolia	H	•	•	•	•	+	•	•	•
Cymbopogon tortilis var. goeringii	H	•	•	•	•	+	•	•	•
Calamagrostis arundinacea (L.) Roth	H	•	•	•	•	•	+	•	•
Artemisia princeps var. orientalis (Pampan.) Hara	H	•	•	•	•	•	+	•	•
Rubus crataegifolius Bunge	H	•	•	•	•	•	+	•	•
Diarrhena japonica Fr. et Sav.	H	•	•	•	•	•	+	•	•
Codonopsis lanceolata	H	•	•	•	•	•	+	•	•
Smilax nipponica Miq.	H	•	•	•	•	•	+	•	•
Angelica decursiva	H	•	•	•	•	•	+	•	•
Adenophora remotiflora	H	•	•	•	•	•	+	+	1.2
Stephanandra incisa (Thunb.) Zabel	S	•	•	•	•	•	•	•	+
	H	•	•	•	•	•	•	+	•
Commelina communis	H	•	•	•	•	•	•	+	•
Phryma leptostachya var. asiatica	H	•	•	•	•	•	•	+	•
Matteuccia orientalis	H	•	•	•	•	•	•	+	•
Rubus oldhamii var.	H	•	•	•	•	•	•	+	•
Cirsium japonicum var. maackii (Maxim.) Matsum.,	H	•	•	•	•	•	•	•	+
Rubus corchorifolius	H	•	•	•	•	•	•	•	+

T tree, *T2* subtree, *S* shrub, *H* herb

Species distributed by class within *Q. mongolica* community are *Q. mongolica* in shrub layer, Aster scaber Thunb and Carex humilis var. nana (H. Lev. and Vaniot) Ohwi in herbaceous layer. The coverage rate is 65–75 % in shrub layer and 30–40 % in herbaceous layer. Species distributed by class within *Q. variabilis* Blume community are *Q. mongolica* in tree layer, *Q. mongolica*, oak in subtree layer, oak in shrub layer, and Aster scaber Thunb in herbaceous layer. The coverage rate is 60–70 % in tree layer, 10–15 % in semi-high tree layer, 5–15 % in shrub layer, and 65 % in herbaceous layer. Species distributed by class within *Q. acutissima-Q. mongolica* community are *Q. acutissima* CARRUTH in tree layer, Quercus aliena Blume in subtree layer, *Smilax china* L. in shrub layer, and ladybell in herbaceous layer. The coverage rate is 25 % in tree layer, 5 % in subtree layer, 55 % in shrub layer, and 60 % in herbaceous layer.

Wetland vegetation in DMZ

Detailed survey on the plant communities at the wetlands is limited since access is prohibited especially in areas declared as "danger areas" where landmines are purposely planted near military facilities. Hence, the survey targeted plant communities that are visible through telescope and those in permitted areas. There were twelve major plant communities investigated in the wetland vegetation and these are the following communities of; (1) Scirpus karuizawensis, (2) *Scirpus fluviatilis*, (3) *J. effusus* var. decipiens, (4) *J. effusus* var. decipiens-*Typha orientalis*, (5) *Juncus alatus*, (6) *Miscanthus sinensis*, (7) *Phragmites communis-Miscanthus sinensis*, (8) *Phragmites communis*, (9) *Phragmites communis-Alnus japonica*, (10) *A. japonica*, (11) *Phragmites japonica*, and (12) *Amorpha fruticosa* community.

Species composition of major communities are *Scirpus maritimus* L., *J. effusus* var. decipiens Buchenau, *Lespedeza bicolor* Turcz. in Scirpus karuizawensis Makino community, Scirpus karuizawensis Makino, *L. bicolor* Turcz., in *S. maritimus* L. community, *S. maritimus* L., *Persicaria sagittata* (L.) H.Gross ex Nakai in *J. effusus* var. decipiens Buchenau community, *S. maritimus* L., *P. japonica* Steud, *P. sagittata* (L.) H.Gross ex Nakai, Cyperus microiria Steud., *J. effusus* var. decipiens Buchenau in *J. effusus* var. decipiens Buchenau community, oak for subtree layer, *L. bicolor* Turcz. and *S. koreensis* for shrub layer, *Miscanthus sacchariflorus* var. gracilis Y.N.Lee, *S. maritimus* L., water pepper, *Artemisia princeps* var. orientalis (Pampan.) Hara, *Artemisia keiskeana* Miq. Hara in *P. australis* community, also in *P. japonica* Steud community, *L. bicolor* Turcz., a wild rosebush flower, *Robinia pseudoacacia* L., *S. koreensis* for shrub layer, *J. effusus* var. decipiens Buchenau, *S. maritimus* L., water pepper, *A. princeps* var. orientalis (Pampan.) Hara, *Erigeron annuus* (L.) Pers., *Typha orientalis* C.Presl, *Phalaris arundinacea* L., etc., mostly appear for herbaceous layer.

Existing vegetation map in DMZ and CCZ

Existing Vegetation Map in Donghae Railroad and Road Area

After the environmental survey of Donghae railroad, and the road connection project area in the CCZ, mapping of the existing vegetation was conducted through an on-site survey. However, as this area is not fully accessible due to regional and military restrictions, the existing vegetation map drafted may not be accurate.

Based on the vegetation survey of the area as evidenced by the photos, *Q. variabilis* Blume community under 40 % of cover degree was formed at the widest in terms of tree layer. *Q. variabilis* Blume community under 60 % of cover degree in tree layer, *Q. variabilis* Blume community, cultivated fields and coastal sand dunes were also investigated as major vegetation.

In the wetland, *A. japonica* Steud., *S. koreensis*, *P. australis* are the major species that spread over the areas, and exotic species are also observed such as; *R. pseudoacacia* L., *L. bicolor* Turcz. are scattered inside and outside.

The reason why the cover degree of tree layer is below 60 % is due to logging activities, military characteristics, and forest fire. Having a cover degree of less than 60 % in tree layer would mean that the forest vegetation in DMZ and CCZ is severely damaged.

The wetland was formed from what used to be cultivated fields that were neglected for 60 years after the war. Near the wetland, *Q. acutissima* CARRUTH that used to be the main tree in the cultivated field and in the abandoned town, is now forming a community of *A. japonica* Steud., *S. koreensis*, *R. pseudoacacia* L. that dominate the forest vegetation in the wetland. *P. australis* is the most dominant species in some parts of the wetland vegetation. Zizania latifolia, *P. japonica* Steud, and *S. maritimus* L. are the main inhabitant. *R. rugosa* var. rugosa and others formed a community in the coastal sand dunes. Inside the coastal sand dune, oak, and pine trees widely formed a community. In the forest area, across the railroad and the road, *Q. mongolica* Fisch. ex Ledeb. and *Q. variabilis* Blume formed a community, while *A. japonica* Steud., *S. koreensis*, *Prunus padus* L. for. padus and *Quercus xurticaefolia* Blume thrived in the lower wetland. The wetland was once a cultivated field that had undergone ecological succession and now has various wetland plants such as *P. australis*, *P. japonica* Steud, *S. maritimus* L., *Typha orientalis* C.Presl, *Typha angustifolia* L., living in. *P. japonica* Steud and *P. australis* are the typical wetland plants in this area. In the sandy dunes, *M. sacchariflorus* var. gracilis Y.N.Lee was observed flourishing mainly in those areas.

However, the damaged forest, vegetation requires urgent restoration of its original and potential vegetation. *L. bicolor* Turcz. and *R. pseudoacacia* L. have invaded the wetland vegetation which disturbed wetland vegetation and also required immediate countermeasures (Fig. 5.39).

Existing wetland vegetation map in Northern Donghae railroad line

DMZ Area

The distinguishable main community of wetland vegetation at the DMZ is composed of; *P. australis*, *P. japonica* Steud, *S. maritimus* L., *M. sacchariflorus* var. gracilis Y.N.Lee, Scirpus karuizawensis Makino, *J. effusus* var. decipiens Buchenau. Although *A. japonica* Steud. mainly forms a community in fringe forest vegetation near wetland, *S. koreensis*, and *P. padus* L. for. padus are also scattered as accompanying species. Especially where a sand dune is developed within the wetland, *P. japonica* Steud community occupied most of those areas.

S. maritimus L. and *P. australis* communities are also found in the wetland areas. The *J. effusus* var. decipiens Buchenau and *E. annuus* (L.) Pers. communities are naturalized plants formed in damaged areas, and *Conyza canadensis* (L.) Cronquist community in the lowland wetland connected to DMZ railroad. *Q. mongolica* Fisch. ex Ledeb., pine, *A. japonica* Steud., *Q. xurticaefolia* Blume, *P. padus* L. for. padus, *S. koreensis* Andersson, and *Quercus dentata* Thunb. ex Murray are scattered near the wetland while the *R. pseudoacacia* L. and *L. bicolor* Turcz. are observed in the drying-up wetland.

Aside from the communities of plants distributed in the wetland, other wetland plants such as; *P. japonica* Steud, *L. bicolor* Turcz.v, Calamagrostis epigeios (L.) Roth are also growing there. But terrestrial plants frequently appear such as; *Equisetum palustre* L., *Hypericum ascyron* L., *Senecio vulgaris* L., Humulus japonicus Sieboid & Zucc., *A. princeps* var. orientalis (Pampan.) Hara,

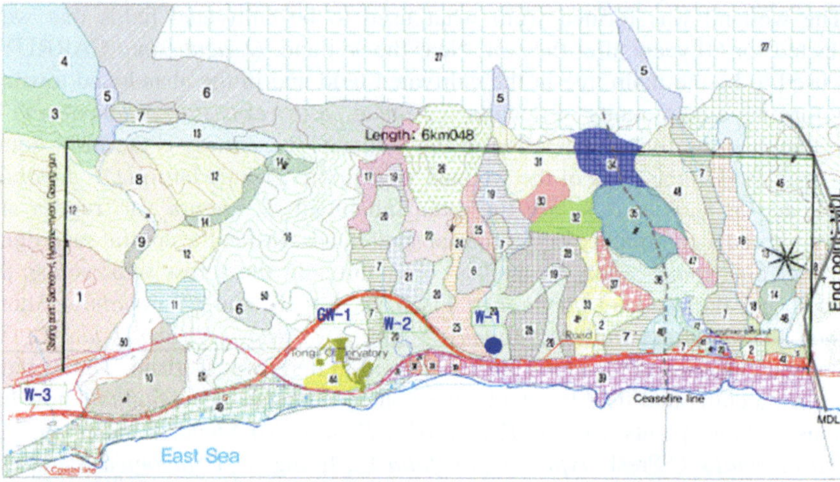

	Community		Community		Community
1	Pine–Cork oak(K)	20	Reed(F)	39	Daimyo oak, Pine, Rosa rugosa(F), Carex pumila, Elymus mollis Trinius
2	Willow–Alder tree	21	Silk worm oak(B)	40	Pine–Silk worm oak(C)
3	Konara oak–Cork oak(B)	22	Cork oak(A)	41	Pine–Silk worm oak(E)
4	Konara oak–Cork oak(D)	23	Silk worm oak(K)	42	Mongolian oak–Pine(F)
5	Mongolian oak–Konara oak(F)	24	Pine(C)	43	Reed–Common reed(F)
6	Cork oak(D)	25	Mongolian oak(D)	44	kuzu vine(F)
7	Alder tree(F)	26	Silk worm oak–Cork oak–Mongolian oak(E)	45	Cork oak–Mongolian oak(B)
8	Cork oak(E)	27	Mongolian oak	46	Willow(shrub)–Manchurian Alder(shrub)–Flame Grass
9	Mongolian oak–Daimyo oak(F)	28	Silk worm oak(C)	47	Pine(B)
10	Mongolian oak(shrub)	29	Stream, Bare land(wetland plant)	48	Cork oak–Silk worm oak(C)
11	Cork oak(C)	30	Willow–False acasia(D)	49	Rosa rugosa(F), Carex pumila, Elymus mollis Trinius
12	Cork oak(B)	31	Silk worm oak–Cork oak(E)	50	Farmland
13	Cork oak–Pine(C)	32	Silk worm oak–Cork oak(D)	A	Cover degree 10%
14	Pine(D)	33	Mongolian oak–Silk worm oak(F)	B	Cover degree 20%
15	Willow(F)	34	Silk worm oak(D)	C	Cover degree 30%
16	Cork oak(K)	35	Silk worm oak–Cork oak(F)	D	Cover degree 40%
17	False acasia–Silk worm oak	36	Silk worm oak(E)	E	Cover degree 50%
18	Willow–Alder tree(F)	37	Willow–Alder tree(F)	F	More than cover degree 60%
19	Willow–False acasia(F)	38	Bastard indigo(F)	K	Below cover degree 60%

Fig. 5.39 Survey area in Northern Donghae railroad

Crepidiastrum sonchifolium, *Oenothera biennis* L., Crepidiastrum sonchifolium, *Petasites japonicus* (Siebold & Zucc.) Maxim., Vitis amurensis Rupr., Hemerocallis dumortieri Morren, Rubia akane Nakai, Viola kapsanensis Nakai, etc. However, the wetland vegetations in sites 4, 5, and 13 of the DMZ have been preserved relatively well (Fig. 5.40). The lowland forest wetland that is situated toward the interior of the railroad near the gate of DMZ is considered the entrance to DMZ. The wetland vegetation in site 4 was the best in the DMZ railroad area. Forested wetland formed between sites 4 and 5 toward the ridge as its boundary. Some areas in this wetland are dried up because of the railroad construction. In addition, there are a lot of depressions developing near the wetland in site 13, which are still within the DMZ area.

P. australis community, *S. koreensis* Andersson community, *A. japonica* Steud.-*P. australis* community, *L. bicolor* Turcz. community, *M. sacchariflorus* var. gracilis Y.N.Lee community and *A. japonica* Steud.-*S. koreensis* Andersson community are forming in lowland wetland near DMZ site 13. *P. japonica* Steud is forming in a large community along the sandy soil of the railroad near site 13.

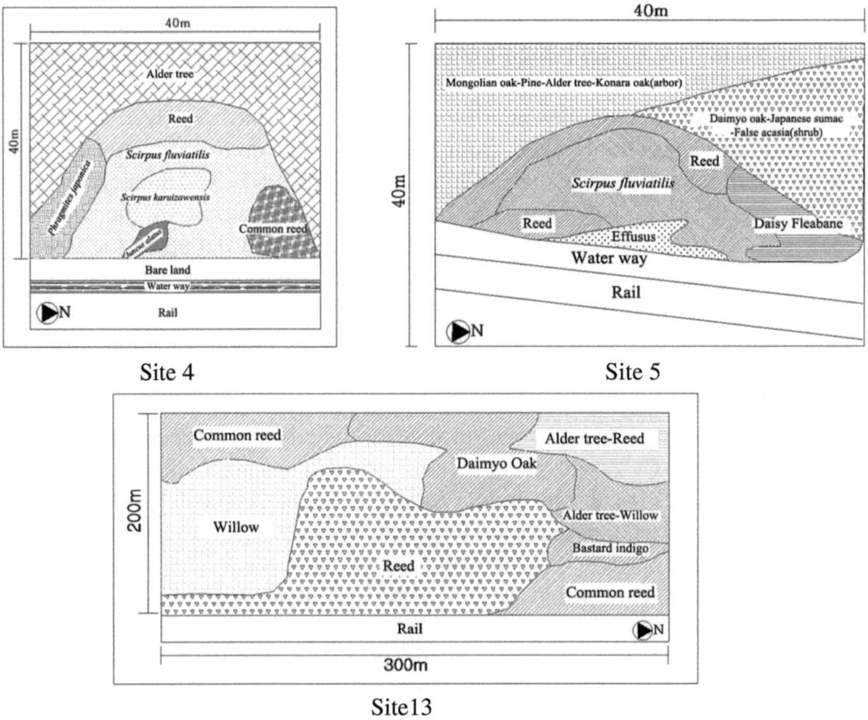

Fig. 5.40 Major existing wetland vegetation map near Northern Donghae railroad within DMZ

Along the forest area near the wetland, *Q. xurticaefolia* Blume, *Q. variabilis* Blume, *Q. mongolica* Fisch. ex Ledeb. are easily seen. Vegetations near this area are *S. maritimus* L., *L. bicolor* Turcz., *E. palustre* L., *T. angustifolia* L., Zizania latifolia and the others. However, in the wetland where *P. australis* is dominant, *S. koreensis* Andersson, *A. japonica* Steud., and *L. bicolor* Turcz. are scattered, which accelerated the drying-up and destruction of the vegetation in the wetland. In particular, the drying-up of the wetland seems to be apprehended quickly in some areas because of the changes in drainage consequent to the construction of the railroad and roads which resulted to succession of the growth of *Q. dentata* Thunb. ex Murray, *Q. xurticaefolia* Blume, and *Q. mongolica* Fisch. ex Ledeb (Fig. 5.40).

Coastal sand dunes

In the coastal areas, various sedimented landforms such as sand beach, lagoon, coastal sand dune, sea cliff, land-tied island, coastal terrace had been developed along the seaside of Korea. In the eastern coast, large-scale sand beaches were continuously formed due to the movement of massive earth and sand towards the sea outline brought about by heavy rain from the steep slope. There are 32 large and small coastal sand dunes near Gangwon province which is of great ecological value in DMZ and CCZ. Vegetation is well developed especially within the coastal sand dunes and had been preserved without any human disturbance for 60 years.

An ecological ecotone connects the marine ecosystem of the East Sea to the coastal sand dunes ecosystem, then to the inland wetland and forest ecosystem of the DMZ and CCZ. It now acts as a buffer and a strong barrier to any climate change-related catastrophe. Seeing that coastal sand dunes are usually formed where there are large tidal range from the west and south coast due to topographical conditions, the coastal sand dunes near DMZ and CCZ area are of great value for preservation. However, the large-scale construction of the Donghae railroad and the road connection near DMZ and CCZ area damaged the lowland wetlands near the coastal sand dunes' vegetation. It is not only a very threatening factor for a large number of animals and plants living in the coastal sand dunes and lowland wetland but also leads to total destruction of the precious habitat. After the construction, it is highly recommended that environmental restoration, conservation, and maintenance initiatives must be implemented.

The characteristics of the vegetation that are widely developed in the coastal sand dunes near the CCZ, are those of tall herbs, short shrubs, herbs with well-developed root system that can endure strong sunlight. Typical dune plants are *E. mollis* Trinius, Aster meyendorfii (Regel & Maack) Voss, and *R. rugosa* var. rugosa and the others.

The vegetations that are widely formed at the low inland wetland near the coastal sand dunes and hill areas which are expected to form coastal forest composed of *Q. dentata* Thunb. ex Murray, and pine trees. *Q. mongolica* Fisch. ex Ledeb., *Q. variabilis* Blume and *Q. xurticaefolia* Blume are the main species identified in the lowland wetland and hill areas across the railroad and the roads. It is also the area where *P. australis*, *P. japonica* Steud, *S. maritimus* L., *A. japonica* Steud. and *S. koreensis* Andersson would form the forest vegetation in lowland wetland. *P. japonica* Steud, and the *M. sacchariflorus* Benth are expected to dominate.

The existing vegetation map and the vegetation transect mimic of the coastal sand dunes and sandy areas near CCZ and DMZ are shown in Figs. 5.41 and 5.42. *R. railroad* var. rugosa community is scattered in the stable belt with pine-shrub forest behind it. However, the railroad and road construction made an axis across

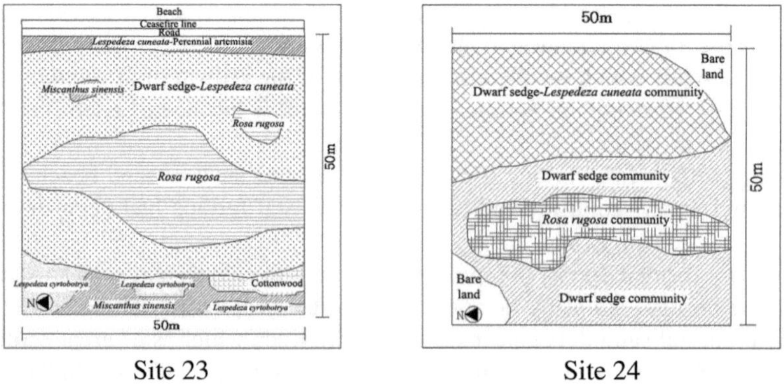

Fig. 5.41 The existing vegetation map of major wetland near coastal sand dunes in DMZ

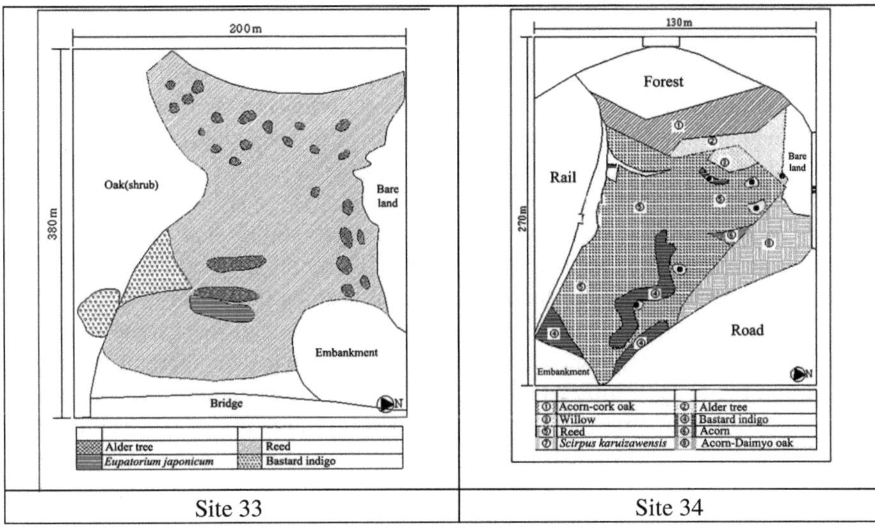

Fig. 5.42 The existing vegetation map of major wetland near Donghae railroad in CCZ

the ecotone between pine-shrub forest and the vegetation breaking away the ecological corridor of marine, forest, and lowland wetland ecosystems.

In addition, since the naturally developed ecotone was obliterated by the construction, that which should have been an effective sand retaining barrier, it is now expected that the coastal sand dunes would rapidly collapse, damage its own ecosystem and, eventually damage the marine ecosystem. Then, tidal waves will inevitably reach the inland vegetation because of the absence of a barrier, destroying all its natural beauty and significant purpose for human and ecological security.

CCZ Area

As one of the most developed wetlands near CCZ area, sites 33 and 34 in Fig. 5.42 have various species composition, wide wetland, and relatively stable wetland vegetation. The wetlands in sites 33 and 34 are connected to each other. Sites 33 formed a forest valley in the upper part of the roads and, in site 34 there are widely formed lowland wetlands between the road and the railroad. The dominant species of these wetlands are *P. australis*, *S. koreensis* Andersson, *A. japonica* Steud., *P. padus* L. for. padus, Fraxinus rhynchophylla Hance, *R. pseudoacacia* L., *Q. acutissima* CARRUTH, *Q. variabilis* Blume and *Q. mongolica* Fisch. ex Ledeb. *L. bicolor* Turcz., *A. japonica* Steud. shrub, *Q. acutissima* CARRUTH shrub, and *S. koreensis* Andersson shrub have invaded and are scattered in the *P. australis* community in wetland. *L. bicolor* Turcz is widely scattered, forming a community which accelerates the drying-up of wetland.

Recently, an ecological bridge has been constructed between site 33 and 34 to eliminate the spatial block. But there are still many other factors and variables to consider in order to maintain the various environments and wetland vegetations. These factors include the vegetative transformation along the drainage built in the construction site, the topographic/landscape changes that disrupted the vegetation of specific species and the artificial disturbance.

Although DMZ wetlands maintain the best wetland ecosystem in the Korean peninsula because of the absence of artificial interference or disturbance, some wetlands had been blocked, some are partially dried up, and some are totally damaged due to the South–North Donghae railroad and road project. Ecological restoration has been made since then, nevertheless, the distribution status of the existing vegetation near the project area is not yet stable. The status of major vegetation communities within the wetlands is crucial in predicting the future changes of vegetation due to the emergence of species that disturbed the natural succession of important plant species in this area. Although the variety of certain special plants and rare plants are very highly populated near Northern Donghae railroad and road, the military facilities, related activities, landmines, and limited access made it impossible to conduct a detailed investigation of the wetland vegetation. However, rapid increase of plants that has naturally adapted to the construction site has made a huge effect on DMZ wetland. Forest vegetation appeared in the wetlands near the railroads and roads and is increasing its population.

In case, individual plants or groups are in danger, special protective measures should also be taken outside the conservation site. It is essential that the rare and endangered species living in this wetland be understood and preserved. In particular, in situ conservation shall be done for targeted plants.

5.2.5.2 Insects

Insects in DMZ and CCZ are mostly found in the sunny wild grasslands, forest areas, and wetland areas. They also have rich diversity due to its connectivity to the neighboring natural environment. Insects in the DMZ and CCZ areas had not been accurately defined because no detailed investigation has been done. It is expected that many more species could be added to those that had been listed in this book (Table 5.14).

Lycaena dispar

Features of Lycaena dispar habitat in DMZ and CCZ area

Neglected for a long time, wild grassland has been formed and herb vegetation has naturally settled in DMZ and CCZ. Most of them are annual herbaceous plants namely; *Rumex crispus* L., *E. annuus* (L.) Pers. and *C. canadensis* (L.) Cronquist, which are the habitat of *L. dispar*. Abandoned fields near the forest are very suitable for butterflies to live in as the grassland which consists of a succession of flowering plants.

In Fig. 5.43, abandoned rice paddy fields transformed into wild grassland in many areas including Paju DMZ and CCZ. With its neighboring environment connected to the forest and the wetland, such areas provide habitats to various insects. *L. dispar* is found here often and the ecological food web is operated well by predaceous insects or birds who feed on butterflies.

Table 5.14 Classification and characteristics of insects that are described in this section

Classification	Name	Characteristics	Distributed area
Lepidoptera	*Lycaena dispar*	Butterfly that has the typical characteristics of palearctic region living in northern Gyeonggi province and Gangwon province. Its number has decreased due to the damage and disappearance of habitat and recently found in northern Gyeonggi province nd Gangwon province. Gathers to *Rumex crispus* L. and flowers such as *Erigeron annuus* (L.) Pers.	Amur, Europe
	Clossiana selene	24 mm of fore-wing, yellow belt in the middle of backside of hind wing, silver pattern on the outside edge. Easily seen in March–September, lives in base of a mountain or grass of a wild mountain, flies slow and low	Korea · China · Sakhalin · Amur
Coleoptera	*Prosopocoilus inclinatus*	Widely distributed throughout the country, mostly found in forest where *Quercus xurticaefolia* Blume and *Quercus mongolica* Fisch. ex Ledeb. are dominant. The most impatient among Lucanidae, threatens enemies with chin or bites to lift up to protect themselves	Korea · Japan · China (Northeastern)
Odonata	*Nannophya pygmaea* Rambur	The smallest dragonfly living in Korea, found in spring in warm areas, in wetland or swamp in summer in cold areas. Weak flying ability, lifts stomach up in the air as in handstand for body temperature control during the day	Korea · Japan · China (Central·Southern) Taiwan
	Coenagrion lanceolatun	No fight for territoriality as large number of individuals copulate in groups in limited habitats. Male and female not separated after mating or only female spawns by putting ovipositor onto the stem of a aquatic plant. The length of an imago is 12–13 mm, light brown color, and lives in flat wetland	Korea · Japan · Sakhalin Island · China (Northeastern) · Siberia

Fig. 5.43 Wild grassland in Inje CCZ: collecting butterflies

Considerations for securing the habitat

The wild grassland functions as a supplier of food for predatory birds but it is difficult to preserve it as the environment is unstable and constantly changing every year. Grasslands can be seen near well-preserved forests, where insects including *L. dispar* are active and many other living things can live and find food. Thus, grassland near forests should be preserved.

Clossiana Selene

Features of C. selene habitat in DMZ and CCZ

There is grassland at the valley of a mountain in DMZ and CCZ that was an abandoned farmland. In this field which is along the riverside, there are herbaceous plants that are normally seen in wild grasslands such as Arenaria serpyllifolia, *E. annuus* (L.) Pers., and Viola kapsanensis Nakai, etc. In an isolated area from the cultivated field, extensive growth of impatiens textori is observed there, along with the cranesbill. Such a habitat is favorable for clossiana slelen to live.

The grasslands that are formed in the DMZ and CCZ are very typical. Especially the gill from neighboring mountain and the streams provide good habitats for butterflies since there are various kinds of mixed vegetation around (Figs. 5.44, 5.45).

Fig. 5.44 Wild grassland in Hwacheon CCZ

Fig. 5.45 *Clossiana Selene*

Considerations for securing the habitat

Various butterfly species are observed in the grasslands connected with surrounding forests, streams, and wetlands. However, some areas of their habitat had been damaged when the riverside was concreted and the grasslands were increasingly cultivated. If these developments are allowed to continue, the butterflies will become extinct. Therefore, the grassland habitat should be preserved and maintained too. Grasses and bushes along streams and rivers, wetlands and forests, open lands and hills should be restored and allowed to flourish for the butterflies to sustain.

Prosopocoilus inclinatus

Features of P. inclinatus habitat in DMZ and CCZ

Trees and shrubs in the montane zone of Inje-county and Goseong-county, in the Yanggu-county (east), and in the Cheorwon-county (center) of DMZ and CCZ are the primary habitats of the lucanidae insects. Because the access to DMZ and CCZ has been limited, these habitats have naturally succeeded to its becoming a good haven for *S. koreensis* Andersson-*R. pseudoacacia* L.-*Q. acutissima* CARRUTH and *Q. mongolica* Fisch. ex Ledeb. Cheorwon in Fig. 5.48 and Goseong in Fig. 5.47 show the areas that have similar characteristics and where *P. inclinatus* lives (Fig. 5.46).

P. inclinatus are distributed throughout the country and live in trees and shrubs where there are *Q. xurticaefolia* Blume and *Q. mongolica* Fisch. ex Ledeb. Being the most impatient insect among lucanidae, *P. inclinatus* protects itself by threatening its enemies with its chin and bites them.

Considerations for securing the habitat

The shrubs and trees in DMZ and CCZ are ecologically excellent as an ecotype. It is mostly covered with minefields thus, people's access is prohibited. However, landmines in some areas had been successfully removed in order to be cultivated.

Fig. 5.46 Woodland patch near Gyewoong mountain in Cheorwon CCZ

Fig. 5.47 Trees and shrubs in Goseong CCZ

Fig. 5.48 Lowland wetland in Paju DMZ

Fig. 5.49 *Nannophya pygmaea* Rambur

This development threatens the existence of the insects. Such areas should revive its forests to be a conducive habitat for entomofauna. It should allow trees and shrubs to grow and be properly maintained or find immediate alternative habitats when damaged.

Nannophya pygmaea

Features of N. pygmaea habitat in DMZ and CCZ

In DMZ and CCZ, the lowland wetlands are well developed under hilly areas and mountains. Some areas used to be farmlands in the past and became wetlands after prolonged exclusion from cultivation. Specifically, in the western DMZ (Fig. 5.48) flood plains are formed. In the area where the estuary of the Han River and Sacheon River meet, there is a widely formed wetland. A huge *P. australis* community occupied most of the wetland area and became a sanctuary of birds, reptiles, amphibians, and insects (Fig. 5.49).

Here, the *N. pygmaea* Rambur lives in the rice paddies that turned into a marsh in the Paju portion of the DMZ. It turned out to be a very good habitat, same as all the other transformed lowland wetlands and stream wetlands. *N. pygmaea* Rambur usually feed on dayflies or Tipula KUb insects living in small wetlands such as; flies that live in ponds and reservoirs that have open surfaces.

Considerations for securing the habitat

A few years ago, *N. pygmaea* Rambur are usually seen in large groups throughout the country. However, as forest lands are cultivated and wetland disappeared in favor of various developments, *N. pygmaea* Rambur also disappeared and nowhere to be found. DMZ and CCZ area may be a better and safe habitat now for *N. pygmaea* Rambur. But it is not also free from the pressures of development. In truth, development seems to be eyeing on this place.

Currently, *N. pygmaea* Rambur is designated as a natural heritage and considered as endangered species grade II by the Ministry of Environment. In order to preserve *N. pygmaea* Rambur, it is most important to preserve their existing habitat and restore those that are damaged.

Fig. 5.50 Lowland wetland in a story shooting range in Paju CCZ

Coenagrion lanceolatun

Features of C. lanceolatun habitat in DMZ and CCZ

Since there is little interference from human beings, DMZ and CCZ wetland ecosystem remained relatively stable. An ample number of organisms had been able to adapt and live here especially water birds, aquatic plants, and insects. They found a suitable habitat in the wetland ecosystem of the DMZ and CCZ.

The Story shooting range in Paju CCZ (Fig. 5.50) is used by the UN Forces. Access to this area is prohibited except for a few weeks when used as a shooting range. In the lowland below the hilly areas, which is an abandoned rice paddy field, various aquatic plants such as; *J. effusus* var. decipiens Buchenau, *P. australis* and Zizania latifolia live in this area. Despite the fact that a detailed investigation was not possible, a number of birds and insects were found during the on-site survey, including the sea bindweed (Fig. 5.51).

Considerations for securing the habitat

The wetland in DMZ and CCZ has been well preserved for a variety of aquatic plants to live. The water gate is well connected and its structures and functions are well maintained, especially in the wetland area which is tightly connected to the forests and grasslands, with high biodiversity and active movement of ecological genealogy. The wetland also is the habitat where the rare sea bindweed lives, so it has to be preserved and maintained.

The wetland in DMZ is not a separate ecosystem that can be divided into South and North, but one ecological unit that requires collaborative effort of South and

Fig. 5.51 *Coenagrion lanceolatun*

North. It is of prime importance that this place be designated as a Ramsar site for global interest and collaboration. This collaboration should not only focus on the preservation of South–North ecosystem but must also explore other areas of common interest like peace of its people and a healthy neighborhood.

5.2.5.3 Fish

Table 5.15.

Hemibarbus mylodon

Figure 5.52.

Features of H. mylodon habitat in DMZ and CCZ

According to Environment Ecosystem Research Group, the slopes of most northern Han river area are 60–100 %, while, in the valley area, the slopes are 30–60 %. The streams in these areas flow down from North to South Korea since the tributaries in the mountains are extensions of the northern Han river. The soil along the streams of northern Han river has very small argillaceous particles while the soil annexing the watersides of the streams consists of pebbles and organic soil. This ecosystem extended to the well-formed sandy plains down to the wetland and inland ecosystems, all are naturally connected and served as a natural bank protection. Due to these environmental characteristics, it evolved as a very suitable site for *H. mylodon* (Berg) to live (Figs. 5.53, 5.54).

Considerations for securing the habitat

The *H. mylodon* (Berg) has been designated as a Natural Monument. It is sensitive to the changes of environment which limited its habitat and the threat for extinction is a clear and eminent danger for this fish. In order to prevent its extinction, it is very important to look at its unique behavior. Its population has been decreased by overfishing and by the water pollution of the streams that survive it. For this reason, the Ministry of Environment has designated it as a specially protected fish in January, 1996 and prohibited its collection, capture, process, and distribution without permission. In addition, a unified stream ecosystem shall be made from the open surface stream, to the waterside and wetland areas, on to riparian buffer areas and into inland habitats. This unified area is reserved for *H. mylodon* fish.

Microphysogobio longidorsalis

Figure 5.55.

Features of habitat in DMZ and CCZ

Surrounded by forests, the northern Han river area in DMZ and CCZ has an excellent natural environment and high quality of water where plenty of wildlife animals are revealed by the survey conducted by the Environment Ecosystem Research Group. In fact, the wetland vegetation in northern Han river in Hwacheon-county (Fig. 5.56) is covered with; *S. koreensis* Andersson, Salix koriyanagi Kimura for. koriyanagi, Salix gracilistyla Miq., Salix integra Thunb., Alnus firma, Fraxinus rhynchophylla Hance, Acer tataricum subsp. ginnala (Maxim.) Wesm. and Phalaris arundinacca.

Table 5.15 Classification and characteristics of fish described in this section

Classification	Name of species (scientific name)	Characteristics	Distributed Area
Carp	*Hemibarbus mylodon*	First order freshwater fish, Korean endemic species, 238th Natural Monument, During the period from April to May, male and female *Hemibarbus mylodon* (Berg) seek for where to spawn. Make oval puddle and spawn. After laying 1,200–2,300 eggs, they carry gemlike pebble to pile a spawn tower, *Hemibarbus mylodon* (Berg) who does not stop piling the tower even with blooding mouth reminds us of maternal affection	Hwacheon Northern Han river
	Microphysogobio longidorsalis	First freshwater fish, Korean endemic species, as carp fish, it lives on the bottom of clean and sandy stream.	Hwacheon Northern Han river
	Microphysogobio jeoni	First freshwater fish, Korean endemic species	Hwacheon Northern Han river
Cobitis hankugensis	*Koreocobitis rotundicaudata*	First freshwater fish, Korean endemic species, living on a bottom of clean stream with fast velocity	Hwacheon Northern Han river
	kosookimia koreensis	First freshwater fish, Korean endemic species	Hwacheon Northern Han river
Catfish	*Silurus microdorsalis*	First freshwater fish, Korean endemic species	Hwacheon Northern Han river
Coreoperca herzi (Herzenstein)	*Coreoperca herzi*	First freshwater fish, Korean endemic species	Hwacheon Northern Han river
Mandarin fish	*Siniperca scherzeri*	First freshwater fish	Hwacheon Northern Han river
Arctic lamprey	*Lampetra japonica*	Endangered wild critters II designated by Ministry of Environment	Estuary of Anhocheon in Goseong
Short ninespine stickleback	*Pungitius kaibarae*	Endangered wild critters II designated by Ministry of Environment	Estuary of Anhocheon in Goseong

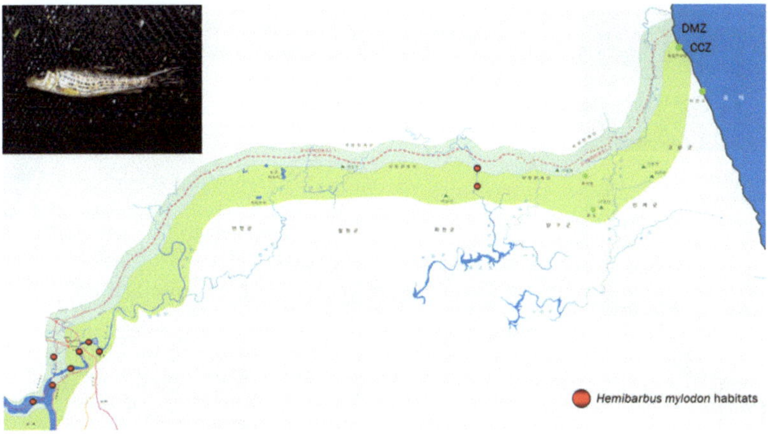

Fig. 5.52 Identified *Hemibarbus mylodon* (Berg) habitats in DMZ and CCZ. ※ The 259th Natural Monument, Korean endemic species

Fig. 5.53 Northern Han river in Hwacheon CCZ

Fig. 5.54 *Hemibarbus mylodon* (Berg)

Fig. 5.55 Identified *Microphysogobio longidorsalis* habitats in DMZ and CCZ. ※ Korean endemic species

Grasslands naturally formed in the buffer zones, between mountains, streams and tributaries that go down from all over the mountains. These stream-grassland-mountain habitats increased biodiversity for it has become a haven of various organisms, including the *M. longidorsalis* species (Fig. 5.57).

Considerations for securing the habitat

C. canadensis (L.) Cronquist and *Ambrosia trifida* L. live in the dry area near the transition area to northern Han river. The rampant spread of these weeds poses as a problem to the fish, and a countermeasure is required. Such exotic plant species have to be removed and should be directly pulled manually or by a machine to uproot deeply buried parts. The best way to do it is to deliberately change its environment by decreasing the concentration of nutrients and change in hydrological system to hinder the growth and spread of these exotic species.

Microphysogobio jeoni

Figure 5.58.

Features of M. jeoni habitat in DMZ and CCZ

As it was presented earlier, the eastern DMZ and CCZ consists of mountains and the valleys. From this area, a big river flows forming wetlands along the northern Han river. The upstream of this river flows from the Ojak bridge to the Peace dam with a distance of 19 km. This area is supplied by the northern Han river where many tributaries flow in ecosystem and environments that have been well preserved for many years. It is the best habitat for *Lutra lutra*. With its good quality water and environment, a variety of fish live in this area and it is also the best habitat for *M. jeoni*.

Figure 5.59 illustrates the research group investigating fish in northern Han river basin in Hwacheon and Fig. 5.60 shows the *M. jeoni* observed in this area.

Fig. 5.56 Upper northern Han river in Hwacheon CCZ

Fig. 5.57 *Microphysogobio longidorsalis*

Considerations for securing the habitat

The eastern area of DMZ and CCZ has been given a lot of external pressures. Military facilities and roads had been constructed and planned for more developments recently which are already becoming a threatening element for the fish habitat. The worst threats are the water pollution due to the installation of dams and the increasing growth of unwanted exotic species that blocks fish passage. The fish habitat needs to be maintained. These facilities have to be removed to allow fish's movement or create a route (reroute) for fish movement, if removal is inevitable.

White-nose loach (*Koreocobitis rotundicaudata*)
Figure 5.61.
Features of White-nose loach habitat in DMZ and CCZ

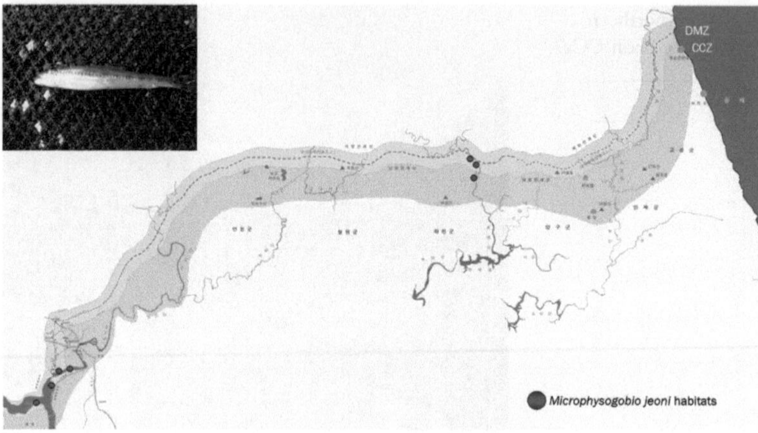

Fig. 5.58 Identified *Microphysogobio jeoni* habitats in DMZ and CCZ. ※ Korean endemic species

Fig. 5.59 Investigating fish in northern Han river in Hwacheon CCZ

Fig. 5.60 *Microphysogobio jeoni*

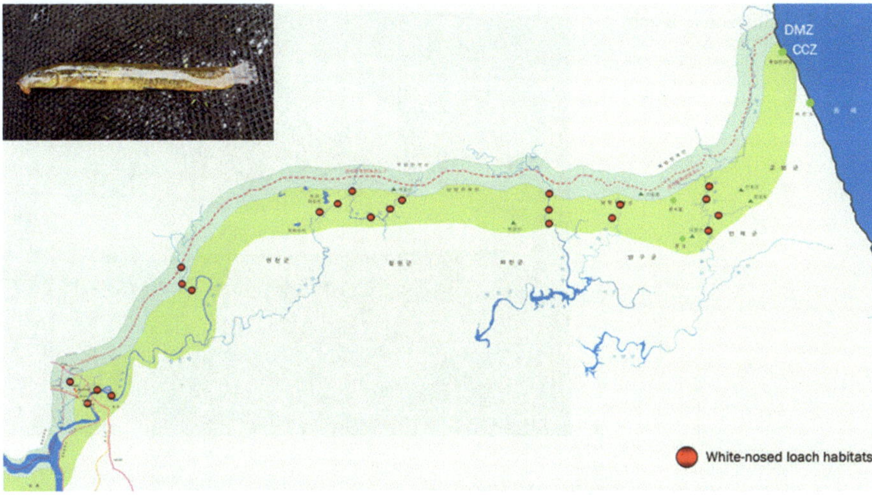

Fig. 5.61 Identified White-nose loach habitats in DMZ and CCZ. ※ Korean endemic species, rare species

The Cheorwon is located at the central part of DMZ and CCZ right in the middle of the Korean peninsula. It forms a boundary with Pocheon-county in Gyeonggi province to the south and there is the Southern Limit Line to the north. There is a Myungsung-dan at an altitude of 923 m in the east of the area, which forms topography of hills in the mountain. To the west, there is a vast plain where residents can cultivate rice farms and the Hantan river serves as the boundary. Such environmental characteristics favor wildlife animals, like the *G. japonensis* and the White-naped crane that pass over in groups during winter in the Cheorwon plain. The Hantan river is the biggest body of water in Cheorwon. Hantan riverbed consists of sand with 15–20 m in width. The *S. koreensis* Andersson community is formed at the water's edge. With these environmental characteristics, Hantan river area is very suitable for White-nose loach species to live (Figs. 5.62, 5.63).

Fig. 5.62 Hantan river in Cheorwon CCZ

Fig. 5.63 White-nose loach

Considerations for securing the habitat

The open fields near the Hantan river had been cultivated and the many tributaries connected to it consist of artificial banks or reservoirs. Some of the tributaries were polluted by rust. In addition, diverse exotic plant species such as *O. biennis* L., *C. canadensis* (L.) Cronquist and Bidens tripartita L. had flourished near the water's edge. Fundamental causes of the environmental problems need to be identified so that appropriate countermeasures can be implemented. Since most of the areas in the Cheorwon are owned by the residents, community-based countermeasures must be implemented with their cooperation. For example, the understanding and cooperation of residents should be solicited especially in terms of their active participation in the maintenance of the streams and tributaries that they also use for agriculture and as a main source of water for household use. The residents should cooperate for the implementation of an eco-friendly agriculture and refrain from using chemical fertilizers and pesticides. The Government plays an important role in this cooperation by imposing protective laws, considering financial incentives, recommending wetland maintenance procedures and methods of preserving the river basin, nationalizing important habitat areas of waterside wetlands connecting fragmented habitats to other habitats.

Iksookimia koreensis

Figure 5.64.

Features of I. koreensis habitat in DMZ and CCZ

Various wild animals live in streams since the wetlands are well developed at the water's edges and a natural environment is created for this ecosystem network. More fish species are found in this area than in other areas because the *L. lutra* is living here. *L. lutra* is an endangered species considered as one of the high consumers in the food web, and its presence in this area could only mean that food is abundant.

The Environment Ecosystem Research Group described critically in their report the aforementioned characteristics of the streams and identified the endemic species that live in the Northern stream, of Goseong CCZ (Fig. 5.65). The streams

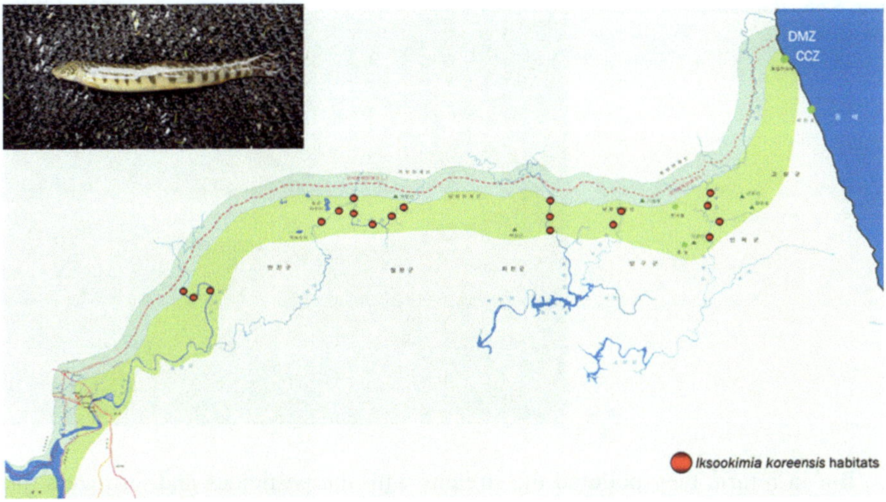

Fig. 5.64 Identified *Iksookimia koreensis* habitats in DMZ and CCZ. ※ Korean endemic species

Fig. 5.65 Northern stream in Goseong CCZ

emanate from Hyangrobong and Geon-bong mountain, and along the course of Geum-gang mountain. The qualities of water, the streams produce are excellent to sustain life. In the mid-downstream, fast velocity of stream is noticed and the waterbed is composed of pebbles which is a perfect habitat for *I. koreensis* (Fig. 5.66).

Considerations for securing the habitat

Mid-downstreams in the DMZ and CCZ smoothly flood neighboring areas and develop into artificial banks, which in turn, are also cultivated as rice paddies by the residents. Meaning to say, the streams also play an important contribution to the human settlers as the main irrigating system of their farms.

Fig. 5.66 *Iksookimia koreensis*

But in return, they polluted the streams with the pesticides and fertilizers that they use in farming. These pollutants flow through the nearby wetlands thereby, polluting its ecosystems and threatens the habitats of the residing wild animals and fish.

In order to resolve such problems, the stream width should be preserved so that it could function as a stream ecosystem. Pollutants should be blocked and riparian buffer areas should be created near the water's edges.

Silurus microdorsalis

Features of S. microdorsalis habitat in DMZ and CCZ

Although most streams in the DMZ and CCZ run from the north to the south, the Nam river is the only one that runs across the DMZ from the south to the north, making U-turn and flows out to the East Sea. This nature or characteristic of the Nam river made it difficult for researchers.

The river basin is based on Hyangrobong on the south and divided into Nam river tributaries, which created a river basin on the north and the Seohwacheon river basin on the south. *S. microdorsalis* was observed in both areas. The Nam river tributary near Hyangrobong is shown in Fig. 5.67. The water system of Nam river tributary secures vast mountains in the northern Hyangrobong. It flows abundant influx of water in fast velocity on the steep slopes along the valleys in the mountain. The tributaries are surrounded by mountains, a suitable habitat for *S. microdorsalis* (Fig. 5.68).

Considerations for securing the habitat

The areas surrounding the Nam river tributaries and the Seohwacheon river basin maintained and developed its natural environment and ecosystems since human interference is very negligible. But again, the presence of military facilities, armories, and access to roads may equally provide access to these natural environments. Human activities had been known to be the worst destroyer of the world's environments. Human use and movements disturb the existing wildlife and its ecology, specifically the *C. canadensis* (L.) Cronquist and *O. biennis* L. that

Fig. 5.67 Nam river tributary valley wetland running from the south to the north near Hyangrobong

Fig. 5.68 *Silurus microdorsalis*

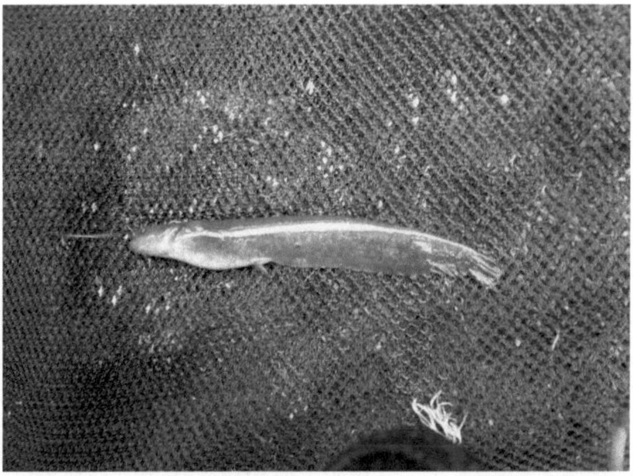

were found in the shooting range. Nam river is one of the primary source of water in north Korea, it is a very important then that these streams should be preserved. This strengthened the recommendation that the area should be designated as wetland protected area and Ecological and Landscape Protection Area.

Coreoperca herzi

Figure 5.69.

Features of C. herzi habitat in DMZ and CCZ

There are numerous rivers and streams formed in the DMZ and CCZ, most of which have good quality of water and well-developed water edges for a variety of species to live. There are many species and individuals of fish that lives here when compared to other areas.

The Namdae stream, where *C. herzi* (Herzenstein) is found (Fig. 5.70) emanated from the top of the Oseong mountain in North Korea and runs about 50 km toward the

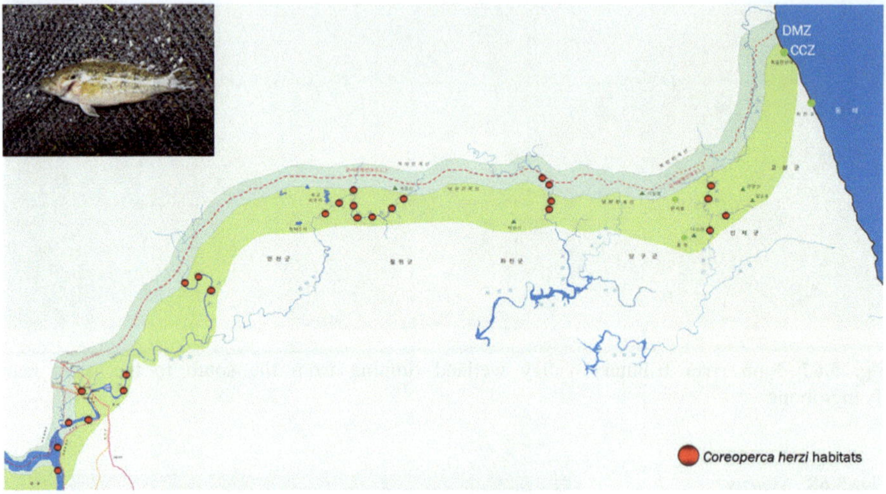

Fig. 5.69 Idnetified *Coreoperca herzi* (Herzenstein) habitats in DMZ and CCZ. ※ Korean endemic species

Fig. 5.70 Downstream of Namdae stream in Cheorwon CCZ

south in Hantan river across the DMZ. There are various vegetations distributed in the vast forest areas where the midstream of Namdae runs through and this provided a friendly environment to wild animals. Ponds are formed all over the flowing water system which also provided habitat for endemic fishes (Figs. 5.70, 5.71).

Considerations for securing the habitat

The Namdae midstream comes down from the CCZ and the distribution of cultivated land and residential areas increased. The connection of habitats is cut due to the formation of artificial banks that are transformed into agricultural lands. Again, water pollution is the greatest concern. In order to maintain the streams, artificial banks should be prevented along riversides in the lower lands. The stream's ecosystem should be restored for aquatic plants and fish to live.

Fig. 5.71 Coreoperca herzi

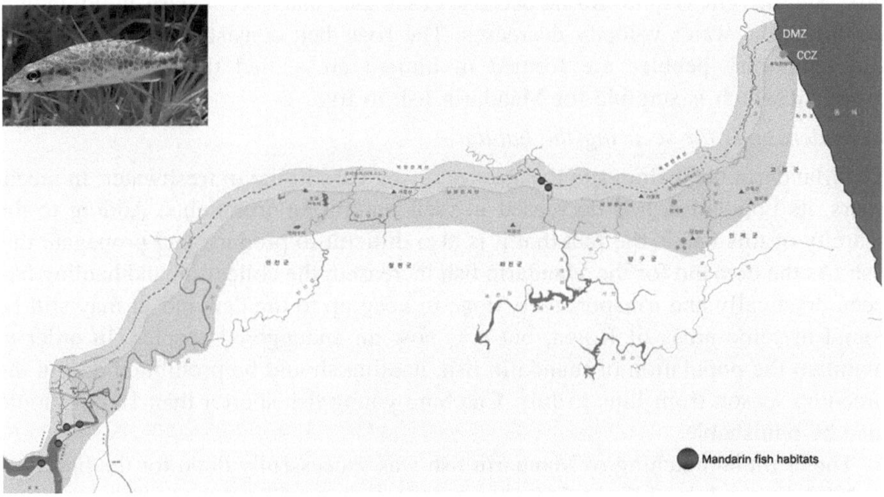

Fig. 5.72 Identified Mandarin fish habitats in DMZ and CCZ

Mandarin fish (*Siniperca scherzeri*)
Figure 5.72.

Features of Mandarin fish habitat in DMZ and CCZ

The Han river, Imjin river, Hantan river, northern Han river, Nam river run from the DMZ and CCZ. These rivers are water sources important for Koreans and for the organisms, plants, and animals that live there. Only the DMZ and CCZ areas maintain the rivers as natural as they are. Mandarin fish lives in the big rivers such

Fig. 5.73 Northern Han river in Hwacheon CCZ

as in northern Han river, Hwacheon CCZ (Fig. 5.73), where the water quality is good. There are other Mandarine fish habitat such as in Aprok river, Han river and Daedong river.

Although the northern Han river in Hwacheon CCZ is the main stream, it actually originated from Geumgang dam in North Korea. It does not have a lot of flux. But the flow is quite strong because of the dam and once it starts building up a reservoir, the water velocity decreases. The river bed consists of sand and mud, and temporary pebbles are formed in limited areas, and the water quality is excellent which is suitable for Mandarin fish to live.

Considerations for securing the habitat

The Mandarin fish belongs to the species of sea-bass living in freshwater. In recent years, its population had decreased in their natural environments. Adding to the scarcity of this fish is the fact that it is also difficult to produce and propagate this fish. As the demand for the Mandarin fish increased, the collection and hauling had been drastically and irresponsively done to keep up to the demand. It may still be found in some areas of Korea, but it is now an endangered species. In order to maintain the population of mandarin fish, hauling should be prohibited during the breeding season from June to July. Catching young fish shorter than 18 cm should also be punishable.

The artificial hatching of Mandarin fish was successfully done for the first time in 1995, by the Inland Water Development Institute in Chungcheongbuk province. But nothing can be compared to its hatching in the natural environments. It is important to preserve the stream where Mandarin fish can live in.

Arctic lamprey (*Lampetra japonica*)

Figure 5.74.

Features of Arctic lamprey habitat in DMZ and CCZ

The habitats of Arctic lamprey are mostly as estuaries of the Han river, Ganghwa, Gimpo, and Paju which are connected to the West Sea and Goseong estuary which is connected to the East Sea. Downstream wetlands are formed in the DMZ and CCZ. A pearl-shaped downstream wetland is formed on the west. The downstream

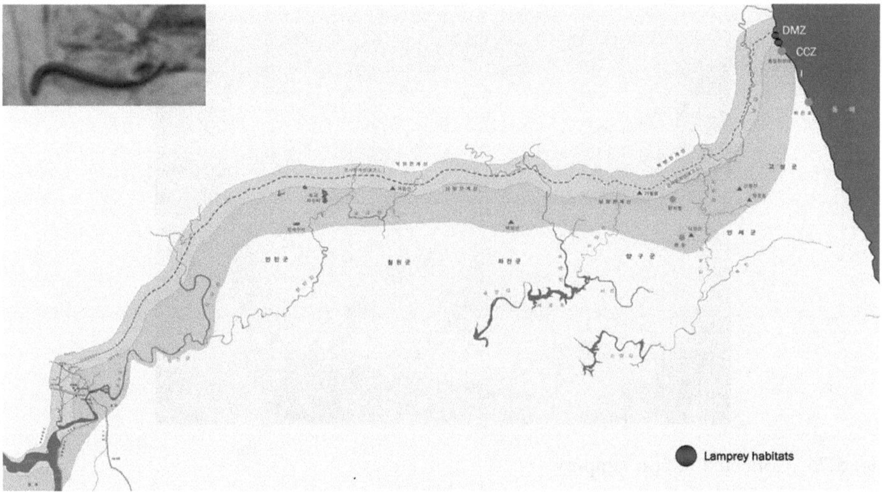

Fig. 5.74 Identified Arctic lamprey habitats in DMZ and CCZ. Endangered species (Grade II) designated by the Ministry of Environment

on the east is connected to the beach of the east coast that forms sand dunes and lagoons along the sea sides.

Due to several courses of the streams running down from the mountains, few downstream wetlands are also formed in Goseong area, especially as an artificial waterway to substitute damaged streams in Gyeonggi railroad line. Waterside plants are well-developed such as; *J. effusus* var. decipiens Buchenau and *P. japonica* Steud in Anhocheon downstream wetland. Such vegetation areas have very good quality water with little interference from human beings. Since it is an intersection of salt water and freshwater, it is nutritious and has well-developed waterside plants which makes it very suitable for Arctic lamprey to live in (Figs. 5.75, 5.76).

Fig. 5.75 Downstream wetland area of Anhocheon in Goseong CCZ

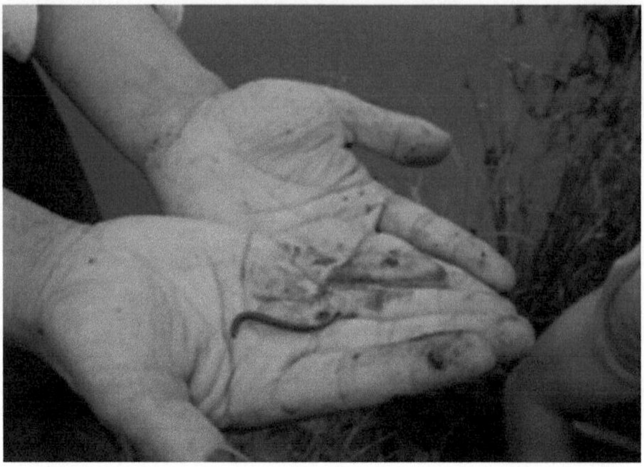

Fig. 5.76 Collected Arctic lamprey

Considerations for securing the habitat

As a rare species, Arctic lamprey should be led from sea to freshwater, but the number of individuals had decreased a lot due to the construction of the dam and bank that blocked the river. Aggravating this circumstance, pollutants are flowing into the streams and artificial banks which made Arctic lamprey more difficult to live. Thus, ecological movement corridors should be secured for the organisms and aquatic plants to flow into the edge of stream where the fish could live and multiply. In addition, the construction of beach facilities in sand dunes that are directly connected downstream should be limited, in order to circulate seawater and freshwater smoothly in the intersections.

Short ninespine stickleback (*Pungitius kaibarae*)

Figure 5.77.

Features of Short ninespine stickleback habitat in DMZ and CCZ

The eastern areas including Inje, Yanggu, and Goseong, are mountains surrounded mostly by forests that are denser compared to other areas. Because of this, the water system of streams running down from the mountains produces pure and a variety of fish lives in these clean waters(Grade I).

Unlike other eastern areas, Goseong in DMZ area has sea on the east, and flat and depression areas in the valleys that formed the alluvial wetland. According to the survey conducted in Goseong area, the alluvial wetland is formed in the valley that is connected to Anhocheon, which leads to the sea and formed downstream wetlands. It is dominated by *P. australis*. Pungitius kaibarae is observed in areas regarded as well-developed waterside due to the slow velocity of the stream. This provides a suitable environment for Short ninespine stickleback to live (Figs. 5.78, 5.79).

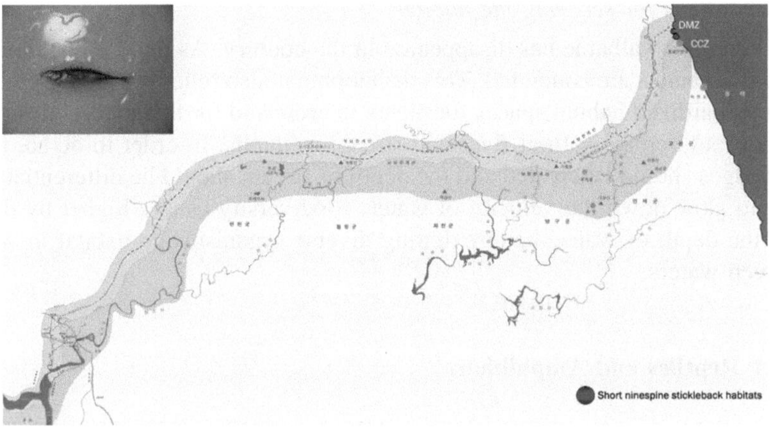

Fig. 5.77 Identified Pungitius kaibarae habitats in DMZ and CCZ. Endangered species (Grade I) designated by the Ministry of Environment

Fig. 5.78 Anhocheon downstream wetland in Goseong CCZ

Fig. 5.79 Short ninespine stickleback

Considerations for securing the habitat

Most Pungitius kaibarae has disappeared in the country. As flood protection, most of the river banks are concreted. This development also relieves parking problems. But it jeopardized habitat spaces for plants to grow and for fish to live. Restoration efforts should create natural river and stream wetlands. In order to do so, natural water edges should be created and the depth of stream should be differentiated and curvy to slow down the velocity of water. Biodiversity can be higher by diversifying the depth of water thereby, letting diverse organisms co-habitat in shallow and deep waters.

5.2.5.4 Reptiles and Amphibians

Twenty eight species—reptiles and amphibians—had been investigated so far. In this section, the characteristics of the various habitats of larger animals are described and recommendations on how to preserve them are also provided.

Reptiles and amphibians are found in many areas in DMZ and CCZ and some of them live in specific habitats. They can live in environments where their proper habitat is created. Since reptiles and amphibians are an important feed source for rapacious birds and large predators, the areas in the DMZ and CCZ with a large number of them have a high biodiversity (Table 5.16).

Hynobidae

Salamander (*H. leechii*)

Figure 5.80.

Features of salamander habitat in DMZ and CCZ

As wetland is well developed in DMZ and CCZ, it is suitable for salamanders especially at the base of a mountain with densed forests, streams, and clean valleys. The natural waterways in the cultivated lands also ideal for salamanders to reproduce.

A salamander lives in areas where clean water is stagnant and where fallen leaves accumulate. Since it does not make any sound, it can be easily found through its eggs (Figs. 5.81, 5.82).

Considerations for securing the habitat

Even if the favored habitat is in mountains and valleys, the salamander is sensitive to water pollution. It requires a lot of dissolved oxygen from birth to maturity and tests the air with its tongue. The inflow of soil and construction wastes into the water system should be completely blocked when building roads. Spraying of agricultural pesticides also caused the decrease of the population of salamanders. Collection of eggs for science experiments in spring by elementary schools should also be prevented.

Bufonidae

Water toad (*Bufo stejnegeri*)

Table 5.16 Classification and characteristics of major reptiles and amphibian species in DMZ and CCZ

Classification	Name of species (scientific name)	Characteristics	Distributer Area
Hynobidae	*Hynobius leechii*	Narrow and long body, long tail and two sets of legs with similar length. As moisture escapes from the outside skin, cannot stand heat and dryness. Lives in moisturized and damp area	Temperate regions
Bufonidae	*Bufo stejnegeri*	Resembles a toad but much smaller and flat, legs look relatively long. Mostly eats insects and earthworm and lives in clean valley of mountains. As a rarely found species, decreases except some mountainous region in Gangwon province	Gangwon province in Korea
Frog	*Rana nigromaculata*	6–9 cm of length, also referred to black-spoted pond frog, big black irregular pattern on the white background on the back of female individuals, male usually brown and little black pattern. Commonly seen in Korea and spawns in pond or rice paddy	Eastern Asia such as Korea·Japan·China
	Rana amurensis coreana	Dark brown body, light brown legs, long hair and a bit sharp mouth. Irregular black dots scattered between eyes and nose. Lives in groups, eats flies, mosquitoes, earthworms, grasshopper, etc. Lives in wetland and pond of mountains or flatland	Korea (North and West)
	Rana temporaria dybowskii	Goes into the deep mountain in summer, stays under stones in water in winter. Used to be caught for food in winter but it is prohibited by law. Residents and environmental organizations of the areas where *Rana temporaria dybowskii* lives in are making effort to protect frogs	Korea
Microhylidae	*Kaloula borealis*	Expanded body in left and right, different form from frogs. Short and small mouth, dull tip and protruded sharpen, vocal sac under chin, cries aloud in forest when rainy. Lives in puddle, ditch, rice paddy, the number of individuals has rapidly decreased as habitat is polluted by agricultural pesticides. Designated as endangered species II by Ministry of Environment	Korea(Seoul city, Gyeonggi, Gyeongnam) and Northeastern China

(continued)

Table 5.16 (continued)

Classification	Name of species (scientific name)	Characteristics	Distributer Area
Lacertid	*Takydromus wolteri*	Length of tail is 2.5 times longer than the length of body, body is about 46 mm long, flat body, sharp mouth tip, the length of hind legs is almost the same as the length of head. Crawls up trees well, lives in grassland of low mountains	Korea · China
Korean mamushi	*Agkistrodon blomhoffii*	Overall short and thick whole body, 40–60 cm of length. Arrow-headed and wide head, upside down U shape in light brown color on the light brown or gray background on the back. Lives in grassland in valley, pile of stones and cultivated field	Korea, Japan, Northeastern China
Colubridae	*Elaphe schrenckii*	Clear scales in the middle of body with mastodon bones of *Elaphe schrenckii* becomes dim toward stomach, black and yellow colored species. The commonly found *Elaphe schrenckii* in Korea lives in stone walls, banks, stones of an embankment around the end of a field. Recently designated as endangered species by Ministry of Environment due to overhunting	Korea · China · Siberia

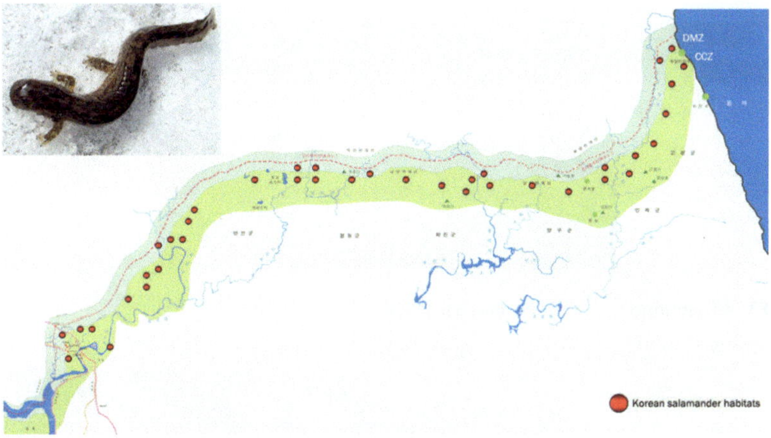

Fig. 5.80 Identified Korean Salamander habitats in DMZ and CCZ

Fig. 5.81 *Hynobius leechii* habitat in Paju DMZ: stream tributary runs from mountains

Fig. 5.82 *Hynobius leechii* habitat in Goseong CCZ: mountainous valley

Features of water toad habitat in DMZ and CCZ

B. stejnegeri is seen in the eastern mountainous valleys of the DMZ and CCZ (Figs. 5.83, 5.84). It is one of the species of amphibians that is difficult to find because it has various colors such as; red, brown, and black on body that blends

Fig. 5.83 Mountainous region on eastern DMZ

Fig. 5.84 Mountainous valley in eastern CCZ

with the surrounding environment. It feeds on insects and earthworms and spawns its eggs in brooks, in small stagnant ponds and paddies.

Due to the destruction of their habitat caused by developments, it is not easily seen but sometimes appears in relatively less damaged areas of DMZ and CCZ. It hides itself in the water and under rocks if an enemy or any threat appears.

Considerations for securing the habitat

B. stejnegeri is known to live mostly in mountains, valleys, and brooks. Its habitat is affected if the forest is damaged or destroyed. A lot of mountainous regions are being destroyed by the construction of roads and railroads in Korea. The DMZ and CCZ are no exceptions. Aside from the Gyeonggi railroad, it is expected that several other railroads and roads would continue to be built in the DMZ and CCZ areas as preparations for the unification of North and South Korea. Mountainous regions and valleys are then expected to be destroyed, there will be water pollution everywhere and eventually destroy the ecosystems and landscapes. It is not only the *B. stejnegeri* habitat that will be scarce. Species will be forced to move and find favorable habitats elsewhere, some may desperately expose itself to humans while in search for food and then, to their eventual extinction. The DMZ and CCZ areas should be protected. Modern engineering had proven state-of-the-art constructions that were impossible in the past. They could surely find ways and means to incorporate environmental sustainability in their developmental plans if they are sincere.

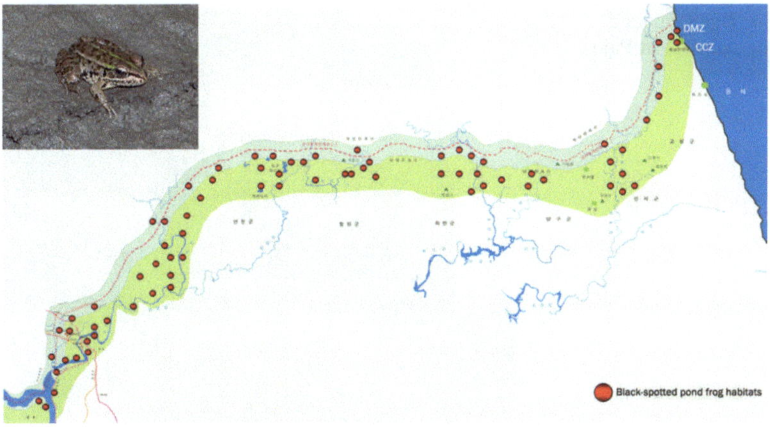

Fig. 5.85 Identified Black-spotted pond frog habitats in DMZ and CCZ

Ranidae

Black-spotted pond frog (*Rana nigromaculata*)

Figure 5.85.

Features of Black-spotted pond frog habitats in DMZ and CCZ

DMZ and CCZ provide wetlands, farmlands, and streams, environments where *R. nigromaculata* can live. This species was seen in the farmlands of CCZ.

R. nigromaculata is scattered in Gimpo, Paju, Yeoncheon, and Cheorwon in CCZ. They are still in large numbers and their presence in the area attracts species of snakes and amphibians that prey on them (Figs. 5.86, 5.87).

Considerations for securing the habitat

Ranidae species mainly lives in water or in rice paddies near water. It used to be the species that had the most number of individuals among anuran before people started using agricultural pesticides. Now, their population is decreasing because of the increasing agricultural developments in the DMZ and CCZ. The use of agricultural pesticides should be controlled, if not be totally prohibited in Korea. Farmers should indulge in organic farming and technologies should be provided by the government.

Korean redfrog (*Rana amurensis corneana*)

Figure 5.88.

Features of Korean redfrog habitats in DMZ and CCZ

As one of the Korean endemic species, *Rana amurensis* can be seen in wetlands and ponds throughout the DMZ and CCZ. This species usually lives in moist and damp places of the mountains, and also migrates down to the flatlands such as; ditches, rice paddies, depressions, and ponds to spawn its eggs. The eggs can be seen floating on shallow waters (Figs. 5.89, 5.90).

Fig. 5.86 Black-spotted pond frog habitats in Paju DMZ

Fig. 5.87 Reptiles and amphibians escapeway for blocked habitat by Gyeonggi railroad

Considerations for securing the habitat

The main habitat of *R. amurensis* are the rice paddies and low wetland. It is very sensitive to the changes in the environment. They only spawn their eggs in waters that are clean and can provide nutrients to its young tadpoles. Maintenance of low wetland is essential if this species be allowed to prevail. It is important to stress that this is native to Korea and deserves to be preserved.
Dybowski's brown frog (*Rana temporaria Dybowskii*)

Figure 5.91.

Features of Dybowski's Brown Frog habitats in DMZ and CCZ
Brown Frogs are also observed in hills, wetlands, and cultivated lands in DMZ and CCZ. Its eggs can be easily seen in the waterways of cultivated lands, brooks, and

Fig. 5.88 Identified Korean redfrog habitats in DMZ and CCZ. ※Korean endemic species

Fig. 5.89 Korean redfrog habitats in Paju CCZ

Fig. 5.90 Korean redfrog observed by the research group

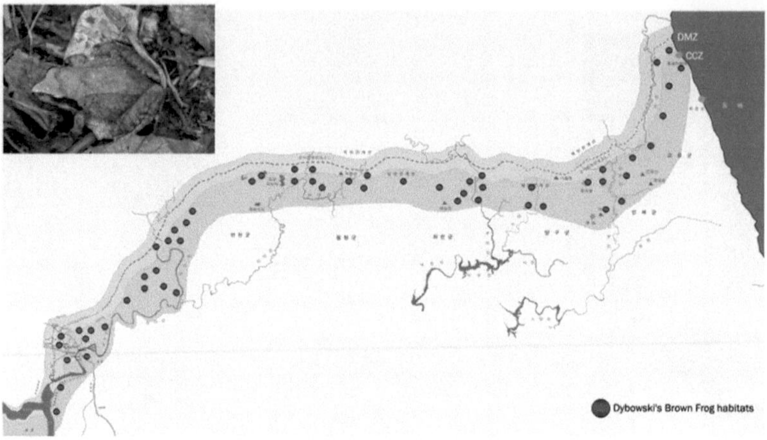

Fig. 5.91 Identified Dybowski's brown frog habitats in DMZ and CCZ

Fig. 5.92 Pond within Paju DMZ

Fig. 5.93 Mountainous stream in Goseong CCZ

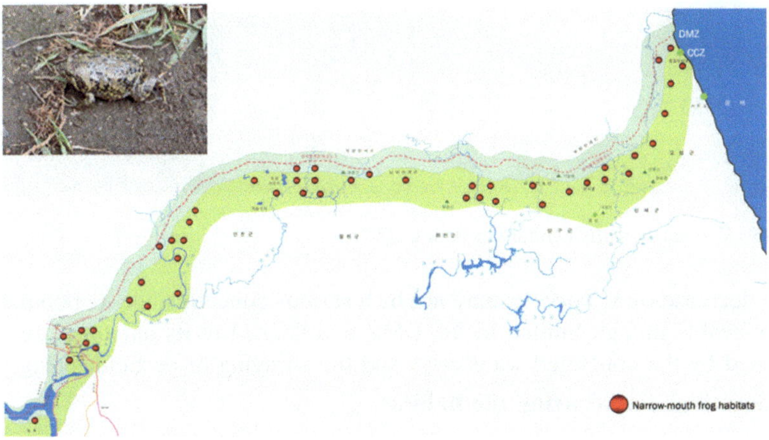

Fig. 5.94 Identified *Kaloula borealis* habitats in DMZ and CCZ. ※ Endangered wild animal and plant II designated by Ministry of Environment

ponds in early spring. It eats insects such as; spiders, ants, and earthworms living in mountains and grasslands. The population of brown frogs is still rapidly increasing in the DMZ and CCZ, even as it is the favorite food of rapacious birds and predatory mammals just like other frogs, which is helpful for the balance of ecosystem. But humans have already included the brown frog in their food list, as an exotic food (Figs. 5.92, 5.93).

Considerations for securing the habitat

Even as the Dybowski has adapted in cultivated lands, monitoring its population is required especially because humans had learned to eat them. This acquired human behavior had already caused the decrease of the brown frogs' population in the past. Nevertheless, this is also a good reason for humans to help in preserving the habitat of these edible brown frogs. Damaging their habitat is a greater threatening element. The pesticides and fertilizers that may be slowly saturating the brown frog's adaptive physiology may not only threaten the existence of rapacious birds and predatory mammals but also to the humans who eat them. The habitat required by Dybowski's brown frog existence should be preserved as much as possible and damaged areas should be restored and ensured to be safe.

Microhylidae

Narrow-mouth frog (*Kaloula borealis*)

Figure 5.94.

Features of Narrow-mouth frog habitats in DMZ and CCZ

K. borealis are found living near houses in DMZ and CCZ and can be distinguished from other frogs by its appearance. It spawns in an entrance to a village or in the footpath of rice field, ditch or depression and disappears through the ground after egg deposition.

Fig. 5.95 *Kaloula borealis* habitats in Paju CCZ

The decrease of *K. borealis* may not be a serious concern since its population is still noticeable though limited in the DMZ and CCZ. But its habitat is obviously threatened by the concreted waterways and the spraying of pesticides (Fig. 5.95).

Considerations for securing the habitat

In the past, “Maengkkong” cries were normally heard during rainy seasons in suburbs and farm villages. Now that urbanization and farmland improvements are increasing, it is no longer heard in ridges between rice fields where ponds and depressions had disappeared. The number of individuals has decreased due to pollution brought about by the use of pesticides. The narrow-mouthed frog has been designated as a protected wild animal by the Ministry of Environment. It is almost not found recently and can only be seen in limited areas. As it lives in groups in the low wetlands, it is absolutely necessary to protect and maintain the low wetland.

Lacertidae

Wolter lizard (*Takydromus wolteri*)

Figure 5.96.

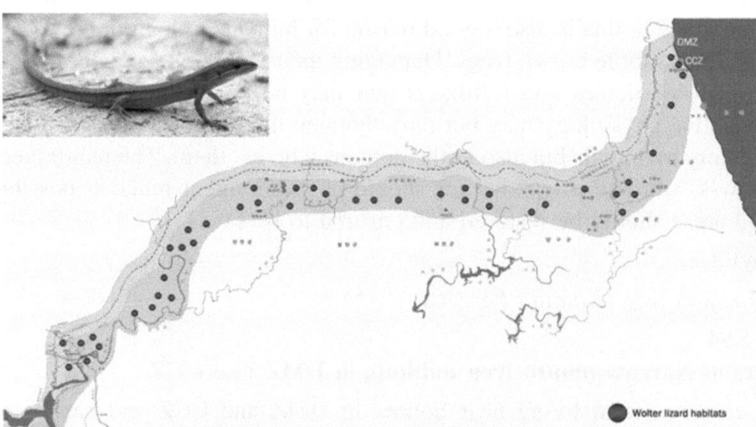

Fig. 5.96 Identified Wolter lizard habitat in DMZ and CCZ

Fig. 5.97 Grassland near Gyeonggi railroad and road

Features of Wolter lizard habitats in DMZ and CCZ

In the edges of the forest and sites where military facilities used to be in DMZ and CCZ, Wolter lizards are seen exposing themselves to the sun. They live in old fields and in edges of the lowland forests near old abandoned military facilities and minefields in DMZ and CCZ (Fig. 5.97).

Considerations for securing the habitat

The primary habitat of these lizards is the irregular forest, weed-grown fallows, and grassland. These habitats should be protected and maintained in order to protect the Wolter lizard. The diversity of insects found in these areas sustains this lizard, the same as to other species that feeds on insects like the frogs.

Crotalidae

Mamushi (*Agkistrodon brevicaudus*)

Figure 5.98.

Features of Mamushi habitats in DMZ and CCZ

The Mamushi lives in rice paddies connected to the forest, specifically in an embankment around the end of a field with shrubs, scrubs, and grasses in the DMZ and CCZ. It is frequently seen in rocky places of the mountains. It is a predator of rats and frogs. The mamushi enjoys sunshine and frequently standby on hot rocks

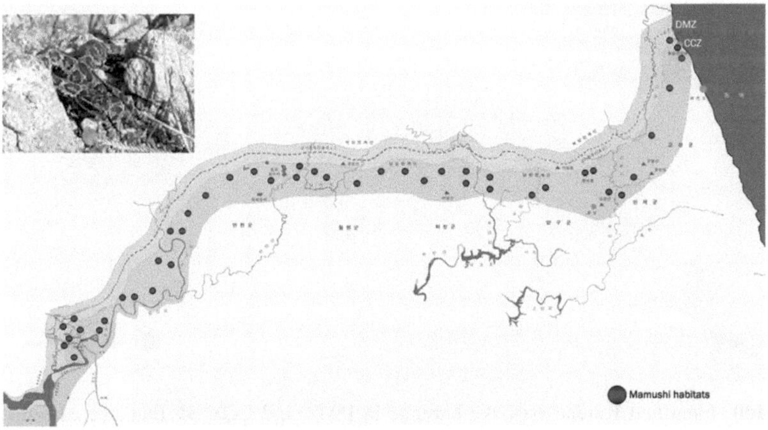

Fig. 5.98 Identified Mamushi habitats in DMZ and CCZ

Fig. 5.99 Grassland and shrub and scrub forest connected to mountains in Paju DMZ: Mamushi habitat

to warm it. Its bite is poisonous and because of this, problems come up from time to time in DMZ and CCZ (Fig. 5.99).

Considerations for securing the habitat

Fields lushed with shrubs and bushes are the habitats of the mamushi. Recently, most of these fields have been damaged. Added to this sad reality, the population of mamushi has rapidly decreased because it is also hunted by humans for its medicinal attributes, and believed traditionally, to build up a healthy body. Hunting of mamushi should be prohibited and its habitat should be restored and maintained.

Colubridae

Korean Ratsnake (*Elaphe schrenckii*)

Figure 5.100.

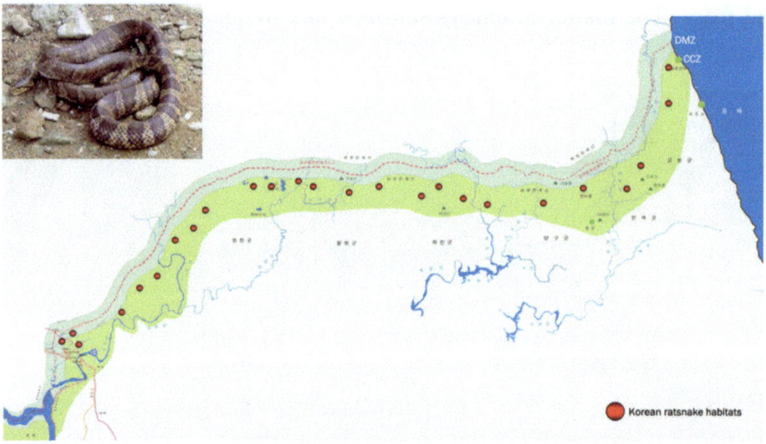

Fig. 5.100 Identified Korean ratsnake habitats in DMZ and CCZ. ※ Endangered wild animal and plant I designated by Ministry of Environment

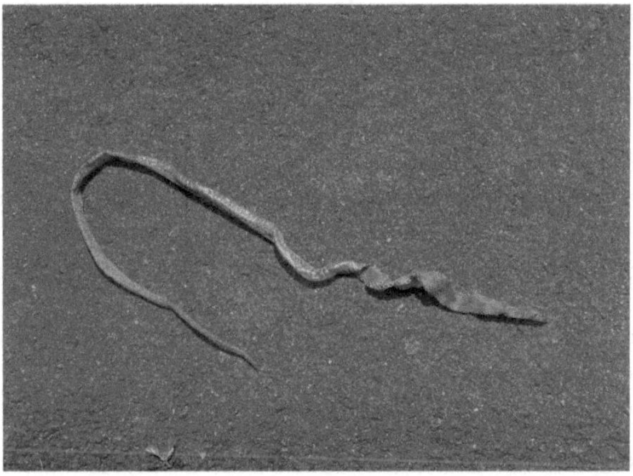

Fig. 5.101 Casted skin of Korean Ratsnake on a road in Paju DMZ

Fig. 5.102 Artificial Korean ratsnake habitats created on ecological bridge in Paju

Features of Korean ratsnake habitats in DMZ and CCZ

The Korean Ratsnake lives in the hills of the DMZ and CCZ. *E. schrenckii* eats small mammals, small birds, and amphibians. Its presence in such areas connotes that there is a high biodiversity of these preys around.

The *E. schrenckii* is the biggest snake species among Colubridae in Korea. Residents used to report that they appear in fences and roofs of their houses. Recently, the population of *E. schrenckii* has rapidly decreased (Figs. 5.101, 5.102).

Considerations for securing the habitat

The *E. schrenckii* is one of the protected species designated by Ministry of Environment. Its habitat is mainly near houses and roads. Its number had been rapidly decreased due to overhunting and is now endangered species. Thus, overhunting should be avoided in order to preserve the species.

5.2.5.5 Birds

According to the compiled data of the by the Environment Ecosystem Research Group, the bird community has 6 departments, 18 sections, and 51 species. Birds living in DMZ and CCZ have more species and larger in population as compared to other animal groups. DMZ and CCZ are less interfered by humans and have various habitats suitable for birds such as forests, wetlands and grasslands. *E. schrenckii* plays an important role in the survival of migratory birds that prey on them. Summer and winter migratory birds of typical species in DMZ and CCZ will be examined in detail in the succeeding sections (Table 5.17).

Summer migratory birds
Scops Owl (*Otus scops*)

Figure 5.103.

Features of Scops owl habitat in DMZ and CCZ

Living in forest wetlands and mountains in the DMZ and CCZ, *O. scops* are making holes for nesting in branches of hardwoods such as; *A. japonica* Steud., *S. koreensis* Andersson, and *Q. mongolica* Fisch. ex Ledeb. It used to live throughout the Korean peninsula. But recently its food has disappeared in other places due to the widespread use of fertilizers and pesticides in farmlands where they feed, which further leads to the destruction of the ecosystem of the mammals that they also eat. The number of *O. scops* has rapidly decreased.

Dominated by *S. koreensis* Andersson and *A. japonica* Steud. communities, the flatland's and wetland's forest such as the hardwood-forest in Inje county CCZ (Fig. 5.104), not only *O. scops* thrived in these forests but also other endemic and rare birds. Large number of *O. scops* is seen specifically in forests near the mountains (Fig. 5.104).

Considerations for securing the habitat

O. scops is a nocturnal animal so it only moves at night after nesting in natural tree holes not only in forests but also in trees near houses. The forest should be protected as it is an important habitat for *O. scops*, holds underground water sources, and serve as barriers of storms and strong winds that destroyed many communities and caused human miseries.

Chinese Sparrow Hawk (*Accipiter soloensis*)

Figure 5.106.

Features of Chinese sparrow hawk habitats in DMZ and CCZ

Having a vast turf, Chinese sparrow hawk is also seen in wetlands, farmlands, and in the forests of DMZ and CCZ. It usually looks for food in the wetlands and farmlands but reproduces and nests in the forests. It comes from China and spends summer in DMZ and CCZ.

The identified *A. soloensis* habitats are in Cheorwon DMZ as shown in Fig. 5.107. This bird flies in the skies over farmlands and prey on frogs, reptiles, amphibians, and rodents in the waterways near the farmland. As same as other

Table 5.17 Classification and characteristics of major birds in DMZ and CCZ

Classification	Name of species (scientific name)	Characteristics	Distributed area
Summer migratory bird	*Otus scops*	Summer bird in northern Korea, some groups are passage migrant birds living in mountains or flat forest, lays eggs into tree holes. Used to symbolize hometown in Korea	Northeast Asia such as Korea · China
	Accipiter soloensis	Summer bird in Korea visiting in May and leaving in September, the number has decreased as it has eaten polluted feed by agricultural pesticides and chemical fertilizer in farm villages. Nests on branches of oak, chestnut tree, pine, *Alnus japonica* Steud., lays 3–4 eggs and raises them	Northeast Asia such as Korea · China
	Platalea minor	Registered as a severe endangered species in Red Data Book of International Union for the Conservation of Nature in 1990, changed its grade to endangered species in 2000. Spends summer in Korea, observed in an uninhabitated island on the west DMZ and Jeju island	China, Korea, Japan, Hong Kong, Taiwan Island, Indo-China Peninsula
	Halcyon pileata	Summer bird that rarely reproduces throughout Korea, mainly lives in cultivated land or waterside of mountains or riverside. Shiny blue color with black head, under chin and gullet in white, the rest in orange and very colorful. Nests in the middle of mountains or cliff making holes or uses tree holes	East Asia, India, Sri Lanka
	Eurystomus orientalis	Rare summer bird in Korea, lives in coniferous forest where tall old trees are occasionally scattered or mixed stand forest or parks or cultivated land. Nests in tree holes and is able to make holes anywhere with hard beak. Usually lives on trees, eats insects while staying on top of trees	Korea, India, Borneo Island, Australia
	Egretta alba modesta	Common summer bird that reproduces throughout Korea, lives on the ground except breeding season, walks along rice paddy, brook, stream, and wetland to find feed. Alone or 2–8 individuals are easily seen in farmland, stays in farmland and eats feed	Korea, China, Japan, Indo-China, the Philippines, etc.
	Motacilla cinerea robusta	Summer migratory bird that spends April–October in Korea, lives in low mountain stream, eats insects and spiders on the waterside. Dark gray color from the crown to waist of male summer plumage, light brown, under the chin in black, edges of neckband in white, chest and stomach in yellow. Nests like a rice bowl in hills near cultivated field or a rent in the rock	Europe, Africa, Asia

(continued)

Table 5.17 (continued)

Classification	Name of species (scientific name)	Characteristics	Distributed area
Winter migratory bird	*Phalacrocorax filamentosus*	Winter migratory water bird in Korea, black body, dark brown bent beak, black legs, webbed feet between 4 toes. Lives alone or in group in downstream of river or coast, lays 4–5 eggs in a nest in dented part of rock or cliff	Asia, South America, West Coast, Galápagos Islands , New Zealand
	Grus vipio, Grus japonensis	Winter migratory bird in Korea, white body, red skin of crown, black forehead, neck, inside the wings, and white blade. Registered as the 46th international protected bird in Red Data Book of International Union for Conservation of Nature and Natural Resources, and designated as endangered species by Natural Monument and Ministry of Environment	Korea, Japan, China, Siberia
	Anser cygnoides	Winter migratory bird spending winter time from October to April in Korea. A bit smaller female species than male but in the same color, rubiginous side of the head, back hair, crown, back of the head, nape, oxblood beneath its chin, white neck, cheek and side of the neck. Dozens of species living in group in lake, polderland, grassland, wetland, farmland in winter, reproduces at riverside, on grassland, in a small island of a lake, nests on a dry grass in a sink area	Korea, Japan, China, Mongolia, Siberia, the Kamchatka Peninsula
	Haliaeetus albicilla	Winter migratory bird in Korea, huge and heavy bird in Accipitridae, equally brown colored except light brown to luggage tan head and neck, white tail. Lives on rocks in the coast, mud flat, marshland, lake, stream, downstream, and open terrain and forest but does not live in a dense forest.	Northeast Asia such as Korea, Siberia
	Aegypius monachus	Winter migratory bird in Korea, likes open area and lives in riparian buffer areas, downstream and coast. Eats animals or dead bodies. Nests on branches or rocks, lays 1 egg in February–April. Recently the number of species has decreased by station as it lost its feed to Corvus corone Carrion Crow or Pica pica	Western Mediterranean, Eastern Asia
	Circus cyaneus	Winter bird seen throughout Korea, lives in wetland where *Salix koreensis* Andersson thinly grows. From forehead to waist colored in gray, white bottom, gray beak, yellow iris. Lays 4–5 eggs in April–May, does not cry except for breeding season and eats passeriformes, anatine, and rodents	Siberia, Mongolia, China (northeast), Europe
	Anser albifrons	Winter bird seen in Korea, groups about 100–1,000 individuals are easily found in farmland. Found more in the south than in the central district, lives in open area such as farmland, wetland, bay, polderland and downstream. Nests in an island of a stream or grassland of wetland and lays 3–7 eggs	Europe, Asia, North America

(continued)

Table 5.17 (continued)

Classification	Name of species (scientific name)	Characteristics	Distributed area
Resident bird	*Paradoxornis webbiana*	Thick break, fast and tends to shake its long tale from side to side when moving. Lives in group of 30–50 individuals except breeding season. Nests in fences of a farm but usually at about 1 m high in forest or grass. Nests with dry grass, grass ears, grass roots, and lays eggs	Korea, China, Myanmar
	Anas poecilorhyncha	The only resident summer duck and bird that often reproduces throughout the country, lives in group in farmland in winter. Huge duck with the length of 61 cm, dark brown body, light brown head and neck, dark brown stomach. Eats leave or stem of water plant, sprouts, hayseed, fruits as well as animals such as insects and invertebrates	Korea, Amur, Ostrov Sakhalin (South), China
	Falco tinnunculus interstinctus	Belongs to Falconidae, the length of 30–33 cm, brown dot on the back of male, scattered big black dots at the bottom of light brown part. Flies in a straight line with wings spread seeking feed on the ground, eats rodents (rats), moles, small birds, insects, and reptiles. Lives in mountains, houses, sits on wires, electroforming, trees, and buildings	All over the world
	Larus crassirostris	Mid-sized seagull, white head, chest and stomach, gray wings and back, is distinguished from other seagulls having a black belt at the tip of tail. Usually lives in grassland in uninhabitated islands, nests with dry grass in groups and lays 4–5 eggs. Eats fish, insects and water plants, lives with *Platalea minor* in uninhabited islands	Korea · Japan · China (Northeast) · Ostrov Sakhalin · Kuril Islands · Maritime Territory
	Aix galericulata	Designated as the 327th Natural Monument, has beautiful feathers in many different colors. Reproduces in mountains, mountain stream, lake, reservoir all over the country in Korea, rests or sleeps under the trees, rocks, on branches on the water escaping people's gaze. Eats tree nuts, snails and small fish	
	Bubo bubo kiautschensis	Big bird in Strigidae, has ear-shaped feathers on head. Dark brown feathers with black vertical stripes on red eyes. Rare in Korea but scattered throughout the country, lives in rocky mountain rather than forests. Young usually eats Phasianus colchicus, hare and rats, mother bird eats frogs, snakes, lizards, insects, etc. Designated as the 324th Natural Monument in Korea	Korea, China
	Cyanopica cyana	Resident bird that reproduces in forests in the north of central area, black head and upper neck, sky blue wings and tail. Nests on branches, lays 6–9 eggs in May–July. Omnivorous, eats not only animals such as amphibians, fish, crustacean, mollusc but also grains and fruits	Korea, Japan

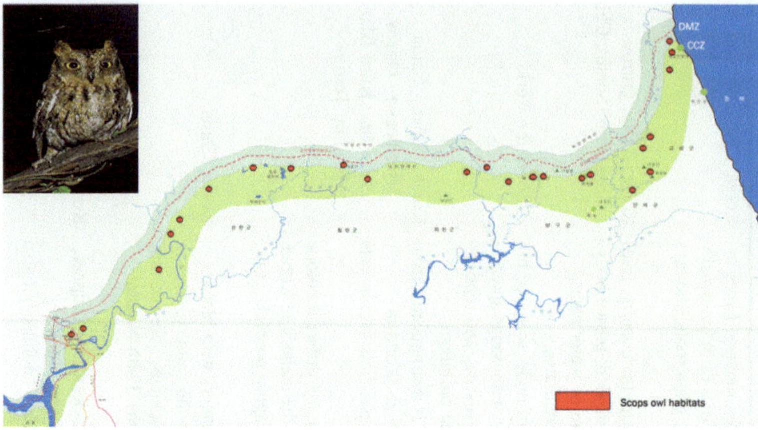

Fig. 5.103 Identified Scops owl habitats in DMZ and CCZ

Fig. 5.104 *Otus scops* habitat in hardwood-forest in Inje-county CCZ

Fig. 5.105 *Otus scops*

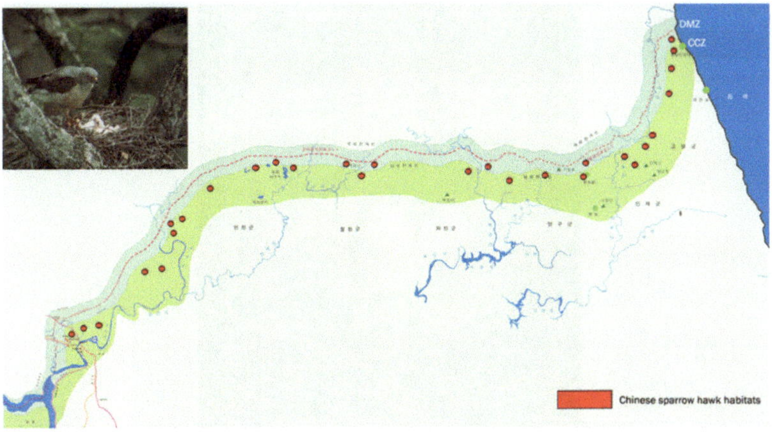

Fig. 5.106 Identified Chinese sparrow hawk habitats in DMZ and CCZ. ※ Natural Monument No. 323

Fig. 5.107 Chinese sparrow hawk habitats in Cheorwon DMZ: wetland and flatland forest are mixed together

birds, it rests, reproduces, and nests in forests and hills near the farmlands of DMZ and CCZ (Fig. 5.108).

Considerations for securing the habitat

A. soloensis used to live throughout Korea but as its feeding grounds decreased, the number of *A. soloensis* equally decreased. Thus, organic farming should be imposed and the use of agricultural pesticides should be punishable.

Black-faced spoonbill (*Platalea minor*)
Figure 5.109.

Features of Black-faced spoonbill habitat in DMZ and CCZ

P. minor is seen living in an uninhabitated island in the west coast of DMZ and CCZ. It is a rare bird species that only lives in this area, but also reported to be seen

Fig. 5.108 Eggs of Chinese sparrow hawk

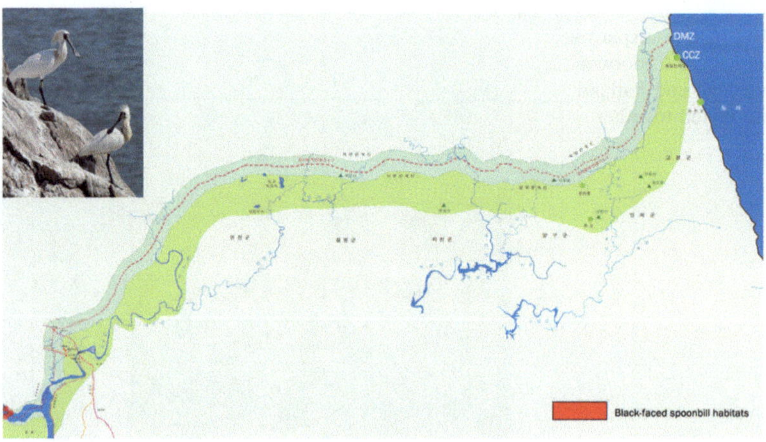

Fig. 5.109 Identified *Platalea minor* habitat in DMZ and CCZ. ※ Natural Monument No. 205-1, Endangered wild animal and plant Grade I designated by Ministry of Environment

in some uninhabited parts of Jeju island in Korea. *P. minor* reproduces and nests in uninhabited islands where there are lots of rocks and eats insects and small fish.

The typical habitat of *P. minor* in the DMZ and CCZ is basically an uninhabited island. This island has an estuary of the Han river and also has rocky mountains in its vicinity that nobody could access to. The neighboring farmland is also the *P. minor*'s source of food, but it is not totally dependent on them, since the island where it lives is also abundant of insects and small fish.

Figure 5.110 shows *P. minor* habitat in Yudo, Fig. 5.111 is *P. minor* actually living in the island.

Fig. 5.110 Yudo within DMZ: *Platalea minor* habitat

Fig. 5.111 *Platalea minor*

Considerations for securing the habitat

The neighboring areas of the estuary of the Han river are also ideal habitats for the *P. minor*. It is required, therefore, such areas be designated protected areas for the birds. A special plan should be developed specifically for *P. minor* which is so selective in terms of its habitat. The uninhabited island where the *P. minor* lives should remain uninhabited and protected as a valued ecological park and where people can observe at a distance and educate themselves so that they can co-exist with this rare bird.

Black-capped kingfisher (*Halcyon pileata*)

Figure 5.112.

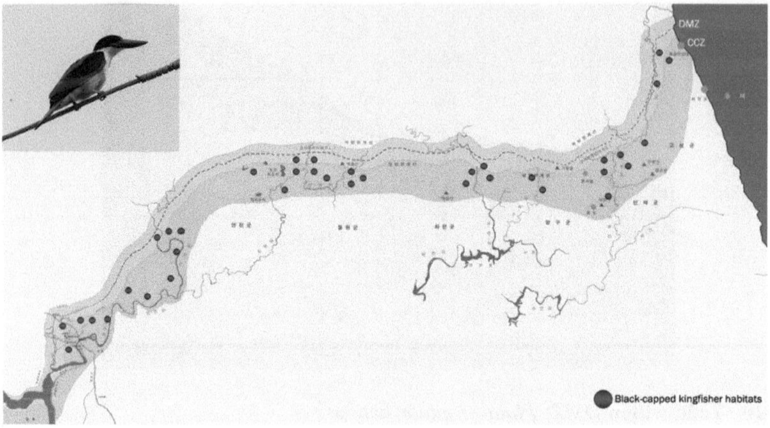

Fig. 5.112 Identified Black-capped kingfisher habitat in DMZ and CCZ

Features of Black-capped kingfisher habitat in DMZ and CCZ

The Black-capped kingfisher also lives in the mountains, wetlands, and riversides of the DMZ and CCZ, It preys on small fishes, frogs, roots at watersides. Although this kingfisher is distributed throughout the Korean peninsula, its preferred wetland habitat is only developed in DMZ and CCZ, hence, the number of Black-capped kingfisher individuals have increased due to well-developed wetlands, and the presence of farmlands where it can feed on seeds and roots, plus the mountains nearby where it could nest.

In fact, the Environment Ecosystem Research Group could observe Black-capped kingfisher at Sacheon reservoir in Panmunbul area in Paju DMZ as shown in Fig. 5.113. The Black-capped kingfishers observed in the DMZ and CCZ were living in a vicinity of farmlands and neighboring wetlands but nesting in holes at the cliffs of the nearby mountains and hills (Fig. 5.114).

Fig. 5.113 Black-capped kingfisher habitat in Sacheon reservoir influent channel, Paju DMZ Panmunbul field

Fig. 5.114 Black-capped kingfisher nests created in soil in Paju DMZ Panmunbul hills

Considerations for securing the habitat

The use of agricultural pesticides and the building of concrete waterways are threatening the habitats of the main food sources of the black-capped kingfisher in the DMZ and CCZ. Thus, it is best to reduce the use of agricultural pesticides, to preserve natural waterways and water edges, cliffs and wetlands, in serious consideration of the Black-capped kingfisher's radius of habitation at a distance and educate themselves so that they can co-exist with this rare bird.

Broad-billed Roller (*Eurystomus orientalis*)

Figure 5.115.

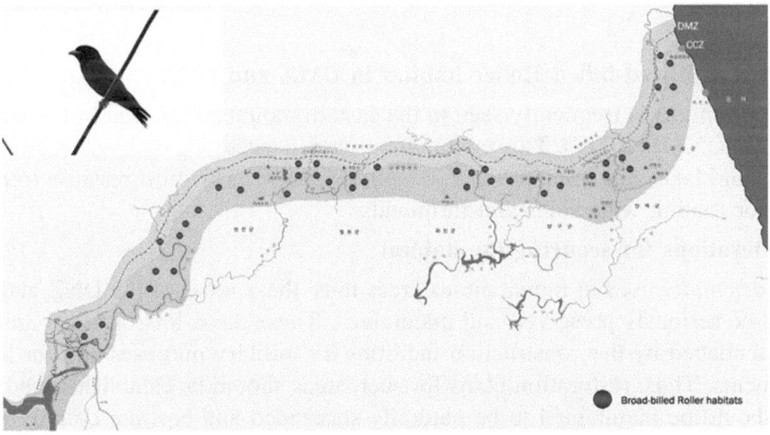

Fig. 5.115 Identified *Eurystomus orientalis* habitat in DMZ and CCZ

Fig. 5.116 Mountainous area in the east DMZ

Fig. 5.117 Mountainous valley area in the east DMZ

Features of Broad-billed Roller habitat in DMZ and CCZ

The *E. orientalis* is frequently seen in the eastern mountainous area of the DMZ or CCZ (Figs. 5.116, 5.117). This mountainous area is occupied by huge trees and the valleys and lakes are suitable for *E. orientalis* to live. This bird rests on trees and hunts for food at watersides and farmlands.

Considerations for securing the habitat

As *E. orientalis* lives in mountainous areas thus, these areas of the DNZ and CCZ should be seriously preserved and maintained. These days, lots of forest areas are being damaged by the construction facilities for military purposes and for human settlements. Thus, restoration plans for such areas should be established and some trees should be maintained to be naturally succeeded and become climax forest.

Large and medium egret (*Egretta alba modesta*)

Figure 5.118.

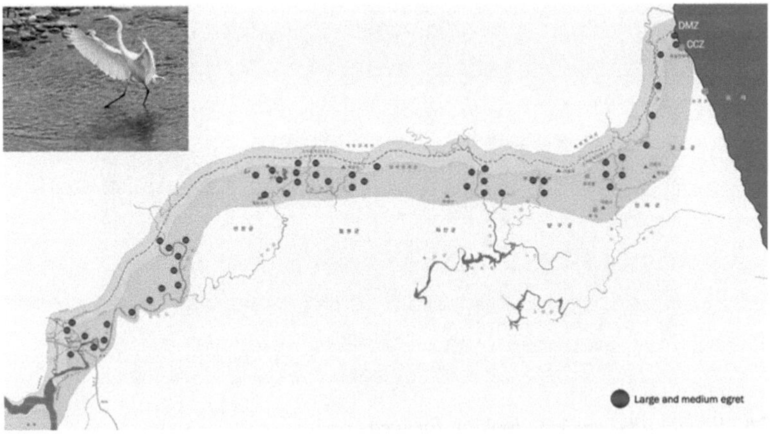

Fig. 5.118 Identified large and medium egret habitat in DMZ and CCZ

Fig. 5.119 Large and medium egrets live on vegetables in farmland

Features of Large and medium egret habitat in DMZ and CCZ

Egretta alba modesta is easily seen in the wetlands and farmlands of DMZ and CCZ, and appears with Ardea cinerea. *Egretta alba modesta* has thin and long beaks and legs, a suitable body to feed in water. It catches reptiles, amphibians, and small fish in farmlands and wetlands with its beak and legs submerged in water while its long neck maneuvers the scanning of the wetland's bed.

It nests on the branches of *S. koreensis* Andersson and *A. japonica* Steud that are abundant in the forested wetlands of DMZ and CCZ. It hunts in the neighboring farmlands and wetlands in flocks, and its number is quite large (Figs. 5.119, 5.120).

Fig. 5.120 *Egretta alba modesta* looking for feed

Considerations for securing the habitat

Ardeids also feed in farmlands and wetlands together with the Egretta alba modesta for they have common preference in food selection. They are both directly threatened by pesticides and fertilizer because they usually feed in farmlands. Thus, it is recommended to prohibit the use of agricultural pesticides and chemical fertilizer in these farmlands, or much better, to use organic agricultural techniques in farming instead. The government must fund the agricultural research on discovering natural pests control like the moths used by vegetable farmers. The forests and wetlands where communities of these birds inhabit should be preserved and maintained.

Gray wagtail (*Motacilla cinerea*)

Figure 5.121.

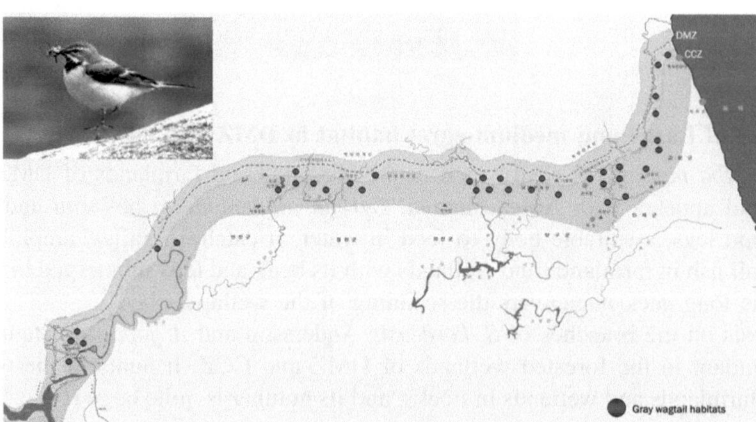

Fig. 5.121 Identified *Motacilla cinerea* habitat in DMZ and CCZ

Fig. 5.122 The whole view of forest in mountainous area of the east CCZ

Fig. 5.123 Valley in a mountainous area in CCZ

Features of Gray wagtail habitat in DMZ and CCZ

M. cinerea lives in deep mountains in the eastern mountainous region of DMZ and CCZ. It is easily found in the valleys. In this region, the mountains and valleys are well preserved compared to other mountainous areas of DMZ and CCZ as there has been little interference by humans. It has clear and very clean waters and lots of fresh water fish which made it a major habitat for such birds.

These birds nest in rocks of the cliffs in the mountains where their offsprings are safe while they hunt for food. Figures 5.122 and 5.123 show areas where *M. cinerea* inhabit in the eastern parts of CCZ.

Considerations for securing the habitat

As of now though, the habitat of the *M. cinerea* may be undisturbed. But at the advent of developments planned to be built in these habitats, and with the influx of human activities, these birds may lose their feeding grounds and nesting cliffs. The mountains and valleys of the DMZ provide one of its kind water processing system

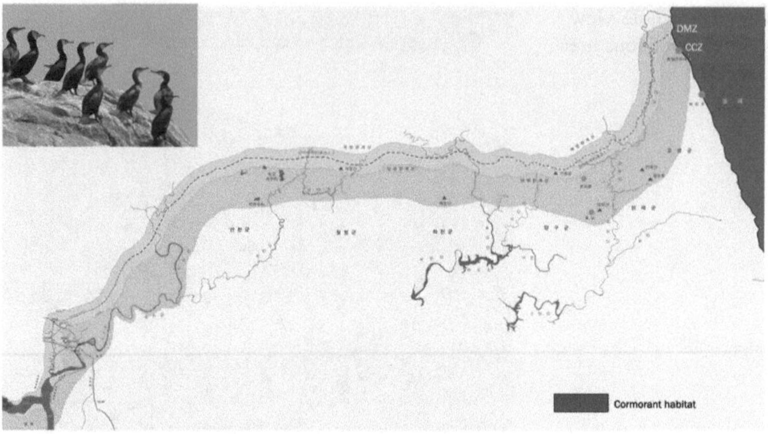

Fig. 5.124 Identified *Phalacrocorax capillatus* habitat in DMZ and CCZ

that produces the best quality of water ever. The water runs down in long stretch of winding tracks, so as the pebbled beds help in the process of sanitizing the water. The minerals of the streams incorporate nutrients in the water, while the highland forests along rivers hold and process underground and surface waters. The waters that come down from these mountains and valleys are of highest quality, best source for humans and wildlife alike. This is a natural process unique to DMZ and CCZ. This natural environment preserves and sustains itself and unknowingly sustains human life, because water is life. Without water, every living things die, including humans. Therefore, this unique ecosystem that nature has created must be sincerely and wittingly preserved and sustained.
Winter migratory birds

Cormorant (*Phalacrocorax capillatus*)

Figure 5.124.

Features of Cormorant habitat in DMZ and CCZ

P. capillatus are winter migratory birds reproducing in an uninhabited island west and east of DMZ and CCZ. It migrates to this area to spend winter. It lives alone or in groups eating small freshwater fish and crustaceans in the estuary of the Han river and those found in mud flats. The *P. capillatus* was observed in an island located in an estuary of the Han river as shown in Fig. 5.126. It is all covered with black feathers, which makes it invisible at night. It has webbed feet for swimming while they feed in the estuaries and even for just resting in the surface of the water (Fig. 5.125).

Considerations for securing the habitat

Although the west coast has not been developed yet, it was observed that some areas of mud flats had been cultivated as farm lands. Of course, where there are farmlands, pesticides, and fertilizers are the worst enemies of the environment and

Fig. 5.125 *Phalacrocorax capillatus* in an island of the west coast

Fig. 5.126 Mud flat in the west coast

its wildlife. The *P. capillatus* feeds on these mudflats and most possibly could feed on surviving and contaminated fish and crustaceans, not to mention its direct exposure to the chemicals and it is drinking the contaminated waters in the estuary. These facts add to the significance of the preservation and maintenance of the DMZ and CCZ's estuaries.

Red-crowned Crane, White-naped Crane (G. japonensis, Grus vipio)

Figure 5.127.

Features of Red-crowned Crane, White-naped Crane habitat in DMZ and CCZ

The wetlands and farmlands in the DMZ and CCZ provide the best habitat to Red-crowned crane and White-naped crane. These birds sleep and eat in different areas, to avoid enemies. They are migratory birds, so they are used in hoping from one feeding ground to another. They seldom stay in one place for a long time.

Habitats in the DMZ and CCZ that are frequently visited by these cranes are; Gimpo, Paju, Yeoncheon, and Cheorwon. Specifically, the Cheorwon plain is the most favored habitat of the red-crowned crane and white-naped crane during winter because it has a vast farmland. The abandoned paddy field wetland along the Yeokgok stream is another place where these cranes spend their entire winter (Figs. 5.128, 5.129).

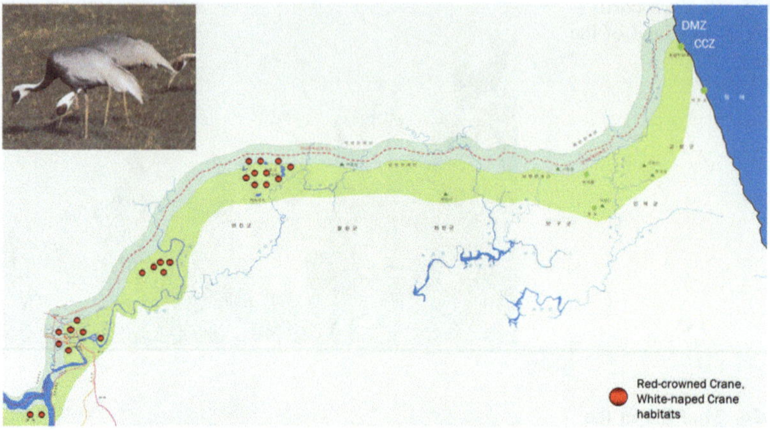

Fig. 5.127 Identified Red-crowned crane and White-naped crane habitat in DMZ and CCZ. ※ Natural Monument No. 202 and 203. International Union for Conservation of Nature and Natural Resources (IUCN) Red Data the 46th International Protected Bird

Fig. 5.128 *Grus japonensis* family in farmland in Cheorwon CCZ

Fig. 5.129 Red-crowned crane and White-naped crane living on vegetables in farmland

Considerations for securing the habitat

As it spends winter in mud flats, mouth of rivers and in frozen grasslands and farmlands, it is crucial to preserve these habitats for the cranes. These birds are very specific and selective in choosing their stop-over places because there are plenty of mudflats and vast farmlands in the entire Asia, but the cranes chose the DMZ and CCZ. Very few cultivated lands implement organic farming and the use of pesticides in farms is very threatening to these rare and beautiful birds.

Cultivated lands near the towns are turned into housing complexes, decreasing the feeding grounds of the cranes. Modernization is slowly crawling toward the margin of its geographic boundaries, jeopardizing natural environments. With these trends and patterns, the situation for cranes and wildlife animals will be dreadful. The future children of the world may not be able to see these cranes alive, and may just content themselves with pictures like those found in this book, while blaming their ancestors for destroying nature.

Thus, the birds should be protected and their habitats should be preserved and maintained.

Swan Goose (*Anser cygnoides*)

Figure 5.130.

Features of Swan goose habitat in DMZ and CCZ

A. cygnoides are also living in the farmlands, estuaries and wetlands of DMZ and CCZ in groups. They eat crop ears in farmlands, crustaceans in the estuary, small freshwater fish, reptiles and amphibians in the wetlands. They feed day and night.

A. cygnoides were observed in Paju, Cheorwon, Goseong areas of the DMZ and CCZ. They feed and rest in groups, and in farmlands, live on vegetables, and nest in grasslands beside rivers and small island lakes to reproduce (Figs. 5.131, 5.132).

Considerations for securing the habitat

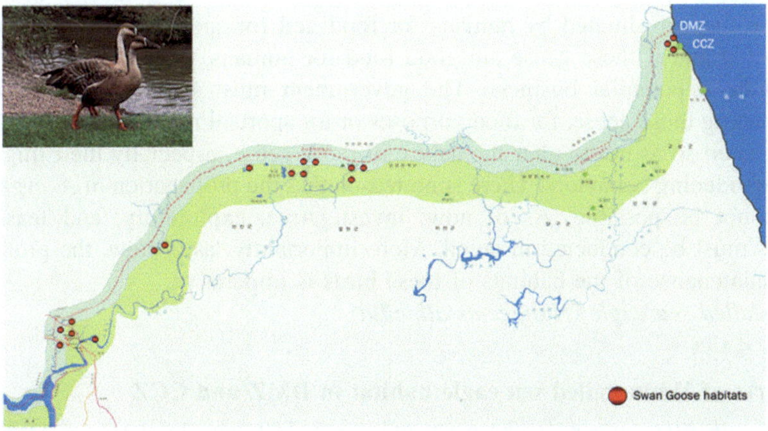

Fig. 5.130 Identified *Anser cygnoides* habitat in DMZ and CCZ. ※ Natural Monument No. 325

Fig. 5.131 Unfrozen wetland in DMZ: Swan goose

Fig. 5.132 *Anser cygnoides* living in wetland

These geese used to be a common species, but they are gradually decreased because they are hunted by humans for food and for sports. Hunting should be prohibited and if these geese are good food for humans, thus, reproducing them would be a potential business. The government must study the feasibility of reproducing these geese, for food purposes or for sports. Proper technology should be explored so as not to alter the nature of these geese, especially their migrating and reproducing behaviors. There is no reason that the propagation of *A. cygnoides* would not be possible. As of now, investigative, exploratory, and feasibility studies must be conducted in detail. More importantly, as of now, the protection and maintenance of the habitats of these birds is imperative.

White-tailed sea eagle (*Haliaeetus albicilla*)

Figure 5.133.

Features of White-tailed sea eagle habitat in DMZ and CCZ

The *H. albicilla* appears in the west and center hills and lowlands of DMZ and CCZ. This bird covers a large radius of flyways hovering over the skies of these

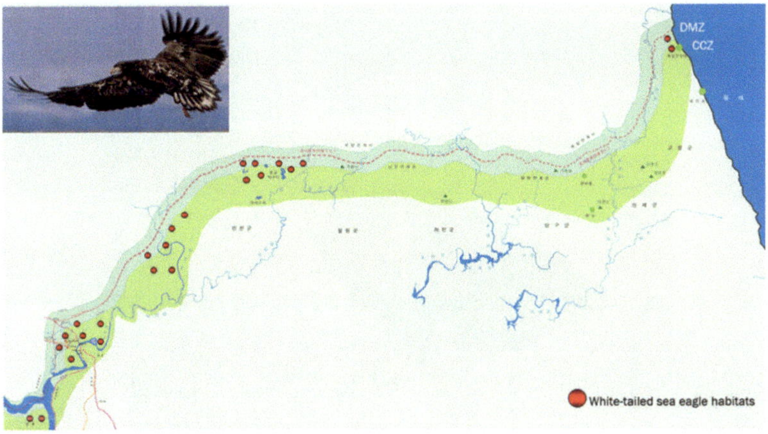

Fig. 5.133 Identified *Haliaeetus albicilla* habitat in DMZ and CCZ. ※ Natural Monument No. 243

Fig. 5.134 Wetland area near estuary of the Han river in Paju CCZ

areas. It prefers to feed on freshwater fish, but it also eats ducks, rodents, plovers and snipes.

It is observed when resting or feeding in the estuary of the Han river and Goseong and, appears mighty, threatening and imposing, a stance typical to eagles. It cannot be found in the eastern mountainous area with dense forest. Figure 5.134 shows a habitat of *H. albicilla* in the wetland area near the estuary of the Han river in Paju CCZ. In Fig. 5.135, the *H. albicilla* was observed by the Environment Ecosystem Research Group.

Fig. 5.135 *Haliaeetus albicilla*

Considerations for securing the habitat

The *H. albicilla* is a protected species all over the world due to its rapidly decreasing population and its significance to the balance of the ecosystem. During winter this bird migrates to Korea and settles in areas where there are streams such as; the Sumjin river, Daeseong-dong area (DMZ), Han river, Nakdong river, estuaries in the east and the west coast and even found in the southern shores of Korea. Since this endangered specie settles in Korea during winter, the key responsibility is in the hands of Korea in this time of the year. The world is expecting that this bird is protected here and its environment as well.

Einereous vulture (*Aegypius monachus*)

Figure 5.136.

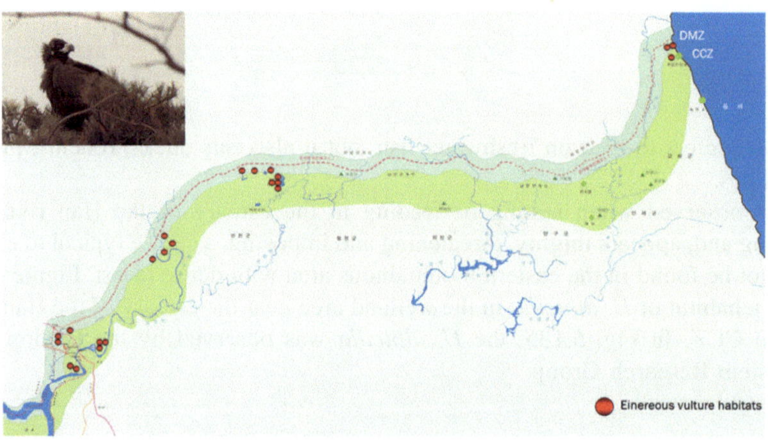

Fig. 5.136 Identified Einereous vulture habitat in DMZ and CCZ. ※ Natural Monument No. 243

Fig. 5.137 Einereous vulture group resting on a bank in Tokyo reservoir in Cheorwon CCZ

Features of Einereous vulture habitat in DMZ and CCZ

Einereous vulture lives in the estuary, stream, hills, and wetlands of DMZ and CCZ. As expected of vultures, it eats dead animals. In these days, however, it starves to death as it is difficult to find dead animals.

Einereous vulture was seen in the Cheorwon Tokyo reservoir and in the Paju Jangdan peninsula (Fig. 5.137). At the time of observation, the vultures were feeding frenzy in groups. These feeding events help this species to survive. Their number is decreasing in wilderness, but they are still patiently waiting for larger animals' leftovers, as well as those accidentally dropped by birds (Fig. 5.138).

Considerations for securing the habitat

Currently, the Einereous vulture is only seen in Daeseong-dong area and in Panmunjeom, DMZ. The preservation areas for this bird keep on changing. The vultures are dying every year, and there is an urgent need that they will be placed in a permanent settlement so that they will not be keep on adapting. The cause for the death of the vultures is presumed to be due to starving. But a study should be conducted to investigate the factors surrounding their deaths including the ecosystem.

Fig. 5.138 Living in farmland

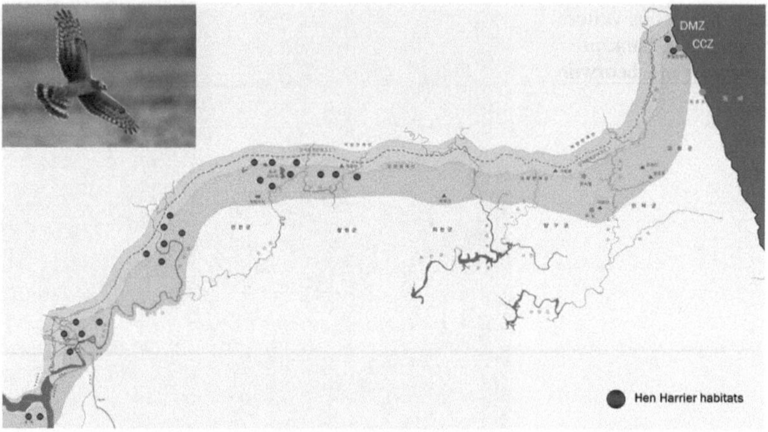

Fig. 5.139 Identified Circus cyaneus habitat in DMZ and CCZ. ※ Natural Monument No. 323-6, Endangered wild animal and plant Grade II designated by Ministry of Environment

Hen Harrier (*Circus cyaneus*)

Figure 5.139.

Features of Hen Harrier habitat in DMZ and CCZ

Except in the mountains and forests, Circus cyaneus can be observed in almost all areas in DMZ and CCZ. This bird finds the wetlands, farmlands, and hills as its habitat. Circus cyaneus feeds on rodents, reptiles, amphibians, and freshwater fish. In the farmlands of DMZ and CCZ, it is often seen feeding. It usually flies alone, with full of energy and appears fierce, as if it dominates the area when it is around (Figs. 5.140, 5.141).

Fig. 5.140 Cheorwon plain in CCZ

Fig. 5.141 Circus cyaneus incubating eggs in nest

Considerations for securing the habitat

The population of Circus cyaneus is also rapidly decreasing. It was once a very common bird that was highly populated in these areas. Now, it is threatened to extinction because their number is decreasing. With the presence of other birds depending on the wetlands and farmlands for food, this bird has to compete in order to survive. Unwittingly, the use of pesticides in farmlands also decreased the population of the animals and fish these birds eat, especially the rodents that is intentionally killed by humans because like pests, they feed on their crops. Pesticides pollute the waters where the fish thrives, delimiting again the sources of food for these species. The wetlands and the grasslands where *P. australis* is abundant, is the best winter habitat for many birds including the Circus cyaneus. Counter measures need to be established soon, as the *P. australis* fields and grasslands are continuously destroyed.

White fronted Goose (Anser albifrons)

Features of White-fronted goose habitat in DMZ and CCZ

The White-fronted goose are always seen in flocks feeding in farmlands during winter. When on air, they look like black clouds and when they gather on the ground, they could paint the fields black. Figure 5.142 shows the White-fronted goose soaring in the air and resting on a vegetable farm. One can just imagine how they could devour a certain area with just one landing. The White-fronted goose lives in farmlands and wetlands. They prefer these areas because it is near water sources like lakes and estuaries. The White-fronted goose is observed in almost every farmland in Paju, Cheorwon, and Goseong. It feeds on vegetables and other edible plants. They are frequently found in farmlands along with *G. japonensis*, the White-naped Crane, and the Anser fabalisserrirostris (Fig. 5.143).

Fig. 5.142 White-fronted goose living on vegetables in farmland in Cheorwon CCZ

Fig. 5.143 Flying in the sky

Considerations for securing the habitat

The White-fronted goose prefers wide and open places like the wetlands. There are still a lot of open spaces in the DMZ and CCZ at present, but it is slowly eaten up by developments. It is important to remember that these areas used to be farmlands but just abandoned by the farmers because of the war. Even after the war, these fields are embedded with mines. Therefore, they are still unable to come back. At the advent of the present program for the removal of these landmines, people are slowly coming back to reclaim the land their forefathers cultivated. In addition, as the DMZ and CCZ is the only remaining barrier between the north and south, moves to reunite the two countries are undergoing and when it prospers, there is no other way to connect the two countries but here. So development is inevitable. All these pressures threatened the status of this wonderful ecosystem.

As the Ministry of Environment keeps on defending the DMZ and CCZ, now brought its peculiar wonder to the attention of the world, we can only hope that the

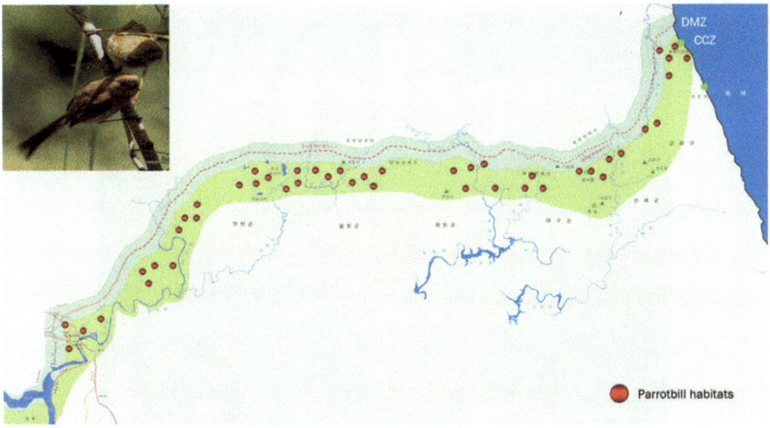

Fig. 5.144 Identified Parrotbill habitat in DMZ and CCZ

world will help advocating for its sustainable management to counterpressure the planned developments on these areas.

Resident birds

Parrotbill (*Paradoxomis webbiana*)

Figure 5.144.

Features of Parrotbill habitat in DMZ and CCZ

P. webbiana spreads throughout Korea and not only in the DMZ and CCZ. But most of its population is in the latter where there is lesser disturbance. It is known locally as the "Korean crow-tit". It lives in the edges of farmlands. This tiny bird nests on the lower branches of the bushes and lays 3–5 eggs at a time. It sometimes unknowingly incubates the eggs of the cuckoo bird, another small resident bird. In the bushes, these birds feed on insects and in the farmlands behind, they feed on grains and seeds of vegetables and other edible plants. This is shown in Figs. 5.145, 5.146, 5.147.

Fig. 5.145 Grassland, shrub and scrub forest in Paju DMZ

Fig. 5.146 *Paradoxomis webbiana* feeding young

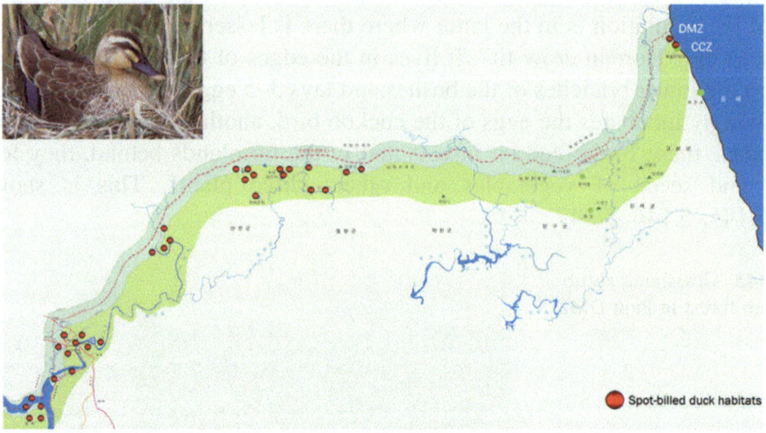

Fig. 5.147 Identified Spot-billed duck habitat in DMZ and CCZ

Considerations for securing the habitat

As earlier said, *P. webbiana* lives at the edges of farmlands but sometimes in places where there are bushes like near houses throughout the country. They are very vulnerable to disturbance especially during cultivation periods, when their nests are accidentally demolished by the clean-up operations. That is why most of them are in the DMZ and CCZ. But as the DMZ and CCZ is the focus of developments, the largest population of this bird is likely threatened as well. Thus, protection and preservation of the habitats of these birds is of prime importance. Care should be given to exotic species, lakes, ponds, marshlands, streams, and even irrigation canals of rice paddy fields. Methods and strategies should be installed in such a way that human and these birds could live and let live as they are both Koreans.

Spot-billed duck (*Anas poecilorhyncha*)

Features of Spot-billed duck habitat in DMZ and CCZ

The Spot-billed ducks live in groups in farmlands during winter. They are always observed floating asleep with their heads under their wings or paddling as they play on the waters of open lakes and calm trapped waters of streams. They usually eat water plants in wetland and crop ears in the farmlands. They are often spotted in Ganghwa, Paju and Goseong areas of the DMZ and CCZ where there are farmlands, lakes, and swamps. They weave dry leaves as their nests in the grasslands and wetlands (Figs. 5.148, 5.149).

Considerations for securing the habitat

The Spot-billed duck has found it difficult to maintain their nests as farmlands expands slowly towards their nesting areas. Nestlings are often disturbed and sometimes die when they are relocated and the mother ducks can no longer locate them. And when this happens, the stressed mother duck make noise and all the rest of the flock join the crackling and wandering. The worst threat of these animals is the concreting of riverbanks and streams. They lose the habitat where they could reproduce. These ducks may be able to adapt in the developing farmlands but eventually become a menace to farmers because they eat their crops.

As of this time, certain areas of the DMZ and CCZ should be designated just for the ducks. And these designated areas should be preserved, maintained, and sustained by law.

Fig. 5.148 Stream connected to estuary of the Han river in Gimpo CCZ: Spot-billed duck stays in winter from time to time

Fig. 5.149 Paddling idly in water

Kestrel (*Falco tinnunculus interstinctus*)

Figure 5.150.

Features of Kestrel habitat in DMZ and CCZ

Falconidae can be found all over DMZ and CCZ. Falco tinnunculus usually lives in houses and neighboring areas. It eats insects, moles, and other rodents lurking around farmlands. It is highly adaptable compared to other falconidae birds. It even appears in the center of the city but Falco tinnunculus prefers to live and reproduce in forests and mountains and hunts in the flatlands of DMZ and CCZ (Figs. 5.151, 5.152).

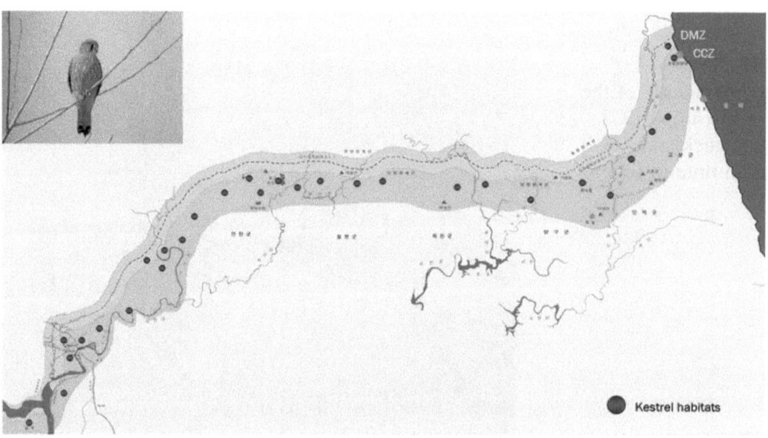

Fig. 5.150 Identified Falco tinnunculus habitat in DMZ and CCZ. ※ Natural Monument No. 323

Fig. 5.151 Falco tinnunculus habitat in Ganghwa CCZ

Fig. 5.152 Falco tinnunculus biting food

Considerations for securing the habitat

Falco tinnunculus nests in cliffs and cracks of mountains, but recently they learned to nest on cracks of high buildings in the city. This bird requires a wider radius of hunting grounds. As development is pushing toward DMZ and CCZ, its habitat is getting smaller and smaller, so the falconidae learned to live in the city.

In addition, rodents and small mammals that are food for the falcon have dwindled in number, not just because so many species prey on them, but because rodents are either deliberately killed by humans or die ingesting pesticides.

Thus, Falco tinnunculus habitat should be preserved and the use of agricultural pesticides should be prohibited in farmlands so as not to put a vacuum or a gap within the food web that is important in maintaining the balance of the ecosystem.

Black-tailed gull (*Larus crassirostris*)

Figure 5.153.

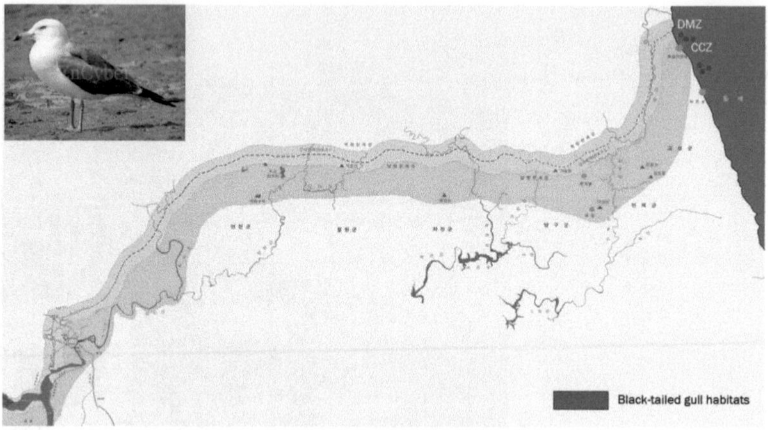

Fig. 5.153 Identified Black-tailed gull habitat in DMZ and CCZ

Features of Black-tailed gull habitat in DMZ and CCZ

Black-tailed gull is observed along the coasts of an uninhabited island of the DMZ. This bird is monogamous and lives with his mate till its death. The female gull lays 4–5 eggs at a time and the nestlings usually survive in this undisturbed environment. They are observed in flocks especially when feeding. They feed on aquatic insects, amphibians and reptiles in the coasts and on rocks in the estuary at Gimpo, Ganghwa, and Goseong areas, and sometimes join other birds in the rice paddies during summer. Figure 5.154 shows the estuary of the Han river in Gimpo DMZ where Black-tailed gull is observed (Fig. 5.155).

Considerations for securing the habitat

The Black-Tailed gull is one of the indicators that the environment where they occupy is healthy. They usually thrive best in areas where there is no human interference. The gulls' population in the entire world is dwindling and therefore

Fig. 5.154 Estuary of the Han river region in Gimpo DMZ: Black-tailed gull eats food here

Fig. 5.155 Black-tailed gull fighting for habitat

calling for protection. They are quite secluded and would like to live in peace in their own community. Efforts should be made to secure their habitats, since these birds are rare and the world could not afford to lose them. During breeding season, these gulls are in its weakest because they seldom leave their eggs and nestlings. When the eggs are hatched, the parents feed their nestlings first. And when food is scarce, they become weak and can only recover when the young birds are able to hunt for themselves. The breeding season therefore is crucial to these birds that needs food supply up to seven times more than their regular hunt. It is the moral responsibility of humans to help protect the habitat of these birds so that they could be healthy during the breeding season to increase their population.

Mandarin duck (*A. galericulata*)

Figure 5.156.

Features of Mandarin duck habitat in DMZ and CCZ

A. galericulata also lives in flocks and are found in lakes, reservoirs, and streams where they are observed resting on the surface of open waters. The *A. galericulata* is sensitive to human interference, flies high and hides immediately in the forest and in the wetlands when approached by humans. They are commonly observed in Paju, Cheorwon, and Goseong areas of the DMZ and CCZ. They eat fish, nuts, and hayseeds. This flock of birds usually stays in hills, valleys, and mountains even in summer time. But the birds leave the mountains during winter and lives in the unfrozen portions of the wetlands, reservoirs, and streams so that they could feed on fish. The Environment and Ecology Research Group observed these birds in the

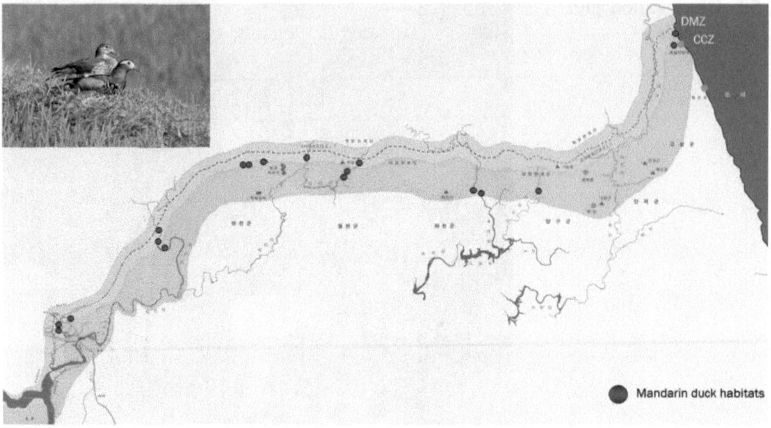

Fig. 5.156 Identified *Aix galericulata* habitat in DMZ and CCZ. ※ Natural Monument No. 327

Fig. 5.157 Upstream of Namdae stream in Cheorwon CCZ

upstream of Namdae stream in Choerwon CCZ as in Fig. 5.157, and in the Anho reservoir in Goseong, still in CCZ as shown in Fig. 5.158.

Considerations for securing the habitat

A. galericulata nests on the sidelines of lakes and reservoirs as well as in the boundaries of trees, but noticeably prefers areas where a large number of esculins are. Thus, preservation efforts should consider propagating esculin communities along the edges of wide and calm waters like lakes and reservoirs that have enough space for these birds to feed and reproduce. The water must be clean and with the proper depths. Second, tree planting of hardwood and maintaining the forests in the hills and mountains should be considered since these birds also reside on them within the three seasons except during winter. Most importantly, these protected areas must be uninhabited by humans who would surely dislocate these sensitive birds at a glance of any human.

Fig. 5.158 Anho reservoir in Goseong CCZ

Eagle owl (*Bubo bubo kiautschensis*)

Figure 5.159.

Features of Eagle owl habitat in DMZ and CCZ

Eagle owls that are found in DMZ and CCZ also occupies the mountains and cliffs. As a nocturnal bird, it rests on its nest during the day and hunts at night. It feeds on small animals such as Phasianus colchicus, rabbit, rats, and on top of that, reptiles, amphibians, and insects.

The Eagle owl is easily seen throughout DMZ and CCZ and it is observed motionless on branches or rocks with wings folded when resting (Fig. 5.160).

Considerations for securing the habitat

Protection of the hunting grounds of the eagle owl should be ensured especially in the woody and rocky places. They should be left undisturbed most especially during the day when they are sleeping and resting on trees. The character of these birds and the characteristics of its habitat must be of prime consideration in planning for their protection and survival. These birds are becoming rare and it is

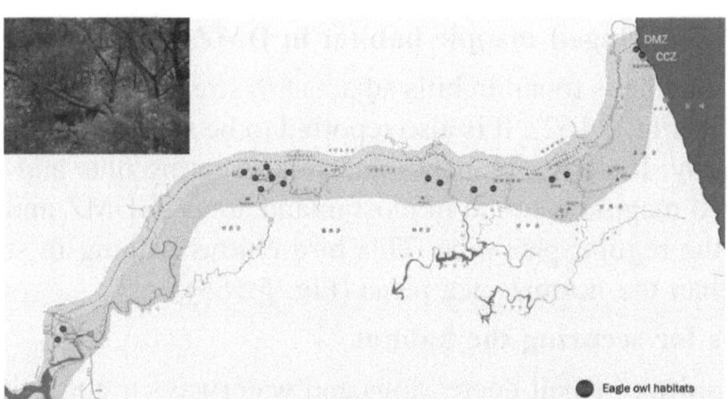

Fig. 5.159 Identified Eagle owl habitat in DMZ and CCZ. ※ Natural Monument No. 324-2

Fig. 5.160 Bubo bubo habitat in Cheorwon CCZ

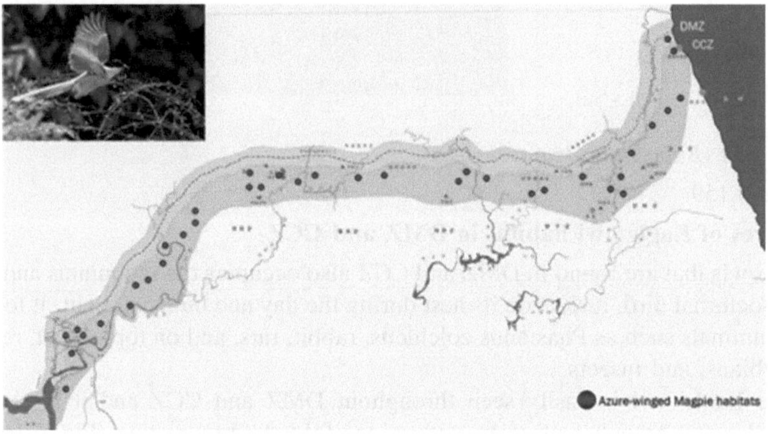

Fig. 5.161 Identified Azure-winged magpie habitat in DMZ and CCZ

one of the most popular birds in the entire world. In fact, before becoming eligible for school, the owl is one of the first animals little children can identify. One can see the excitement on the faces of the children in seeing birds, especially the owl that may be the queerest of them all, but surely the dearest.

Azure-winged Magpie (*Cyanopica cyana*)
Figure 5.161.

Features of Azure-winged magpie habitat in DMZ and CCZ

Azure-winged magpie is found in hills adjacent to streams and also in flatlands that have forests in it (Fig. 5.162). It is also reported to be seen near houses where there is a stream nearby. It sits on branches and fences eating nuts and insects.

Azure-winged magpie is found in most inland areas in DMZ and CCZ. Its color is grayer than the regular pica pica. This bird enjoys bathing in small ponds and streams more than the normal pica picas (Fig. 5.163).

Considerations for securing the habitat

The concreted sides of small depressions and waterways in the cultivated lands at the DMZ and CCZ limited the playgrounds of these bathing birds. In fact, the

Fig. 5.162 Azure-winged magpie habitat in Goseong CCZ

Fig. 5.163 Hardwood-forest near Sewol stream in Paju, Yeoncheon CCZ: prefers forest and stream

cultivated lands itself sequestered the pure waters of pica picas. The farmlands are also primarily situated where there are abundant waters but people inevitably and unwittingly contaminate them with pesticides. When these birds happen to bathe in the contaminated waters, they may be unable to reproduce or be poisoned and die. Farmers should be advised not to concretize watersides where the birds play and feed. Other larger rivers are also concretized to prevent flood. This practice must be changed into more environment friendly methods in preventing flood. The habitats after all help in counteracting global climate change that causes the floods to happen. It is the devastation of our environment that caused the miseries of the world in the first place.

5.2.5.6 Mammals

According to the compiled data investigated by the Environment and Ecology Research Group, there are 6 departments, 18 sections, and 51 species belonging to the mammals. Although mammals are distributed all over DMZ and CCZ, specific species of mammals identified are driven more in the eastern mountainous areas due to the ecological characteristics of these areas. In this section, 6 species are presented as they are the ones recognized as the more important species to be protected in the DMZ and CCZ (Table 5.18).

Chinese water deer (*H. inermis*)

Figure 5.164.

Features of Chinese water deer habitat in DMZ and CCZ

H. inermis is seen throughout DMZ and CCZ. It is because most areas in DMZ and CCZ consist of mountains and wetlands that provide an ideal environment for *H. inermis* to survive. Recently the number of *H. inermis* had increased since it has no enemies or predators in this area and food is abundant. It is only during winter that they migrate down to the farmlands where they are likely to be harmed. The increase of Hydropotes may be good news but the more they increase, their feeding grounds must also increase.

As a herbivore (plant-eating animal), *H. inermis* hides and reproduces in mountains beneath thick forests, in shrubs and bushes. They drink water in the clear springs and streams in the mountains. In the lowlands during winter they feed on vegetables in the farms and drinks water in the wetlands.

Since the habitats are widely developed in the DMZ and CCZ, the *H. inermis* roams from one mountain to another devouring edible herbs and grasses along the way. The movement of this animal is continuous and that requires a wider radius of coverage for them to survive. They are frequently observed in the highland wetlands (Figs. 5.165, 5.166) year round except during winter when they go down to the lowlands.

Considerations for securing the habitat

H. inermis likes water. It swims twice a day and drinks often, especially more during summer when it is hot. Protecting *H. inermis* habitat is very important and that is, preserving the forests and maintaining the purity of the mountain springs and watersides that sustain the grasses and herbs they feed on. But the most important factor that should be considered soon is the widening of the protected habitat for *H. inermis* since their population is growing. The fact that they risk to go down to the farmlands where there are people who might eat them only means that their food is getting scarce and not enough for their present population. This is urgent. Their number should be calculated and equated to the estimated feeding area that these mammals need. There should be massive tree planting and widening of the forest, enough water space and seclusion from humans who are their only predators.

Leopard cat (*Felis bengalensis manchurica*)

Figure 5.167.

Table 5.18 Classification and characteristics of major mammal species with the DMZ and CCZ

Classification	Name of species (scientific name)	Characteristics	Distributed Area
Cervidae	*Hydropotes inermis*	*Hydropotes inermis* has long and bent canine teeth	Base of a mountain all over the country, farmland and wetland
		Rough and thick hair, long hair on neck and waist in light brown color	
		Cowardly and homing that makes it come back to where it used to live in	
Felis catus	*Prionailurus bengalensis euptailurus*	Lives in forest, looks like leopard	Mountains in India and Southeast Asia
		Yellow brown or light brown back, white stomach with lots of black dots and stripes	
		Nocturnal, eats birds or small mammals	
Bovidae	*Nemorhaedus goral raddeanus*	The 217th Natural Monument and endangered species designated by Ministry of Environment	Siberia, China and Mountains near Seorak-san, Odae-san, Daegwan-ryeong, Taebaek mountain
		Only five species exist in limited areas all over the world	
		Grayish brown hair, dark brown hair tip, black horn	
Suldae	*Sus scrofa coreanus*	Brown or black body, lives in hardwood-forests in remote mountains	Center and South of Eurasia
		Comes down to hills when cold and snowy	
Canidae	*Nyctereutes procyonoides*	Very short and round tail covered with extremely fine long hair	All over Korea
		Almost black colored body, the color of belt on the back, spots under the eyes and belt on forelegs is darker	
Mustelidae	*Lutra lutra*	The body of *Lutra lutra* is formed to be suitable to living in water	Widely distributed in Europe, Africa, Asia, Northern and Central stream in Korea
		Flat round head, round nose and eyes are covered by wrinkled skin	
		Nocturnal, rests during the day and eats at night	

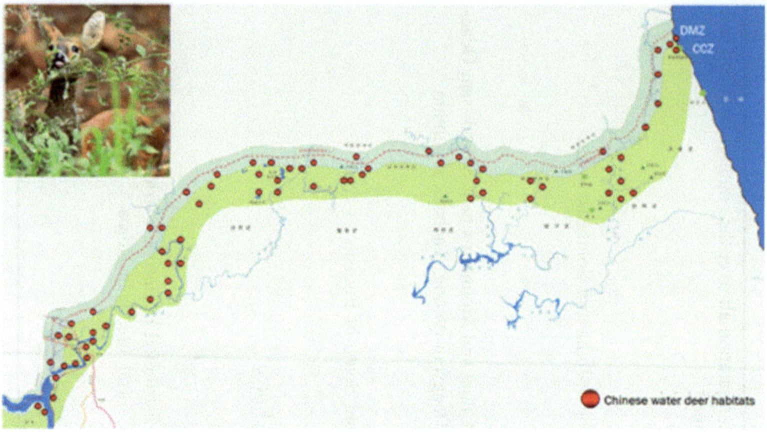

Fig. 5.164 Identified Chinese water deer habitat in DMZ and CCZ

Features of Leopard cat habitat in DMZ and CCZ

The main habitat of leopard cat is hills of DMZ and CCZ. But it is also found in cultivated lands where rodents are plenty. This cat is widely found in the DMZ and CCZ. It belongs to the upper level of the food chain, as a predator of small mammals and rodents. Thus, where leopard cat lives, there has to be high in biodiversity. Phasianus colchicus is added to the leopard cats' diet in order to gain fur. It is the only animal that survives on Felis catus in Korea, so it is very rare and needs protection. The number of leopard cats rapidly decreased since the war ended (Figs. 5.168, 5.169).

Considerations for securing the habitat

Although it is nocturnal in nature, it moves during the day in the woods. It lives in the valleys where there are rock formations and in the mountainous areas. It was also observed near brooks with a façade of lush shrubs where it can hide. It is a rare mammal and poachers hunt them. Poaching and hunting should be prohibited and the forest habitat should not be fragmented. A forest with covered canopy is the best place for this rare cat. Hence, the conservation initiatives should be focused on developing a wide interconnected dense forest just like the *H. inermis*. Wild boar/Wild pig (*Sus scrofa creanus*)

Figure 5.170.

Features of Wild boar/Wild pig habitat in DMZ and CCZ

The wild boar co-exists with the leopard cat and Chinese water deer in the interconnected hills and forest of DMZ. The whole DMZ area is seen as the best habitat for wild pigs as it has high biodiversity. There are also a lot of food wastes from the military facilities which contributed to the increase of their number in recent years. The problem is when it goes down to farm houses scavenging waste food, digging up cultivated lands, and devouring human food stocks during winter.

Fig. 5.165 Wetland area in Paju DMZ

Fig. 5.166 Water of rice paddy wetland area in CCZ

Wild pigs are sensitive and also nocturnal. They enjoy mud baths that is why they dig. They can be traced easily by following their fresh diggings near wetlands, according to the report of the survey conducted. Through the mud bath, the parasites and mites in their hairs will be removed this in turn, prevents the incidence of diseases among the pigs. In addition, mud baths cool down the body of these pigs that does not have sweat glands. Traces of wild pig diggings can be found throughout DMZ and CCZ. Figure 5.171 shows wild pigs eating food waste in front of a military unit in Inje county (Fig. 5.172).

Considerations for securing the habitat

The population of wild pigs increased because it seemed to have adapted well to human food wastes and are more daring to invade the farmlands. People found a good reason to hunt them. The pigs destroy the farmlands by their diggings and consume people's food stock in barns. Monitoring and controlling the number of these pigs in the wild should be done with the full cooperation of the residents in that area and the local government.

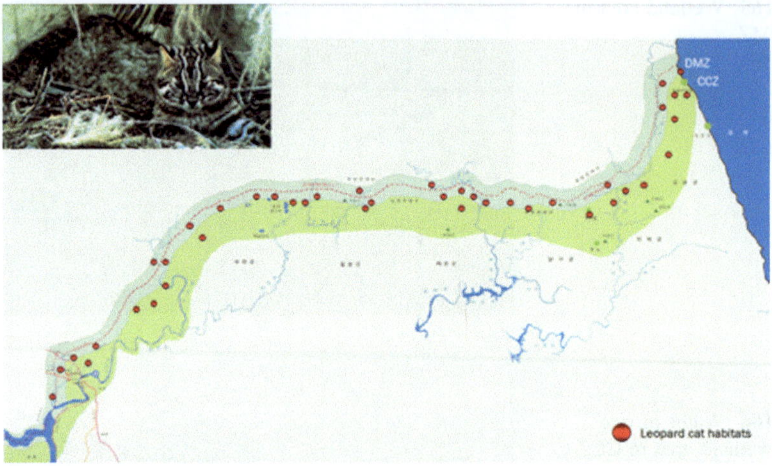

Fig. 5.167 Identified leopard cat habitat in DMZ and CCZ. ※ Endangered wild animal and plant Grade II designated by Ministry of Environment

Fig. 5.168 Stream habitat connected to neighboring forest near Pilseung bridge, Yoncheon DMZ

Fig. 5.169 Wetland area in Paju DMZ

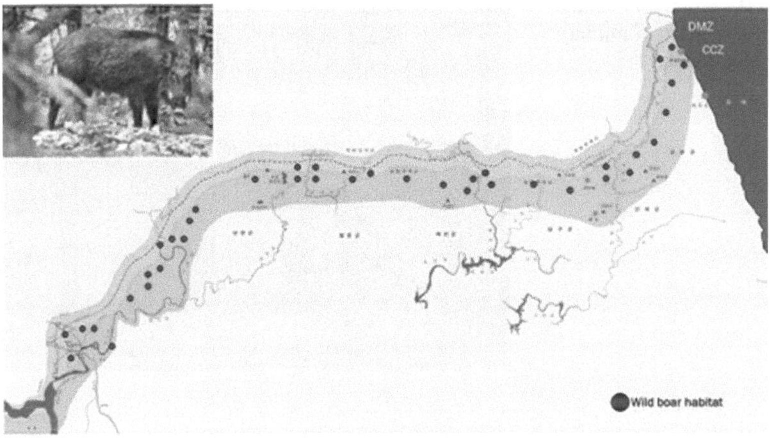

Fig. 5.170 Identified Wild boar/Wild pig habitat in DMZ and CCZ

Raccoon dog (*Nyctereutes procyonoides*)

Features of Raccoon dog habitat in DMZ and CCZ

N. procyonoides is widely distributed in DMZ and CCZ as shown in the map in Fig. 5.173. It adapts well in farmlands near the mountainous areas, especially the wetlands and depressions where they can catch fish and amphibians. In the farms and mountains, they hunt reptiles. It was also reported to have come down to houses and eat eggs of birds nesting in the bushes.

N. procyonoides is the only canidae animal that goes into hibernation. It digs a deep tunnel and nests in the midsection of the tunnel, then makes a few side holes toward the surface. It spreads dry moss and grass into its nest before hibernation.

The number of *N. procyonoides* in DMZ and CCZ is increasing because food is abundant and their predators like the wolves, leopards, and Meles meles had eventually became extinct. If there are any of its predators which survived in this area, their number may be very negligible. Thus, the number of *N. procyonoides* has suddenly increased (Fig. 5.174).

Considerations for securing the habitat

N. procyonoides moves relatively wide, because it lives in various habitats in the DMZ and CCZ. The usual areas where they reproduce are in mountain corridors, near streams, in the forests, and wetlands. Thus, the developments along the corridors to the forests and wetlands are a clear and eminent danger to the *N. procyonoides*. Therefore, the connectivity of the forests to streams and wetlands should be restored and the connectivity should be preserved for this animal to prevail.

Fig. 5.171 *Sus scrofa* eating food waste in front of military unit in Inje county

Fig. 5.172 Footprint of *Sus scrofa* discovered in wetland

European otter (*L. lutra*)

Figure 5.175.

Features of European otter habitat in DMZ and CCZ

L. lutra lives in the inland streams and in an estuary dam with clear water and abundant fish. Habitats for the *L. lutra* are found in many parts of the DMZ and CCZ. Being nocturnal, *L. lutra* feeds in watersides at night and hides behind the aquatic plants. It especially prepares its bed by digging a tunnel near the water, brings forth its young into the tunnel, and raises it.

The best habitat for *L. lutra* habitat in DMZ and CCZ are the natural streams within mountainous forests, and as well as in the lowlands where there are clear water with abundant fish and aquatic plants in the surface water or on the sides of the waterways. Thus, *L. lutra* can be an indicator of the health of the stream. It belongs to the highest level of the ecosystem, and functions as a Keystone Species, that is if one species becomes extinct, the diversity of all the other species are affected. *L. lutra* could be observed in the upstream of northern Han river as shown in Figs. 5.176, 5.177.

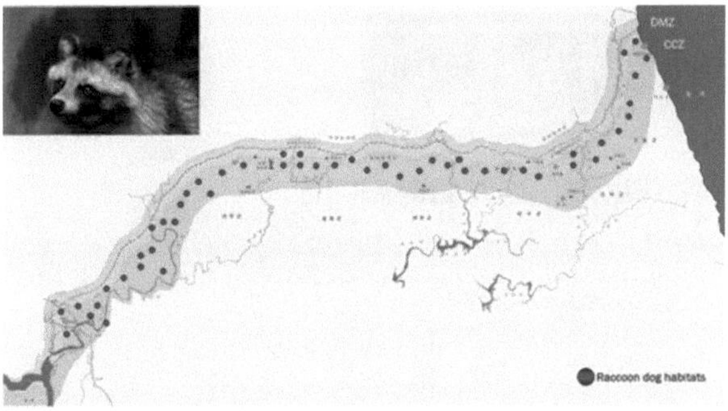

Fig. 5.173 Identified *Nyctereutes procyonoides* habitat in DMZ and CCZ

Fig. 5.174 *Nyctereutes procyonoides* habitat in wetland area in Sacheon river flood plain within Paju DMZ

Considerations for securing the habitat

In Japan, *L. lutra* is almost extinct. In Korea, it used to be distributed widely in the inland, in estuaries and even in the coastlines of Korea. But due to the destruction of habitats and the worsened water quality in lowland springs near cultivated areas, they are marginalized toward the mountains to find fresher waters. Ever since a bank was built in the mouth of the river, and all the other projects like the making of the dam, roads, railroads, and the concretizing of waterways, the lutra–lutra just now appears in some mountainous valleys and estuaries in DMZ and CCZ.

In order to secure the habitat of *L. lutra*, projects along and near waterways should not be allowed. The wetland should be maintained clean to serve as a buffer habitat of the *L. lutra*. Water quality must always be monitored and the causes of pollution must be by all means irradicated. *L. lutra* should be provided with clean water with the formation of sedimented minerals in water beds, including rocks and pebbles that help sanitize the water.

Goral (*Nemorhaedus goral raddeanus*)

Figure 5.178.

Features of Goral habitat in DMZ and CCZ

The Goral is an endangered species living in some parts of the mountainous areas east of DMZ and CCZ. It has thick, strong legs, and feet so as not to slip when jumping in cliffs and rocky areas where it chooses to live. The goral found by the Environmental Research Group which surveyed the area was standing on a cliff over 45°. No person could have been able to do as easily as this mammal could (Figs. 5.179, 5.180).

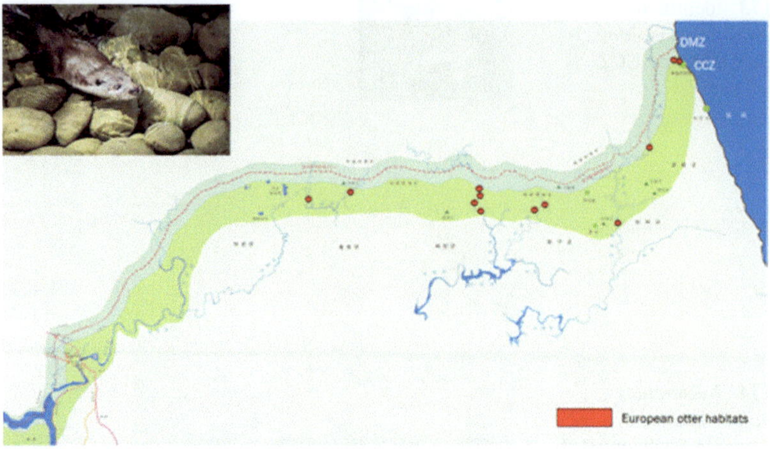

Fig. 5.175 Identified *Lutra lutra* habitat in DMZ and CCZ. ※ Endangered wild animal and plant Grade I designated by Ministry of Environment

Fig. 5.176 *Lutra lutra* habitat in upstream near northern Han river near DMZ

Fig. 5.177 Footprint of *Lutra lutra*

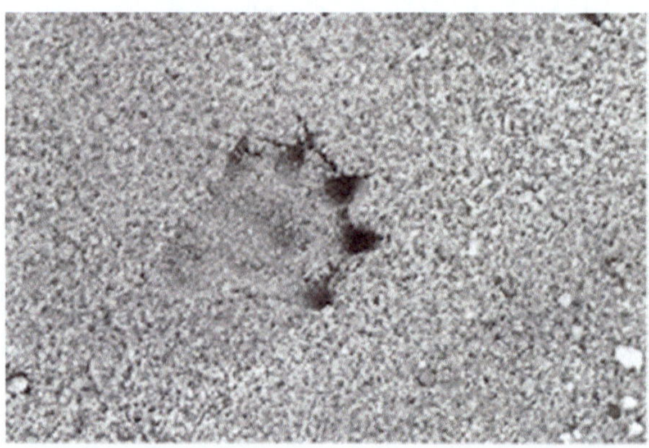

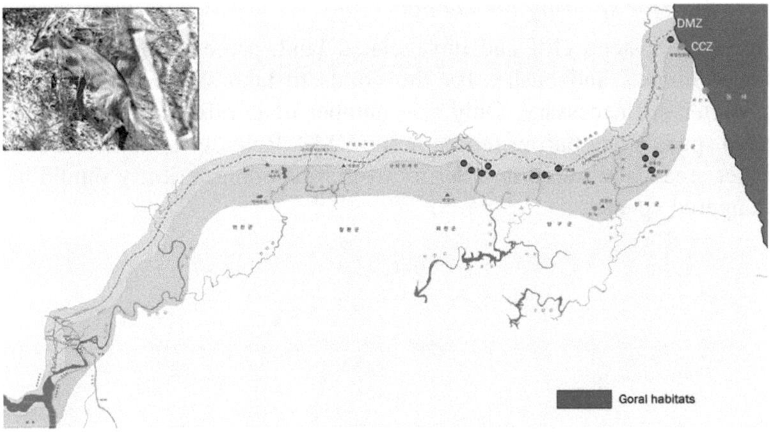

Fig. 5.178 Identified Naemorhaedus goral habitat in DMZ and CCZ. ※ Natural Monument No. 217, Endangered wild animal and plant Grade I designated by Ministry of Environment

Fig. 5.179 Goral habitat in Hwacheon-county DMZ: Cliff in mountain

Fig. 5.180 Naemorhaedus goral family

Considerations for securing the habitat

As it lives on a steep cliff and unvegetated land, preservation, restoration, and creation of tunnels and bushes for the gorals to take shelter from heavy snow during winter are necessary. Only few number of gorals are surviving in a disconnected space along barrier fences in the DMZ. Release of Naemorhaedus goral from other areas is required in order to avoid inbreeding. Priority should be given on endangered species.

Chapter 6
Biodiversity

6.1 Components of Biodiversity

In this book, biodiversity is classified into four categories which include; (1) Diversity of landscape, (2) Diversity of ecosystems or habitats, (3) Diversity of species, and (4) Diversity of genes. A diagram of the classifications is shown in Fig. 6.1.

6.1.1 Classification of Biodiversity

Based on the components of biodiversity shown in Fig. 6.1, in this chapter, the values of biodiversity and ecosystem service are presented. Then, biodiversity strategy and indices are described. Case studies include Cheorwon biodiversity strategy and the DMZ hotspots.

6.1.2 Landscape Biodiversity

The parameters describing landscape biodiversity are shown in Fig. 6.2. The information on the landscape biodiversity provide the basics for describing other components of biodiversity.

6.2 Biodiversity and Ecosystem Values

DMZ biodiversity and the respective values of the ecosystem can be described as follows: Diversity as a wildlife habitat; historical value; recreational opportunities; and education.

K.-G. Kim, *The Demilitarized Zone (DMZ) of Korea*,
DOI: 10.1007/978-3-642-38463-9_6,

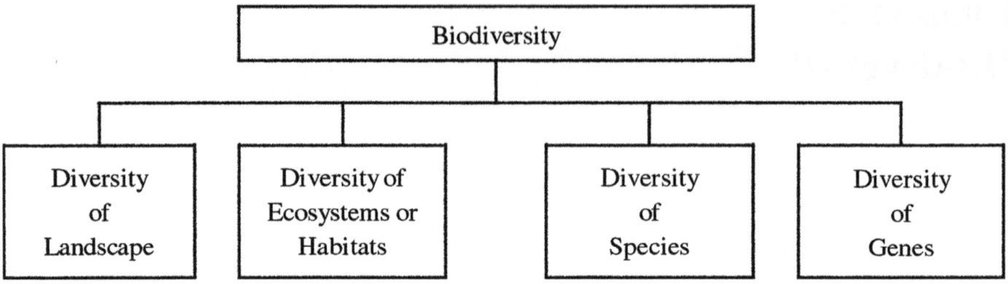

Fig. 6.1 Components of biodiversity

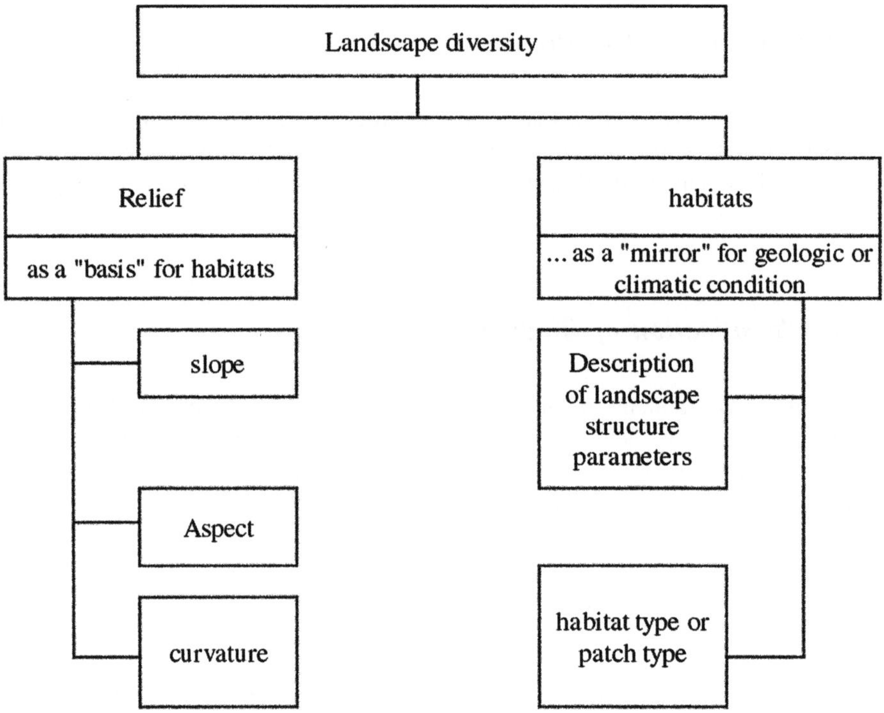

Fig. 6.2 Parameters describing landscape diversity

6.2.1 DMZ Biodiversity as a Wildlife Habitat

DMZ wetlands provide a habitat for many species of amphibians, reptiles, birds, and mammals that are uniquely adapted to wet environments. Upland wildlife such as Chinese water deer commonly uses wetlands for food and shelter. Wetlands are particularly vital to many migratory bird species, especially cranes. The cranes are included in the IUCN Red List as endangered species. There are only 1,700–2,000 cranes left in the world and 200–300 of them are found living in the

DMZ while another 300–350 cranes are living in North Korea. Practically 29–32 % of the surviving cranes are in Korea.

6.2.2 Historical Value

Historical relics found in the DMZ area provide mythical significance, a sentimental meaning, and define as well the regional characteristics within the premises of its tumultuous historical background. The ceasefire line from the east to the west represents the sadness of the separation and the remains of the train showed traces of time past. Unlike the fast changing cities in Korea, the DMZ is a remembrance from the past—the history and culture that symbolize the spirit of the Korean people.

6.2.3 Recreational Opportunities

DMZ and CCZ areas have become more and more popular as people began to recognize its values. Paju city and Cheorwon county areas in particular, have attracted many people with the appearance of rare birds, specifically the cranes and *Aegypius monachus*. The DMZ Experience Tour has become one of the most well-liked tours in Korea.

6.2.4 Education

DMZ provides an important resource for general education to everyone. For instance, the area can be used for an educational program for the enhancement of the understanding and recognition of environmental issues. There are currently various visit programs in the CCZ.

6.3 Biodiversity Strategy

6.3.1 An International Tendency

Though many biodiversity indices are suggested, there is not yet one that is widely accepted and used. The major challenges that one might face in developing biodiversity indices can be summed up into the following aspects:

1. Inadequacy of data,
2. information loss that may occur when scaling down very complex and multi-dimensional concepts into one-dimensional index, and the
3. lack of understanding on the causal link between biodiversity and the feature of ecosystem.

Recently, various efforts are exerted in a statewide and global level as a means of supporting policy decisions related to the sustainable development of the DMZ. Hundreds of indices for the measurement of biodiversity conditions are either suggested or undergoing development. Although there are indices that cover all the levels of biodiversity (diversity of genes, species habitat or ecosystem, and landscape) and aspects of the diversity (component, structure, and feature), the amount of effort put in different areas are not the same. Now let us take a look at the following types of biodiversity indices that have been suggested till date.

6.3.1.1 Gene-Level Indicators

The diversity is measured on the number of varieties within a species. Despite the fast development, there is not yet sufficient number of single gene-level indicators that are being widely used. Until further progress is made in this area, species-based indicators are recommended as the sound scientific basis for the development of indicators.

6.3.1.2 Species-Based Indicators

The richness of species is measured on the number of species within the given security area and its taxonomy. This is the simplest and most commonly used biodiversity index. However when this index is used, one should be careful about the weight being imposed on each species.

6.3.1.3 Other Indicators

Other indicators include population abundance-based indicators; ecosystem level area-based indicators; functional indicators, integrity indicators, and composition indicators

Till date, most of our biodiversity indicators are biased to that based on species. In the future various kinds of indicators including genes, abundance, and ecosystem-based indicators should be developed as a means of assessing the present conditions as well as for achieving international goals and targets in defining biological resources. The establishment of a standard biodiversity D/B and integrated information center should be promoted since experts with high credibility and consistency are essential for the development and utilization of various kinds of indices.

6.4 Cases

6.4.1 Biodiversity Strategy in Cheorwon

6.4.1.1 Land Use and Zoning for the Biodiversity in Cheorwon

The Man and Biosphere (MAB) program developed by UNESCO was applied for the land use division of the biodiversity in Cheorwon. Grade 1 was set as core area, Grade 2 as the buffer zone, and Grade 3 as the transition area. Based on these settings, the biodiversity comprehensive assessment map of Cheorwon, including the land-use division, was defined as shown in Fig. 6.3.

6.4.1.2 Current Biodiversity Strategy in Cheorwon

Biodiversity Strategy at a National Level

Through the Ministry of Environment, the biodiversity strategy implemented at a national level was to expand the "Biodiversity Management Contract Program" to seven migratory bird sites in five cities and counties including Chonsoo Bay in Seosan city. This means that the "Biodiversity Management Contract Program" increased the compensation to local residents from three cities and counties in 2002 to five cities and counties in 2003 who engage in farming around migratory bird arrival sites and participate aggressively in the protection activities such as; providing food and preserving rest areas for the migratory birds.

Fig. 6.3 Land-use zoning for the biodiversity strategy in Cheorwon

In order to protect quality ecosystem areas including migratory bird habitats, the Ministry of Environmental developed a guideline for the implementation of the "Biodiversity Management Contract Program" for 2003 through the active participation and cooperation of local residents. The mandate was delivered to the respective local governments on January 21, 2003.

The "Biodiversity Management Contract Program" is a natural environment conservation program attended by residents, including signing of a contract between the head of the local government and a landowner or manager on change of cultivation method, supply of food to migratory birds, and creation of wetlands to conserve endangered wildlife, migratory bird arrival sites, and quality biodiversity area and compensating residents with loss incurred by fulfilling the contract. This is widely applied in the UK, the Netherlands, and Japan. However, it was first introduced in Korea in 2002 in five migratory bird arrival sites in three cities and counties including Changwon (Joonam reservoir), Kunsan (Kumkang lake), and Haenam (Kochon Am lake, Youngam lake, and Kumho lake).

Expansion of the Biodiversity Management Contract Program seeks protection of migratory bird arrival sites and quality ecosystems through proactive participation and cooperation of residents and offering financial compensation to local residents. This is expected to serve as an opportunity to level up a natural environment conservation policy.

Sites designated for 2003 are migratory bird arrival sites in Seosan (Chonsoo Bay) and Kimjae (Dongjin River and Mankyung River estuaries) that were newly designated in 2003 in addition to migratory bird arrival sites in Changwon (Joonam reservoir), Kunsan (Kumkang Lake), and Haenam (Kochon Am Lake, Youngam Lake, and Kumho Lake), which were pilot project areas in 2002.

Local residents living in the above-described areas are compensated by the head of local governments if they are engaged in migratory bird protection activities such as leaving agricultural crops unharvested to provide food to migratory birds, creating and managing rest areas for migratory birds like, for example, keeping water in rice paddies and among other agreed responsibilities.

The subject of management contracts depends on the type of protection management to be engaged:

In the Lease Management Contract, the head of the local government rents the land within the radius of the migratory bird arrival site, allows local residents to cultivate and plant barley or wheat, and compensates the farmer on the net loss incurred by the migratory birds.

In the Protection Activity Management Contract, local residents are compensated for migratory bird protection activities, including the practice of environment-friendly farming in rice paddies, leaving unharvested agricultural crops as food for migratory birds and, providing rest areas, including water reservation.

The Cultivation Suspension Management Contract protects natural ecosystems such as wetlands by suspending farming temporarily and compensating the farmers on the net loss incurred by the suspension. In Table 6.1, Biodiversity Management Contracts are categorized by types and their features are summarized.

Table 6.1 Features of biodiversity management contracts by types

Item	Lease management contract	Protection activity management contract			Farming suspension contract
		Environment-friendly farming	Remain unharvested	Offer rest area	
Policy objectives	Migratory bird protection and natural ecosystem conservation				Quality biodiversity area conservation including wetland
Contractors	Local residents and head of relevant authority (MOE, central govt, local govt.)				
Target site	Migratory bird arrival sites (barley and wheat cultivation land)	Migratory bird arrival sites (rice paddies)			Quality biodiversity area such as wetlands (cultivation land where farming needs to be suspended temporarily)
Contract area	Area where migratory birds actually cause damage (discuss with local govt, experts, and residents)		Calculate based on migratory bird population and required grain		Area affecting ecosystem destruction seriously
Obligations	Sincere farming; no actions harming birds	Reduce use of agricultural chemicals	Do not harvest agricultural crops	Keep water (3–5 cm deep)	Suspend farming
Compensation	Compensate residents with net loss incurred by fulfilling the contracts				
Compensation calculation	Shortage of average income from agricultural crops	Decline in crop production due to less use of agricultural chemicals	Sales price of unharvested grain	Rest area creation and maintenance cost	Based on standard income from agricultural crops in the previous year
Payment of compensation	10 % at the signing of contract, 90 % after harvest				
	May in the following year	November	November	March	November
Obligation verification	Insincere Cultivation (use Agricultural Technology Center)	Check remaining agricultural chemicals (less than 1/2 of permissible level)	Check whether unharvested crops remain	Whether water depth of 3–5 cm is kept	Whether farming is suspended
Consequence when not fulfilled	• Do not pay out remaining amount and collect contract money • In case of insincere cultivation, pay 20–50 % less than the agreed amount				

Biodiversity Strategy at Local Government Level

Although Cheorwon county office initiated organic farming and various projects, they all failed due to lack of cooperation from farmers. They end up with returning the allocated conservation program budget to the central government. Based on old documents, cranes had been recorded inhabiting the Cheorwon 400 years ago. Flocks of cranes began returning to the area when Cheorwon residents reclaimed the area and began farming in 1970. It is obvious that cranes began coming to the area to feed on the leftovers of harvesting. Once again, the Cheorwon plain became the habitat of cranes. Since it is difficult to make a living with rice farming only, the Cheorwon government considered eco-tourism as an alternative. Although Cheorwon residents are positive about protecting cranes, they are not supportive of any regulation, especially the designation of their farmlands as conservation areas. They believe that this will restrict their main source of livelihood.

In Yangsoo-ri village today, cranes are being fed artificially. This is why they barely enjoy the fame of being the crane's habitat. Their local residents are also not accepting administrative restrictions on biodiversity.

In order to protect crane habitats, the Cheorwon county office introduced the rice paddy farming sabbatical year program. However, not a single farmer applied for the program. Instead, a compromise agreement was ensured and they called it the "Bird Feeding Site Lease Payment." It was agreed that few agricultural lands will be rented and preserved as crane and bird habitats. After thorough discussion with residents, the agreement was signed. Figure 6.4 illustrates the agreement including the location of bird feeding sites, migratory bird arrival sites, and quality natural ecosystem areas in Yangsoo-ri, which are managed for biodiversity conservation by Cheorwon county.

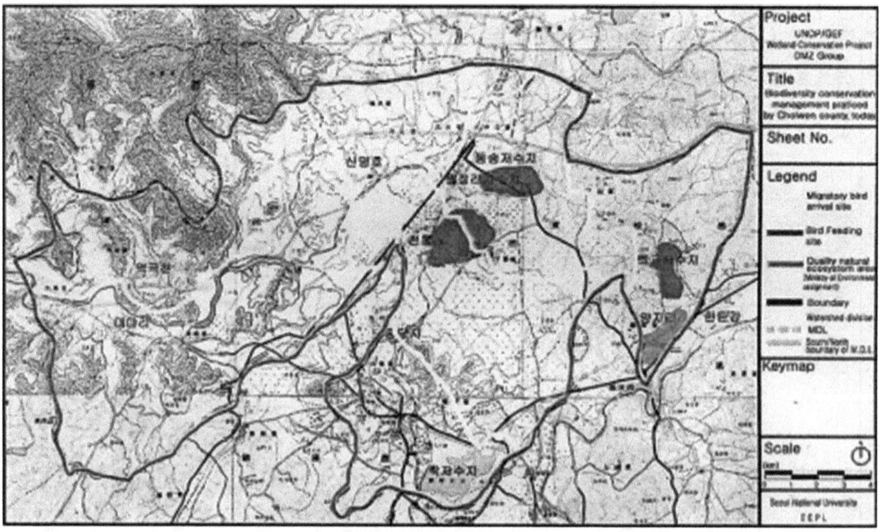

Fig. 6.4 Biodiversity conservation management practiced by Cheorwon county

Biodiversity Strategy at NGO Level

It is fair to acknowledge though that the opposing Cheorwon residents had been supportive of the IEC (Information, Education, and Communication) activities of the conservation program. They include learning camps and bird feeding events organized by crane schools and bird associations. Bird feeding events take place regularly every winter organized by various sectors including government and private sectors like the environment groups, media organizations, and military-based environmental organizations. These events became very frequent to such an extent that local protectionists claim that these events already stress migratory birds, especially the cranes, and jeopardize its very purpose. Because of this alarming consequence, the Cheorwon Chapter of Bird Protection Association planned to relocate the feeding events from food-picking sites to nearby roadsides. To improve further the implementation and to make the Bird Feeding more meaningful and symbolic to participating individuals, they are encouraged to willingly donate food bags. A community meeting was held in Cheorwon county on December 13, 2001 attended by residents from five villages of north CCZ. The assembly suggested unifying bird feeding events and making them more systematic.

The arrival sites of migratory birds in Cheorwon are then developed as eco-tourism program sites. The highlighted migratory birds which were observed to draw most of the attraction to Cheorwon are the *Grus japonensis*, Grus vipio, and *Anser albifrons*. Security sites are also strengthened in Woljeong station, 2nd South–North infiltration tunnel, and Labor Party headquarters and in Paekma heights for the security of the tourists.

Overview of Existing Biodiversity Strategies

The overall picture of the existing biodiversity strategies is shown in Fig. 6.4. They include migratory bird arrival sites at the national level and specifically at Yangsoo-ri village that was recognized as implementing a natural ecological best practice. Such strategies are executed to increase biodiversity and should be managed well to facilitate their execution.

Biodiversity Strategy Newly Proposed in Cheorwon

The New Biodiversity strategy proposed is the development of a Land Management Plan for Cheorwon Watersheds and the Upgrading of Technical Knowledge Base. To this end, the following are presented:

1. Increasing ecological information to conserve biodiversity in Cheorwon

- Conduct descriptive studies and surveys to provide the required information to be used as basis for action planning in the conservation of cranes such as habitat selection and use of food, nesting, minimum resource required, factors affecting survival and lifespan, effects of the interactions between cranes and human activities, and the impact of the disturbances on cranes.

- Intensify education campaigns and public relations (PR) based on appropriate concepts for effective management of biodiversity, conservation and protection.
- Reinforce research capabilities and specialization to increase ecological information in terms of:
  - Increased support to universities, institutions, and companies that study the conservation and the ecologically sound use of biodiversity.
  - Create a social environment by upgrading the activities of specialists working on biodiversity and conservation.
  - Upgrade and expand museums to highlight the environmental heritage and its significance to human and ecological security.
  - Set up a "Biodiversity Institution" composed of specialists in biology, sociology, economics, law, ethics, and other related specialties that shall study national biological resources and develop strategies on its sustainable use and conservation.
  - Introduce an "Ecological Certificate Program" that is recognized nationally to enhance ecological research ability of students and scholars.
- Data collection, research methodology, and procedure should be examined by recognized research authorities in terms of validity and reliability.
- A management information system (MIS) should be installed and the accessibility and effective use of information should be promoted and intensively implemented to trigger people's consciousness and active participation. Suggested mechanisms are as follows:
  - Information on biodiversity should cover environment, distribution, characteristics, value, threat, conservation technology, and use.
  - Centralized management of information to ensure the validity and reliability of the data that are disseminated to the public, specifically managed by government institutions (national or local public subsidiaries) and by "accredited" private research institutions and agencies.
  - Set up Information Centers promoting exchanges of biodiversity information and technology at home and abroad to build a network of environmental institutions.
  - Facilitate information exchange nationally and internationally.
  - Furnish updated data to specialists and professionals who are engaged in environmental research and other research of the same nature and purpose for the continuity of data collection and data triangulation.
  - Public and Private newspapers and publications should be mandated to publish information, conduct their own research, or support research endeavors on the status and use of biodiversity.
- Promote technology exchanges at home and abroad for biodiversity conservation such as:

- Collect information on conservation technology that is available across the country and globally.
- Promote exchange of technology by conducting an annual "ECO-TECH EXPO" at a national level.
- Enhance Civic and People's Cooperative Systems by the inclusion of programs that promote the exchange of conservation technology, ecological policies and research, in collaboration with universities, other agencies, and companies.
- To increase ecological information, Fig. 6.5 describes a conceptual map for building a biodiversity GIS.

Comprehensive Strategy for Cheorwon Biodiversity

The comprehensive biodiversity strategy in Cheorwon consists of land-use plan, the existing and newly proposed biodiversity strategies. In core wetland protection areas (migratory bird arrival sites), ecoparks were set. In the buffer area, specifically the agricultural lands are applying organic farming, and sabbatical years are introduced. Migratory bird observation parks were also designed. Transition area was planned to fully reflect existing land use. A program is also conceptualized linking the strategies of core area and buffer area. For instance, eco-village, security eco-tourism, and the natural ecological best practice area were planned and designed in such a way that the cultural remains will be well preserved. In addition, an ecological corridor was planned along the flyway of *G. japonensis* to facilitate their migration and to sustain the ecological connectivity of these birds to

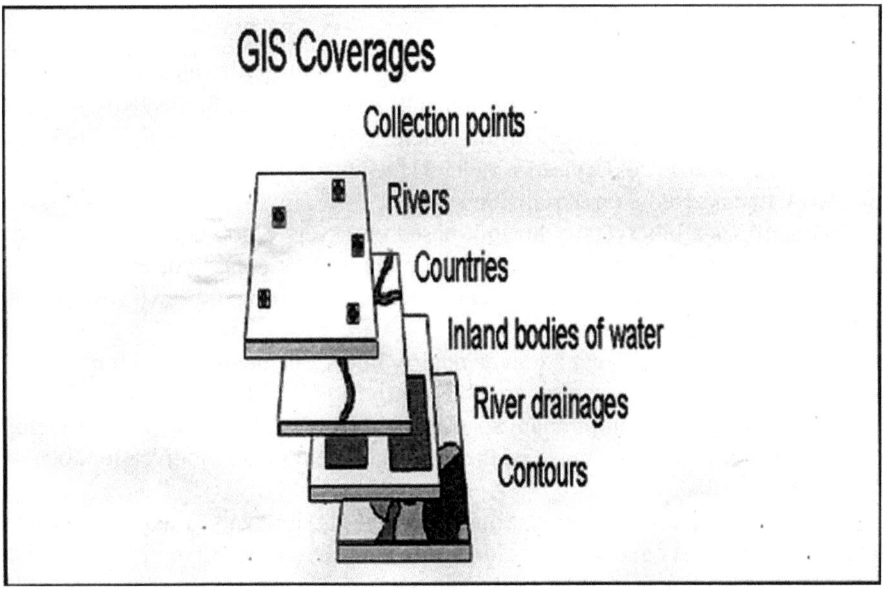

Fig. 6.5 A conceptual map for building biodiversity GIS

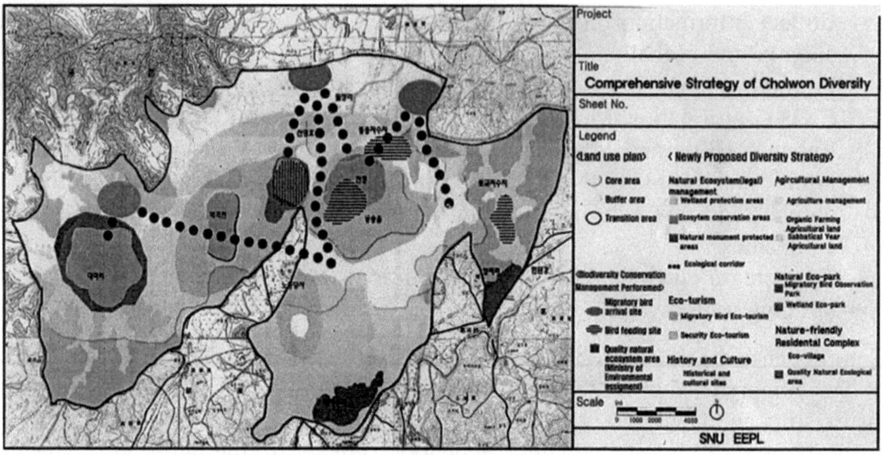

Fig. 6.6 Comprehensive strategy of Cheorwon biodiversity

their habitats. The biodiversity comprehensive strategy in Cheorwon is shown in Fig. 6.6.

6.4.2 DMZ Biodiversity Hotspot

6.4.2.1 Methods

A biodiversity hotspot is a biogeographic region with a significant reservoir of biodiversity that is under threat from humans. The concept of biodiversity hotspots originated from Norman Myers in his article, "The Environmentalist" (1988 [1] and 1990 [2]) and in his revised article, "Hotspots: Earth's Biologically Richest and Most Endangered Terrestrial Ecoregions."

To qualify as a biodiversity hotspot based on Myers (2000 edition) of hotspot-map, a region must meet two strict criteria: (1) it must contain at least 0.5 % or 1,500 species of vascular plants as endemics and (2) it has to have lost at least 70 % of its primary vegetation.

Around the world, at least 25 areas qualify under this definition with nine other possible candidates. These sites support nearly 60 % of the world's plant, bird, mammal, reptile, and amphibian species, with a very high share of endemic species. Only a small percentage of the total land area within biodiversity hotspots is now protected.

Several international organizations are working in many ways to conserve biodiversity hotspots and whose initiatives are all based on scientific criteria and quantitative thresholds namely:

- Critical ecosystem partnership fund (CEPF), a global program that provides funding and technical assistance to nongovernmental organizations in their participation to protect the Earth's richest regions of plant and animal diversity including: biodiversity hotspots, high-biodiversity wilderness areas, and important marine regions.
- Conservation International (CI) works in more than 40 countries on four continents, with headquarters near Washington, D.C.
- The World Wildlife Fund is the international organization that has derived a system called the "Global 200 Ecoregions" which aimed at selecting priority Ecoregions for conservation within each of 14 terrestrial, 3 freshwater, and 4 marine habitat types. They are specifically selected for their species richness, endemism, taxonomic uniqueness, unusual ecological or evolutionary phenomena, and global rarity. All biodiversity hotspots contain at least one Global 200 Ecoregion.
- Birdlife International is the organization that has identified 218 "Endemic Bird Areas" (EBAs), each of which holds two or more bird species that cannot be found elsewhere in the world. Birdlife International has identified more than 11,000 Important Bird Areas all over the world.
- Plantlife International is the one that coordinates several projects around the world aiming to identify Important Plant Areas.
- Alliance for Zero Extinction is an initiative of a large number of scientific organizations and conservation groups which cooperate to focus on the most threatened endemic species of the world. They have identified 595 sites, including a large number of Birdlife's Important Bird Areas.
- The National Geographic Society is the organization that has contributed to the development of a World map of the hotspots, Arc View shape file and compiled the metadata for the Biodiversity Hotspots including details of the individual endangered fauna in each hotspot, which is available from Conservation International.
- The high profile of the biodiversity hotspots approach has resulted in considerable criticism. Papers such as Kareiva and Marvier (2003) have argued that the biodiversity hotspots:
- Do not adequately represent other forms of species richness (e.g. total species richness or threatened species richness).
- Do not adequately represent taxa other than vascular plants (e.g. vertebrates, or fungi).
- Do not protect smaller scale richness hotspots.
- Do not make allowances for changing land-use patterns. Hotspots represent regions that have experienced considerable habitat loss, but this does not mean they are experiencing ongoing habitat loss. On the other hand, regions that are relatively intact (e.g. the Amazon Basin) have experienced relatively little land loss, but are currently losing habitat at tremendous rates.
- Do not protect ecosystem services.
- Do not consider phylogenetic diversity.

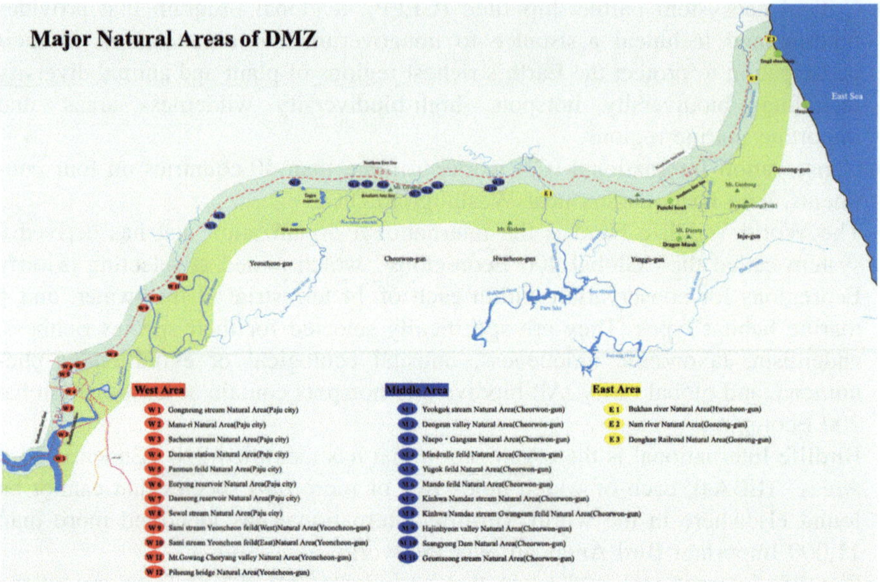

Fig. 6.7 Major natural areas of DMZ

Recent papers have pointed out that biodiversity hotspots do not address the concept of cost. The purpose of biodiversity hotspots is not simply to identify regions that are of high biodiversity value, but to prioritize conservation spending. The regions identified include regions in the developed world (e.g. the California Floristic Province), alongside regions in the developing world (e.g. Madagascar). The cost of land is likely to vary between these regions by an order of magnitude or more, but the biodiversity hotspots do not consider the conservation importance of this difference. It is either called Biodiversity Hotspot protected areas "Eco-regions' "Crisis Ecoregions" or "High-Biodiversity Wilderness Areas."

In this book we have adopted the concept of "Natural Areas" which was first used in the English Nature at UK, and selected the site by the conservative value of biodiversity per ecoregion. After the selection, we described the characteristics in terms of habitat, species richness, unusual ecological or evolutionary phenomenon, and global rarity (Fig. 6.7).

6.4.2.2 Major DMZ Hotspots

Han River Basin

Han River basin, Paju downstream region

This region has a brackish water wetland that lies on the estuary of Han river. The region is formed by a wetland of 200,000 m^2 with *Phragmites australis*. It is an area of well-developed wetland and estuary ecosystems in which both the community of *P. australis* and other wetland vegetation are widespread. Moreover, wetland vegetation is also well developed around the tributaries gathering into the estuary of Han River. Within waterways, various kinds of species are also found.

In this region, the tributary, the mainstream, and Gokneungcheon estuary of the Han River are connected. This area is regarded as having a high connectivity among species at the edge of the water. However, as the ecosystem of inland and forest is disconnected, restoration work is highly recommended.

Present Condition of Habitat

Along the waterways and wetlands, communities of *P. australis* are found in a wider scale that typifies the entire region. The brackish water type of the wetland emanates from the fusion of the Han River waters and the western sea water. It is regarded to be one of the widest brackish water type wetland habitats found in the region extending toward the foreshores of the western sea.

According to the detailed investigation conducted, the vegetation found at the estuary of Han River are quite simple with no scrub and tree vegetation. It is mostly filled with grass which provides the best habitat for insects. Since it is adjacent to farmlands, various kinds of natatorial birds inhabit the place such as *A. albifrons*, Anser fabalisserrirostris, and *Anser cygnoides*. Since these are legally protected species, the region is now regarded as an advantageous and valuable habitat for various kinds of species.

Figure 6.8 shows the view of the waterway and wetland taken during late fall in which grass and wetland vegetation are well developed (Fig. 6.9).

Present Condition of Species

Table 6.2 presents the number of species found in the estuary of Han River. Scrutiny on reptiles and amphibians could not be carried out due to the site condition.

Fig. 6.8 Tributary wetland of Han River downstream

Fig. 6.9 Wet grassland created in the estuary of Han River

Table 6.2 The number of species found in the estuary of Han river

	Vegetation and plants	Mammals	Birds	Reptiles and amphibians	Fish	Inland insects	Aquatic species	Total
No. of species found	35	4	14	–	5	3	3	64

Regarding vegetation and plants, approximately 20 ha of *P. australis* community is developed downstream of Han River. A typical brackish water wetland is found downstream as well. Moreover, the plants observed are: *P. australis, Carex scabrifolia* Steud, *Miscanthus sacchariflorus* var. gracilis Y.N.Lee, *Artemisia scoparia* Waldst. & Kit, *Calamagrostis pseudophragmites* (Haller f.) Koeler, *Themeda triandra* var. japonica, *Phacelurus latifolius* (Steud.) Ohwi, *Robinia pseudoacacia* L., *Sium suave* Walter, *Cyperus difformis* L., *A. scoparia* Waldst. et Kitaib., *Artemisia gmelini* Weber ex Stechm., *Plantago asiatica* L., *Setaria glauca* (L.) *P. Beauv., Xanthium strumarium* L., *Lycopus lucidus* Turcz. ex Benth., *Calamagrostis epigeios* (L.) Roth, *Imperata cylindrica, Ambrosia artemisiifolia.* L., *Erigeron annuus* (L) Pers, *Conyza canadensis, Aster tripolium* L., *Lepidium apetalum* Willd., *Chrysanthemum boreale, Rubus oldhamii* var., *Polypogon fugax, Mosla dianthera* (Buch.-Ham. ex Roxb.) ex Maxim., *Cardamine amaraeformis* Nakai, *Aster subulatus* Michx., *Cnidium japonicum* Miq., *Medicago lupulina* L., *Humulus japonicus* Sieboid and Zucc. and *Trifolium pratense* L. Near the estuary of Han river is where the following plants are widely found: *P. australis, Phragmites australis–Miscanthus sacchariflorus* var. gracilis Y.N.Lee-Carex scabrifolia Steud., *I. cylindrica* var. koenigii, Artemisia lavandulaefolia DC. *–I. cylindrica* var. koenigii-Setaria viridis (L.) P.Beauv. var. viridis., *Aster pilosus* Willd.,*T. triandra* var. japonica (Willd.) Makino., L. *apetalum, M. sacchariflorus* var. gracilis Y.N.Lee, *C. boreale, R. oldhamii* var. and *P. fugax*, etc.

There are no mountains around the region. However, scantily distributed 20-year-old oak populations such as *Quercus acutissima* Carruth, *Quercus aliena*, Blume and Quercus serrata, as well as the pine tree and *R. pseudoacacia* L. are also observed. In the future, we expect that an inspection of the background forest will be carried out in order to determine the connectivity status of the ecosystem.

Fig. 6.10 Phragmites australis, Carex scabrifolia Steud., and Themeda triandra var. japonica Makino observed at the estuary of Han river

In addition, exotic species including *A. artemisiifolia*, *R. pseudoacacia* L., *Erigeron canadensis*, *E. annuus* (L.)Pers., *T. pratense* L., *A. subulatus* Michx. and *A. pilosus* Willd. are found along the road and in close vicinities (Figs. 6.10 and 6.11).

A large-scale shore ecosystem protection is implemented along the waterway of Han River tributary where inland insects such as; dragonfly species, *Colias erate* (Esper), and *Pieris rapae* are highly observed. However, due to the seasonal nature of insect movement in grasslands and the dominance of one vegetation in the area, very few inland insects were detected during our inspection. A research in the near future to determine the movement of inland insects and the appearance of phanerogam per season should be conducted.

The research on fish and aquatic species was conducted in Gokneungcheon. Various kinds of aquatic species including *Acanthogobius hasta* (Temminck et Schlegel) (dominant specie), *Chelon haematocheilus*, *Oryzias sinensis*, *Tridentiger bifasciatus*, *Carassius auratus*, freshwater shrimp, Sesarma dehaani and Ilyoplax deschampsi are found in the community of *P. australis* that covered the vast plain, the wetland, estuary, and stream. *A. hasta* (Temminck et Schlegel) is an anadromous species and its presence in the region proves that it has brackish water. Figures 6.12 and 6.13 show *A. hasta* (Temminck Et Schlegel) and other aquatic species, Sesarma dehaani, found in the region.

A total of 14 bird species with an estimated total population of 522 entities were discovered in the Han river region and they include; *Emberiza rustica.*, *Hypsipetes amaurotis*, *Paradoxornis webbianus.*, *Parus palustris.*, *A. albifrons*, *Streptopelia orientalis orientalis*, *Lanius bucephalus*, *Ardea cinerea*, *Veratrum oxysepalum Turcz.*, *Phoenicurus auroreus*, *Passer montanus*, and *Pica pica*. The appearance of Anser fabalisserrirostris (250 entities) and *A. cygnoides* (50 entities), which are both grade II endangered wild animals, proves the Han River's adequacy as a habitat for birds.

The community of *Anser cygnoides.* was discovered resting at the junction where the mainstream and tributary of Han River meet. Nearby rice paddies were used as the feeding grounds for brent goose. Besides, birds such as *P. webbianus*, and *E. rustica.* were found nesting in communities of *P. australis* but their number

Fig. 6.11 Nearby background forest of Gokneungcheon stream at the estuary of Han river

Fig. 6.12 Acanthogobius hasta (Temminck et Schlegel) observed at the estuary of Han river

did not vary compared to those found in other areas. Furthermore, in the walk that was formed within the *P. australis*, the footprints of *Hydropotes inermis* and *Sus scrofa* were found.

We assume that this is also the habitat of the white-naped crane as there are many kinds of estuary vegetation in the area. But considering the recent wild expansion of *P. australis* and the reduction of other estuary vegetation such as *Scirpus maritimus* L. and *Suaeda japonica* Makino, the number of white-naped crane and *A. cygnoides* is expected to be reduced.

Only common kinds of birds such as Brent goose, duck, and Passeriformes are found nesting in the community of *P. australis*. But still the region is an important habitat and resting place for Brent goose (*A. cygnoides*, Anser fabalisserrirostris, *A. albifrons*). However, the environmental carrying capacity of *A. cygnoides* is found to be decreasing as the prey in the major riverside habitat of *A. cygnoides* is diminishing.

Fig. 6.13 Sesarma dehaani observed at the estuary of Han river

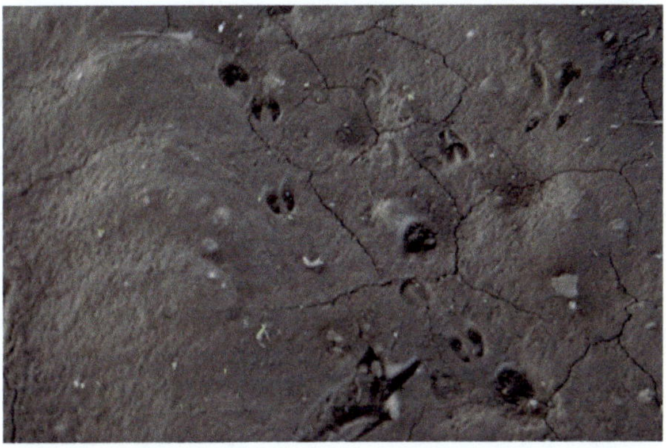

Fig. 6.14 Hydropotes inermis's footprint observed at the estuary of Han River

Regarding mammals, the habitat of some rodents and artiodactyl such as *Apodemus agrarius*, *Rattus norvegicus*. Berkenhou, wild cat and *H. inermis* were confirmed inhabiting the area but their diversity is very poor as the quality of the environment is also poor and the area is too small to sustain a larger population.

Figure 6.14 shows the appearance of *H. inermis*, a type of mammal. It shows that the area is being used as the ecological corridor of *H. inermis*.

Sacheoncheon River Basin

Paju Sacheoncheon: upper stream region

The region has three significant features namely: (1) the flow of Sacheoncheon, (2) its proximity to Panmunbeol region in its north, and (3) the development of the floodplain of Sacheoncheon.

The Panmunbeol wetland ecosystem that is formed by Sacheoncheon is shared by both North and South Korea. In this way, the Sacheoncheon functions as the central axis of the water's edge ecosystem. Moreover, North and South Korea are connected by the forest ecosystem surrounding the Sacheoncheon. As a result, the water's edge and the inland ecosystems are well connected and display a high connectivity with the plain wetland.

Present Condition of Habitat

The area along the Sacheoncheon that was formerly used as farmlands has now turned into a natural or semi-natural wetland. It has remained untouched for a long time. The wetland spreads widely around the Sacheoncheon and formed a wetland vegetation zone that is in good condition. This area is also occupied by *P. australis* and *S. maritimus* L.

Moreover, *H. inermis* was also found at the nearby wetland basin. The forest vegetation on the other hand sustains the *Q. acutissima* Carruth and *Q. aliena*. It is observed to have provided a perfect habitat for both birds and mammals. The grassland vegetation is dominated by *M. sacchariflorus* var. gracilis Y.N.Lee and communities of *Salix koreensis* Andersson and *Alnus japonica* Steud are observed at the nearby area adjacent to the wetland.

The region is covered with wetland vegetation that spreads widely. However, allowing the destruction of the forest vegetation will be preventing the further development of wetland vegetation. A domino effect is actually the significance of the high connectivity of these ecosystems, whether toward destruction or toward sustainable development. Therefore, in order to protect the region from such grim apprehension, urgent and speedy measures must be implemented. Prevention and control of the wide spread of *Amorpha fruticosa* L. in the wetland is also crucial to the conservation of the habitats sustained by these areas. The various kinds of habitats along Sacheoncheon are presented in Figs. 6.15, 6.16, 6.17, 6.18, 6.19, visibly providing at the same time, a beautiful landscape while maintaining the high level of biodiversity.

Figures 6.20 and 6.21 show Sacheoncheon and North Korea viewed from the pavilion located within Daeseong village. Around Sacheoncheon, *S. koreensis* Andersson communities can be observed. Birds including natural monuments such as *Circus cyaneus cyaneus* and *Buteo buteo*, as well as endangered species like the *G. japonensis* were detected in these regions besides the *A. albifrons*, *Coccothraustes coccothraustes*, *S. orientalis*, *E. rustica.*, *A. cinerea* and *Emberiza elegans*. The water regime flowing across Daeseongdong to the tributary of Sacheoncheon is especially regarded as a good habitat for fish as vegetation of *P. australis*, and *M. sacchariflorus* Benth are abundant.

Thus, continuous monitoring of the farmland and the Daeseong village located around Sacheoncheon is highly recommended in order to protect the area from potential threats.

Fig. 6.15 Sacheoncheon river basin and surroundings observed on the highland

Fig. 6.16 Marsh with herbaceous wetland plants

The floodplain developed within the tributary of Sacheoncheon is a highly developed wetland and has a well-formed water gate that provides a very conducive habitat for many species. Communities of *P. australis* and *S. koreensis* Andersson thrive here, as well as *Prionailurus bengalensis*.

The floodplain in Sacheon is one that has both marsh and swamp in which wide open water surfaces are found. Although a detailed inspection was not possible as it is within the military zone, it is assumed to be a good habitat for wild animals as

Fig. 6.17 Swamp with woody wetland plants developed along Sacheoncheon

Fig. 6.18 Wet grassland developed near Sacheoncheon

Fig. 6.19 Bush and scrub forest developed within Sacheoncheon river basin

it is fully equipped with reservoir, farmland, water gate, and wetland. However despite such good habitat conditions, sustainable restoration effort for the natural wetland is still recommended since the farmland was expanded up to the inner part of the bank. Near the wetland, *Anas poecilorhyncha*, *Nycticorax nycticorax*, *Egretta garzetta*, and *Ardea modesta* are seen. Even a trace of *Phasianus colchicus*, *Anguilla Japonica* Temminck & *Schlegel* and *Eriocheir sinensis*. could be found (Figs. 6.22, 6.23).

Fig. 6.20 Sacheoncheon river

Fig. 6.21 Feature of North Korea seen over Sacheoncheon

Present Condition of Species

Species population in Sacheoncheon, Paju, is presented in Table 6.3. Detailed inspection of inland insects was not made possible in the 2008 research due to the condition of the target area, thus data obtained from the 2006 research was used instead as reference in this presentation.

Fig. 6.22 Sacheon reservoir

Fig. 6.23 Agricultural waterway inflowing to Sacheoncheon

Table 6.3 The number of species found in the upper stream of Sacheoncheon

	Vegetation and plants	Mammals	Birds	Reptiles and amphibians	Fish	Inland insects	Aquatic species	Total
No. of	species	27	8	11	2	6	–	5
59								

To completely enumerate and describe the representative communities of wetland vegetation, it is illuminated generally as; the *P. australis* community, Zizania latitolia community, *Scirpus wichurae* var. asiaticus (Beetle) T.Koyama.community, and S. maritimus L. community that are found as most widespread; *Acer tataricum* subsp. ginnala (Maxim.) Wesm.community, *A. japonica* Steud. community, and *S. koreensis* Andersson community that are also found scattered all over; *R. pseudoacacia* L. community that is found dispersed at the lowland and in some forest hill areas; while PhytolaCCZ americana. L. that is found along roads at the boundary of DMZ.

Meanwhile on the mountain side, *Q. acutissima* Carruth is the major coverage of the forest with a few of *Q. aliena* Blume. *R. pseudoacacia* L. and *S. koreensis* Andersson found in the nearby wetland and in areas where resident houses are historically situated. Large scale of *A. japonica* Steud. community is also observed on the way to GP that was used as farmlands in the past. The woody plants that grow in this area are the following: *Q. acutissima* Carruth., *Q. aliena* Blume, *R. pseudoacacia* L., *Quercus mongolica*, Lespedeza, Symplocos chinensis for. pilosa (Nakai) Ohwi, *Morus alba* L., *Rosa multiflora* Thunb. var. *multiflora*, *A. tataricum* subsp. ginnala (Maxim.) Wesm., *S. koreensis* Andersson, *Prunus padus* L. for. padus, *Pinus rigida* Mill., *Elaeagnus umbellata*. Thunb., *A. japonica* Steud., *Corylus heterophylla* var. thunbergii, *Betula davurica* Pall., *Zanthoxylum schinifolium* Siebold & Zucc., *Ailanthus altissima*, and a considerable number of *Aristolochia contorta* entity, a tropical plant strangely found in the open area. Figures 6.24 and 6.25 present and define the vegetation and plants found to have grown in the Sacheoncheon river basin and its surrounding areas.

The mountainous area is dominated by *Q. acutissima* Carruth. The northern farmlands are still maintained while the farmland in the south of the truce line is transforming into a natural wetland.

Large scale of *A. japonica* community in low wetlands and flat areas where farmlands are assumed to be situated in the past is embanking around the end of fields. Its width and length is expected to be approximately 50–100 m and 1 km respectively.

Fig. 6.24 Vegetation developed within Sacheoncheon of DMZ

Fig. 6.25 Alnus japonica community developed around Sacheoncheon

Fig. 6.26 Zizania latifolia—Phragmites australis community and Quercus acutissima Carruth. forest observed around Sacheoncheon

Fig. 6.27 Quercus acutissima Carruth. forest observed around Sacheoncheon

Figures 6.26 and 6.27 show the *Zizania latifolia–P. australis* community in the wetland and the *Q. acutissima* Carruth developed in the hill area. The vegetation is seen generally growing widely at a good and stable pace but in some regions, poor vegetation is seen in the middle and lower layers with remnants of forest fires.

Figure 6.28 shows the dominance of *Q. acutissima* Carruth. In Fig. 6.29, *P. rigida* Mill. forest is shown forming a small community that was assumed to have been planted in the past. Figure 6.30 presents a community of *S. koreensis* Andersson–*A. tataricum* subsp. ginnala (Maxim.) Wesm. formed in a wet area near GP with a circumference ranging from 20 to 50 cm and an average height of 14 m. Figure 6.31 shows huge *P. padus* L. for padus growing in the surrounding area. Figure 6.32 presents the *R. pseudoacacia* L. forest that developed in the hill area near the wetland with a height of 5–7 m. They are shorter in height compared to the expected and the main stems are all dead due to forest fire; nevertheless, new stems are seen sprouting in its trunk.

The density and diameter class of forest trees of this region's vegetation are found to be higher compared to that of Yeoncheon county, but due to the frequent forest fires, they showed poor layer development in terms of species composition

Fig. 6.28 Quercus acutissima Carruth. forest and Populus davidiana Dode (in light color) observed around Sacheoncheon

Fig. 6.29 Pinus rigida Mill. forest observed around Sacheoncheon

and forest stand structure. The middle and lower layers were found to be in simple structures, therefore, weak and still unstable.

The widespread wetland vegetation though was forming a community that is in quite a good condition. But as consistently mentioned in earlier chapters, the destruction of forest vegetation is regarded to be having negative impact on wetland vegetation and the control of *A. fruticosa* L. is necessary to prevent it from turning into a dry land.

The research on fish and aquatic species was conducted at the small stream next to the pathway to DMZ. The water quantity was not plentiful but the ecosystem was maintained in quite a good condition. Fish such as *O. sinensis*, Macropodus chinensis, *C. auratus*, *Misgurnus anguillicaudatus*, and *M. anguillicaudatus* were found here including the *Aphyocypris chinensis* Cirinherthat is dominant. Aquatic species such as *Muljarus japonicus.*, *Notonecta triguttata*, *Ranatra chinensis* Mayr, *Nepa hoffmanni* Esaki, and *Neocaridina denticulata* were also observed.

Typical species are easily found in the wetland area. The *Macropodus chinensis* is observed to be decreasing in number but is expected to increase and repopulate. Figures 6.33 and 6.34 show the *A. chinensis* are Cirinher and Macropodus *chinensis* are observed in the area.

Fig. 6.30 Salix koreensis Andersson-Acer tataricum subsp. ginnala (Maxim.) Wesm. community observed around Sacheoncheon

Fig. 6.31 Prunus padus L. for. padus observed around Sacheoncheon

Fig. 6.32 Robinia pseudoacacia L. forest observed around Sacheoncheon

Fig. 6.33 Aphyocypris chinensis observed around Sacheoncheon

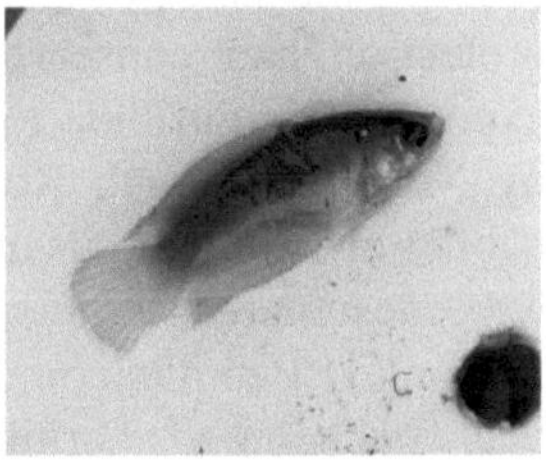

Fig. 6.34 Macropodus chinensis observed around Sacheoncheon

Fig. 6.35 Dropping of Hydropotes inerm is observed around Sacheoncheon

It is a perfect breeding place for amphibians during spring. Reptiles and amphibians including *Rana coreana* and *Glandirana rugosa* were also observed in this region. As for birds, more than 30 entities of *P. webbianus*., 2 entities of *P. colchicus* karpowi, over 20 entities of Emberiza cioides, 10 entities of *P. palustris*., over 17 entities of Corvus macrorhynchos, over 10 entities of *P. pica*, 3 entities of *G. japonensis*, 5 entities of *P. montanus*. Linnaeus, 7 entities of *A. albifrons*, 2 entities of Parus ater, and 9 entities of *A. monachus* were found at a specific survey time frame. As for mammals, the following inhabitants were identified by the droppings and footprints systematically observed in the region namely; *P. bengalensis*, *Mustela sibirica*, *Nyctereutes procyonoides*, *S. scrofa*, *H. inermis*, *Micromys minutus*, *A. agrarius*, and wild cats. Figure 6.35 shows the dropping of *H. inermis* found during the inspection.

Around Sacheon reservoir and nearby farmland, dead *H. inermis* and *Elaphe dione* were observed as well as *Egretta intermedia*. Within the reservoir, *Silurus asotus*, *Carassius cuvieri*, *E. sinensis*, *Rhodeus uyekii*, and *Zacco platypus* were found (Figs. 6.36, 6.37).

Paju Panmunbeol region

For reference purposes, the research conducted in 1996 was limited to Paju Panmunbeol region, which was the first one conducted at the DMZ while under the control of the UN Command (UNC). The UNC assisted the survey and monitoring of the habitats of wild animals in Panmunbeol area which is known to be one of the major habitats for migratory birds. The main objective of this investigation is to describe the status of the present ecosystems in the area and to suggest appropriate protection and conservation management strategies. The investigation is also expected to contribute to the establishment of the ecological data jointly initiated by North and South Korea for the purpose of implementing and mapping the habitat of wild animals and the monitoring of migratory birds. A total of ten spots were designated as hotspots including (1) Eoryong reservoir, (2) Daeseong reservoir, (3) Panmunjeom, (4) North Korean side wetland, (5) Daeseongdong, (6) Sacheoncheon, (7) Farmland within Daeseongdong, (8) abandoned rice paddy field wetland near reservoir, and (9) the pond within farmland.

Based upon the collected opinions from experts of various species field, this investigation recreates the features of Panmunbeol, western part of DMZ on the record for the first time in Korea. Such investigation is very significant since it is able to provide a data-based assessment of the natural value of the regions of DMZ and made it possible for program organizers and implementers to plan and manage the area scientifically through data utilization and processing.

Dead Hydropotes inermis | Ardea cinerea | Egretta alba modesta | E. dione

Fig. 6.36 Species observed around the farmland of Sacheon reservoir

Silurus asotus | Carassius cuvieri | Eriocheir sinensis

Fig. 6.37 Species observed at the Sacheon reservoir

Panmunbeol region is composed of farmlands with surrounding hilly areas. A tributary of Sacheon is flowing within these areas and reservoirs can be seen everywhere. The hydraulic and hydrologic analysis of the region's forested wetland revealed that the flow of water and location of hydrosphere provided ample quantities of water which largely emanated from many reservoirs and tributaries stretching within the area. A landscape scenery was observed with interconnecting ecosystems of; wetland shrubs, farmland, stream, reservoir, forest and, forested wetland in the Panmunbeol region. The dominant wetland vegetation within Panmunbeol region was described to be *Salix gracilistyla* Miq. and the community of *S. koreensis* Andersson, while *P. australis* and *A. japonica* Steud. were found within the wetland of the abandoned paddy field.

The agronomical survey on Panmunbeol forested wetland revealed that the region is generally moist soil. It is an organic soil found to be having excellent qualities favorable for the development of wetland vegetation. In turn, the lush wetland vegetation attracted many kinds of winter migratory birds that are observed in large numbers during the entire duration of the season.

Daeseong reservoir region

The targeted research locale was the area close to MDL and the farmland such as rice paddies and ginseng fields in the vicinity of Daeseongdong village. In this research locale, various kinds of habitats were investigated namely lakes, ponds, forests, and shrubs and forest.

Various shapes and size of islands were found within Daeseong reservoir and *S. maritimus* L. was obviously seen dominating around the islands. Moreover, birds such as *N. nycticorax*, *E. garzetta*, *Dendrocopos leucotos quelpartensis* and *Phalacrocorax filamentosus* were confirmed to be using the islands as their resting place.

Present condition of habitat

The forest of Daeseong reservoir is found to be quite dense compared to other regions with oak population as the most dominant species but pine trees are also found in some parts. Aquatic wetland and nearby forest ecosystems are well connected in the Daeseong reservoir. The wetland is also well developed due to the abundance of water from the reservoir that has well-defined inlet and outlet of water flows supplying the wetlands. This specific research locale provides an opportunity for the investigating team to study the "transition process" of a marshland transforming into a swamp ecosystem (Fig. 6.38).

Fig. 6.38 Daeseong reservoir and surrounding habitat

Present condition of species

S. koreensis Andersson community is highly detected in this area. This feature, characteristic of an area, is regarded as an advantageous condition for maintaining high diversity of species. The forest near Daeseong reservoir was dominated by *Q. acutissima* Carruth and had a considerable number of trees, specifically the pine tree and *P. rigida* Mill. In some regions connected with wetland, *R. pseudoacacia* L. forest was found while *A. japonica* Steud. and *S. koreensis* Andersson were observed in the low wetland areas. The condition of wetland vegetation formed at the lower part of wetland influent channel was found in a good condition.

Regarding the forest vegetation, *Q. acutissima* Carruth., *Q. aliena*, and *Quercus variabilis* were found to be dominating the area and it is easy to assume that the desired natural vegetation is competing with the undesirable communities of major accompanying species. Forest vegetation near Daeseong reservoir was observed with the community of *Q. acutissima* Carruth., *Q. aliena*, and *Q. variabilis* and, the diameter at breast height and maximum height of the major species were found to be 10–30 cm and 10–14 m respectively.

The hydrosphere of forest vegetation is divided into four layers namely (1) the tree layer where *Q. acutissima* Carruth. was found dominant and *Q. aliena*, to include the pine trees, (2) the subtree layer where *Quercus dentata* var. fallax Nakai, *Q. aliena* and *Q. acutissima* Carruth were also found, (3) shrub layer with *Lespedeza bicolor* dominating and *Q. acutissima* Carruthis also presents, *Q. aliena*, *Rhododendron mucronulatum* Turcz. var. mucronulatum were also observed, and (4) the herbaceous layer where *Carex lanceolata* was found dominant, to include, *Artemisia keiskeana*, *Solidago virga-aurea* var. asiatica, *R. mucronulatum* Turcz. var. mucronulatum, *M. sacchariflorus* var. gracilis Y.N.Lee and *Atractylodes ovata* (Thunb.) DC.

Between the low water bed and forest vegetation of Daeseong reservoir, *Spiraea prunifolia* for. simpliciflora. Nakai, *S. virga-aurea* var. asiatica, *Lespedeza virgata* (Thunb.) DC., *R. mucronulatum* Turcz. var. mucronulatum, *Celastrus orbiculatus*, *Eupatorium japonicum* Thunb. ex Murray, *Rhamnus davurica*, *Lonicera japonica*, *Lonicera maackii* (Rupr.) Maxim., *Viburnum opulus* var. calvescens, *Solidago serotina Aiton*, *Rubus hirsutus*, *Prunus sargentii* and, *Styrax japonicus*, *Onopordum cynarocephalum Boiss & Blanche*, *R. multiflora*, *Rhus trichocarpa* were found which are highly regarded as food for birds (Fig. 6.39).

In the wetland area, species such as *M. sacchariflorus* Benth., *M. sacchariflorus* var. gracilis Y.N.Lee, *Isachne globosa* (Thunb.) Kuntze, *Saussurea pulchella*, *S. wichurae* var. asiaticus (Beetle) T.Koyama, *S. koreensis* Andersson, *A. japonica* Steud., *S. prunifolia* for. simpliciflora. Nakai, *Osmunda japonica*, *Acorus calamus* L., *Rubus parvifolius* L. for. parvifolius, *R. hirsutus*, *C. orbiculatus*, *Juncus effusus* var. *decipiens Buchenau.*, *S. suave* Walter, *Z. latifolia*, *Persicaria sieboldii*, *Cyperua microiria*, S. maritimus L., *Trapa japonica*, *Nymphoides peltata*, *Hydrilla verticillata*, *Ceratophyllum demersum* L., *Potamogeton crispus*, and *Potamogeton malaianus* Miq. were found as well. Some species appeared in communities, including *P. australis* community, *Z. latifolia* community, *S. wichurae* var. asiaticus (Beetle)

Fig. 6.39 Ecosystem in eastern part of Daeseong reservoir

Fig. 6.40 Quercus acutissima Carruth. appeared near wetland

T.Koyama community, and *A. calamus* L. community (Figs, 6.40, 6.41, 6.42, 6.43, 6.44, 6.45, 6.46, 6.47, 6.48).

In the 2008 research, the fish species found were the *C. auratus* (dominant), and *O. sinensis*, *Pseudorasbora parva*. While the aquatic species found were the *Muljarus japonicas*, *R. chinensis* Mayr, *Laccotrephes japonensis*, and *N. denticulata*. The Daeseong reservoir region is a perfect spot for spawning aquatic species including swamp fish with various kinds of water plants at the edge of the reservoir.

The investigation conducted in 2006 was held during summer since it was a perfect time to observe various kinds of fish and aquatic species. Fish investigated in this time of the year include *Squalidus gracilis majimae*, Odontobutis intereupta, *Pseudobagrus fulvidraco, C. auratus, Rhinogobius giurinus, Pseudoperilampus notatus*, *Crustacea, shrimp*, and *Cipangopaludina chinensis* malleata. Not as many species were seen during the 2008 investigation compared to that of 2006 because the latter research was conducted during the cold winter. Many species can only be observed during spring and especially in summer (Figs. 6.49, 6.50, 6.51, 6.52, 6.53).

In the research of 2008, there were no insects, reptiles, and amphibians found since these species are not typically active during winter. The 2006 research

Fig. 6.41 Robinia pseudoacacia L. forest located at the piedmont

Fig. 6.42 Settlement of Salix koreensis Andersson and Alnus japonica Steud. at the rice paddy in the past

Fig. 6.43 Wetland occupied with Phragmites australis which was rice paddy in the past

revealed inhabitants such as the *Rhabdophis tigrinus*, *Agriocnemis pygmaea* (Rambur) and *R. chinensis* Mayr. It was also a perfect time for insect inhabitants especially that many kinds of plants flourish in this time of the year and the wetland vegetations are at its best because of the sunshine (Figs. 6.54, 6.55, 6.56).

Fig. 6.44 Damage made to pine tree due to forest fire

Fig. 6.45 Daeseong reservoir migratory bird and nearby forest vegetation

The 2008 research revealed a total of 19 species and 7,457 entities of birds and they include duck and wild goose species, *Ardea alba*, *Tachybaptus ruficollis*, crow, *V. oxysepalum Turcz.*, *P. colchicus*, *Uragus sibiricus*, *P. palustris.*, *P. webbianus*, *E. rustica*, *P. auroreus*, *Cettia squameiceps*, and *l. bucephalus* as well as natural monuments *Aix galericulata* and white-naped crane and Anser fabalisserrirostris designated as the level II endangered wild animals. Most of them were migratory birds as the research was held in the winter. Since it is a reservoir, it had provided clean water for the birds while the farmland nearby provided food. Such features of the Daeseong reservoir made it a perfect habitat for the stopover of Anser fabalis and duck species in both spring and fall. According to the 2008

Fig. 6.46 Salix koreensis Andersson observed in Daeseong reservoir

Fig. 6.47 Scirpus maritimus L. observed in Daeseong reservoir

Fig. 6.48 Acorus calamus L. observed in Daeseong reservoir

Fig. 6.49 Carassius auratus (2008)

Fig. 6.50 Ranatra chinensis Mayr (2008)

Fig. 6.51 Rhodeus uyekii and Cipangopaludina chinensis malleata (2006)

Fig. 6.52 Squalidus gracilis majimae and Odontobutis intereupta(2006)

Fig. 6.53 Scapharca broughtonii (2006)

Fig. 6.54 Rhabdophis tigrinus observed in Daeseong reservoir

Fig. 6.55 Agriocnemis pygmaea (Rambur) observed in Daeseong reservoir

Fig. 6.56 Ranatra chinensis Mayr and shrimp observed in Daeseong reservoir

research, approximately 7,000 entities were using the region as the stopover including Anser fabalis, duck species, and *A. galericulata*. A family of white-naped crane is frequently observed in the nearby rice paddy.

Based upon these facts, the reservoir within DMZ plays a "pole-vaulting" role in the preservation and protection of endangered water birds and winter migratory birds that inhabit the region during winter and fall (Figs. 6.57, 6.58).

During spring and summer on the other hand, different kinds of species appear but some of the selected rare species were observed to remain. The 2006 research conducted during summer revealed the presence of *A. alba*, *Aquila chrysaetos*, white-naped crane, and about 300 *A. albifrons*, *Mergus merganser* and *S. orientalis* were found around the reservoir. As the reservoir was used as the resting place and the farmland nearby the reservoir as the major habitat providing preys for birds, the entire region was regarded as the perfect habitat for all kinds of birds throughout the four seasons (Fig. 6.59).

Fig. 6.57 White-naped crane observed in Daeseong reservoir

Fig. 6.58 Birds flying observed in Daeseong reservoir

In 2006, the footprints of a mammal called the *P. bengalensis* was found and is shown in Fig. 6.60. A follow-up inspection carried out in year 2008 still found traces of *P. bengalensis* to include *M. sibirica*, *S. scrofa*, *N. procyonoides*, and *A. agrarius* in addition to the footprint of *H. inermis*. Many mammals take birds as their prey and stay in the region even in the cold winter because of the abundance of their birds of prey even in the cold weather. For details, please refer to Fig. 6.61.

The biggest threats to Daeseong reservoir are from the nearby artificial bank. It was not only contaminating the water of the natural waterway but also

Fig. 6.59 Birds observed within the reservoir of Daesungdong (2006)

Fig. 6.60 Footprint of Prionailurus bengalensis (2006)

Fig. 6.61 Droppings of Hydropotes inermis (2008)

disconnecting the reservoir from the surrounding natural habitats. More threatening is the cultivation of the mountain areas into farmlands posing a serious problem to the quality of water that may radiate to the entire interconnectivity of ecosystem in the region that is solely dependent on the abundance of quality water. As evidenced by the presence of empty herbicide bottles and traces of agricultural pesticides on waterways, these are the major contaminants that would inevitably destroy the ecosystems and biodiversity of the environment. In addition, the permanent fishnet boxes installed by settlers in the reservoir destruct the aquatic ecosystem. Decaying dead fish contaminate the water. Therefore for the effective water quality management, the cooperation of nearby farmers and residents is a must (Fig. 6.62).

The waterways near Daeseong reservoir are habitats for various kinds of fish and supply water to the stream wetland that was well-developed in the region. In addition, it has a typical plain region where the form and characteristics of the past stream was well maintained. It was functioning as a habitat for birds in daytime and for *P. bengalensis* and feral cats at nighttime. Moreover, a community of *S. koreensis* Andersson was observed around the waterways. The shape of the

Fishing activity | Soil diggings | Chemical fertilizer sack | Herbicide bottle

Fig. 6.62 Threats to Daeseong reservoir (2006)

waterway was taking a natural form where some parts of the tributary connected to Daeseong reservoir, in which the sand bar and the wetland are well developed. It was regarded as an appropriate habitat for wildlife animals with its stream wetland and scrub forest.

Eoryong reservoir

The survey on Eoryong reservoir located within Panmumbeol was performed together with the Daeseong reservoir. A brief review of the survey is presented in the succeeding sections (Fig. 6.63).

On the way to Eoryong reservoir, various kinds of wildflower and vegetation were already observed. This reservoir is also regarded as a perfect habitat for wild animals and plants just like Daeseong reservoir. Because of its common nature and characteristics, the two reservoirs are presumed to have a more or less equivalent vegetation, ecosystems, and of course, the consequent biodiversity. A community of *S. koreensis* Andersson was also found extended along the edge of Eoryong reservoir that has a high level of biodiversity with its round and diverse forms of shore protection. Even ducks were found in the wetland developed around the community of *S. koreensis* Andersson. Besides, from the *M. sibirica* observed, the bird Eurasian Buzzard was found in the region, designated by the Ministry of Environment as a grade II endangered species (Figs. 6.64, 6.65).

However, the ginseng field located in the region has a potential risk of disconnecting the natural habitats. Therefore, urgent preventive measures should be taken for the protection of these species.

Like other regions in DMZ, the scrutiny on certain species were not available in Eoryong reservoir. Therefore, the results of the investigation are just briefly described in Figs. 6.66, 6.67, 6.68, 6.69, 6.70, 6.71 which present the vegetation and plants observed within Eoryong reservoir. As seen in those figures, the reservoir wetland is developed around the reservoir. In this area, Acorus calamus var wetland vegetation was observed. The *A. japonica* Steud was observed inhabiting the swamp areas while the L. *bicolor* community was observed at the nearby forest. The development of L. *bicolor* is largely due to small forest fires and continuous cutting of trees, and the former was identified as the major cause.

Species observed in Eoryong reservoir include Odontobutis intereupta, *Abbottina rivularis*, *Z. platypus*, *stone moroko*, *Asiatic ricefish*, *C. auratus*, and

Fig. 6.63 Species found on the waterway to the Daeseong reservoir (2006)

Fig. 6.64 Habitat on the way to Eoryong reservoir (2006)

Fig. 6.65 The present condition of habitat in Eoryong reservoir (2006)

Fig. 6.66 Alnus japonica Steud. observed in Eoryong reservoir

Microphysogobio yaluensis for fish, *G. rugosa* and *Hyla japonica* for reptiles and amphibians, and *G. japonensis* for birds. As the scrutiny was not available, the result of the investigation might be unsatisfactory. But the species observed in the inspection are presented in Figs. 6.72, 6.73, 6.74, 6.75, 6.76, 6.77.

Fig. 6.67 Bridal wreath observed in Eoryong reservoir

Fig. 6.68 Lespedeza bicolor observed in Eoryong reservoir

Fig. 6.69 Acorus calamus var observed in Eoryong reservoir

Fig. 6.70 Potentilla fragarioides var. major observed in Eoryong reservoir

Fig. 6.71 Baby brier observed in Eoryong reservoir

Panmunjeom Region

Another targeted area for the research was the wetland within Panmunjeom that is adjacent to the Military Demarcation Line and the wetland, formerly an abandoned paddy field where the communities of *P. australis*, and *M. sacchariflorus* Benth

Fig. 6.72 Grus japonensis observed in Eoryong reservoir

Fig. 6.73 Droppings of Mustela sibirica observed in Eoryong reservoir

Fig. 6.74 Droppings of Hydropotes inermis observed in Eoryong reservoir

Fig. 6.75 Stone moroko observed in Eoryong reservoir

Fig. 6.76 Microphysogobio yaluensis observed in Eoryong reservoir

Fig. 6.77 Hyla japonica observed in Eoryong reservoir

were observed. From the habitat of Ardeidae found on the left side of the way to Daeseong-dong village from Panmunjeom, many Ardeidae were observed during summer time. *A. monachus* were also known to have been detected in the region.

The wetland was well developed around the 'No return bridge' located within Panmunjeom.

Although no animal species were detected, communities of *P. australis* and *S. koreensis* Andersson are observed here and the water gate is being well developed. Upon thorough analysis, the habitat of Ardeidae has adequate conditions to be a habitat conducive for reptiles and amphibians. Therefore, it was expected that various kinds of reptiles and amphibians can be observed here if a detailed field investigation is to be conducted in this area (Figs. 6.78, 6.79, 6.80, 6.81, 6.82).

The scrutiny concentrated only on birds since the conduct of investigation was not favorable due to the exceptional conditions in Panmunjeom area. It is a place untouched by humans and communities of Ardeidae readily observed in the

Fig. 6.78 Panmunjeom forested wetland (1998)

Fig. 6.79 The present condition of habitat of Panmunjeom (2006)

Fig. 6.80 'No return bridge' in wetland (2006)

Fig. 6.81 Habitat of Ardeidae (2006)

Fig. 6.82 Habitat of woodpecker (2006)

wetland. The abundance of bird preys are seen in the well-developed water gate that made the region the best habitat for all kinds of birds including Ardeidae.

The old forest are full of trees and vegetations conducive for woodpecker and a total of 1,000 birds including cattle egret, black-crowned night heron, little egret, medium-large egret, and Chinese pond heron were found in the region (Figs. 6.83, 6.84, 6.85, 6.86, 6.87, 6.88, 6.89).

Imjin River Basin

The region lies on the basin of Han river and Imjin River with wetlands having brackish water. The brackish wetland was developed on Imjin river estuary that

Fig. 6.83 Ardeida observed in Panmunjeom

Fig. 6.84 Black-crowned night heron observed in Panmunjeom

Fig. 6.85 Pheasant observed in Panmunjeom

Fig. 6.86 Black-crowned night heron observed in Panmunjeom

Fig. 6.87 Ardeida egg observed in Panmunjeom

Fig. 6.88 Cattle egret observed in Panmunjeom

Fig. 6.89 Ardeida observed in Panmunjeom

has a *P. australis* wetland of approximately 2,000,000 m^2. Although it has a well-connected water fringe ecosystem, it is also regarded as a place where the inland forest ecosystem is disconnected due to the bank road and the farmland.

The Paju Imjin river estuary region

This region enjoined the estuary of the Han and Imjin Rivers. It shows similar features of habitat and species as that of the Han River estuary. The region is found to have wetlands with brackish water as influenced by the brackish water wetland of the Han River estuary. There are no mountains around the area and expectantly, no forest vegetation and shrub and bush forest exist in the region. But it has a vast wet grassland where various kinds of vegetations are developed. Hence, it is regarded as a good habitat for insects (Figs. 6.90, 6.91, 9.92).

Table 6.4 shows the number of species found in Imjin River estuary. No investigation for reptiles and amphibians could be carried out as it was held in late fall.

The plants and vegetations in the area typically characterize brackish water vegetation. *P. latifolius* (Steud.) Ohwiwas found at the edge of the *P. australis* community spreading widely in the area. The *M. sacchariflorus* Benth was detected at the higher water bed of the same *P. australis* community. Besides, from *P. australis*, plants such as *C. epigeios*, *I. cylindrica* var. koenigii (Retz.) Pilg., *C. scabrifolia* Steud., L. *apetalum*, Oxalis corniculata L., *T. pratense* L., *M. sacchariflorus* Benth., *Artemisia princeps*, *C. pseudophragmites* (Haller f.) Koeler, *C. amaraeformis* Nakai, *Rorippa islandica* (Oed.) Borb., *Digitaria ciliaris*, *C. microiria*, *Digitaria sanguinalis*, Green Bristle grass, *P. asiatica* L., *M. lupulina*

Fig. 6.90 Wetland vegetation developed in the Imjin River estuary

Fig. 6.91 Wetland vegetation developed around the tributary waterway of the Imjin River estuary

Fig. 6.92 Wet grassland developed around the Imjin River estuary

Table 6.4 The number of species observed in Imjin River estuary

	Vegetation and plants	Mammals	Birds	Reptiles and amphibians	Fish	Inland insects	Aquatic species	Total
No. of species observed	22	7	9	–	3	3	3	47

L., *A. tripolium* L., and *T. pratense* were observed as well. The discovery of the existence of a rare endangered species—*P. latifolius* (Steud.) Ohwi, has significantly increased the value of the estuaries.

Although it has no forest, some pine trees and *R. pseudoacacia* L. trees are detected besides oak trees which are dominant in the area. Figures 6.93, 6.94, 6.95, and 6.96 show the vegetation and plants found in the region, especially the picture showing the North Korean part across the Imjin river where the *S. koreensis* Andersson is seen dominating the riverside up to the distant cultivated fields in the mountains. The density of the trees in the forest was observed to be low and still at a young age. It is believed that the forest suffers from soil loss.

Inland insects such as *C. erate* and *P. rapae* including dragonfly type were observed. As the investigation was held in late fall, vegetation dominated by one plant species were mostly observed and not many inland insects were detected.

Creatures inhabiting the deep mud waterway such as the hasta (Temminck et Schlegel), *C. auratus*, and *O. sinensis* fish were found here including Acanthogobius and other aquatic species such as; freshwater shrimp, *Sesarma dehaani*, and *Ilyoplax dentimerosa*. This is a region where species of both brackish water and freshwater could be found. For safety reasons, a detailed investigation on fish species was not performed because of the proximity of the area to North Korea. Figure 6.97 shows *O. sinensis* observed in this area.

A total of 17 bird species with a population of 5,189 entities were observed within the region. Due to its geographical features, similar entities as that of the Han River estuary were detected. A large number of wild geese were using this area as

Fig. 6.93 Phragmites australis observed in Manwoo-ri mud flat, Imjin River estuary

Fig. 6.94 Phacelurus latifolius (Steud.) Ohwi observed in Imjin River estuary

Fig. 6.95 Forest and Imjin River estuary of nearby Manwoo-ri observed at the Odusan Observatory of Imjin River estuary

their habitat due to the presence of rice paddies and *P. australis* widely developed around the Imjin River. *Phalacrocorax capillatus*, *A. cinerea*, *A. albifrons*, *P. colchicus*, *Steptopelia orientalis*, crow, *l. bucephalus*, and *Dendrocopos kizuki* were also observed here. Moreover, 4,000 entities of Anser fabalisserrirostris appear in nearby streams, wetland vegetations, and in rice paddies. A grade II endangered

Fig. 6.96 North Korea across Imjin River observed in Imjin River estuary

Fig. 6.97 Oryzias sinensis observed in Imjin River estuary

wild animals were observed frequenting this area, proof that it is maintaining the good qualities of the choice habitat for those birds.

However, since only simple *P. australis* community is found in the area, only the common simple bird species such as wild goose species, duck species, and *P. auroreus* are also observed here. The biodiversity of the region is found to be quite low. Figure 6.98 shows a group of Anser fabalisserrirostris flying over Imjin River estuary proving the value of the estuary as a habitat.

As the region is very wide and with well-developed vegetation, it is considered as a good habitat for mammals such as artiodactyl and some rodents. However, due to its simple environmental factors, the species of mammals inhabiting here is very low. Nevertheless, the density of *H. inermis* is found to be higher in this area compared to other regions, and are even detected to be crossing the river toward the north. Figure 6.99 shows the footprint of *H. inermis* found in the region. In addition, traces of other mammals such as *A. agrarius*, *Microtus fortis pelliceous*, *M. minutus*, *M. sibirica*, *N. procyonoides*, and *S. scrofa* were also detected.

Fig. 6.98 Community of Anser albifrons

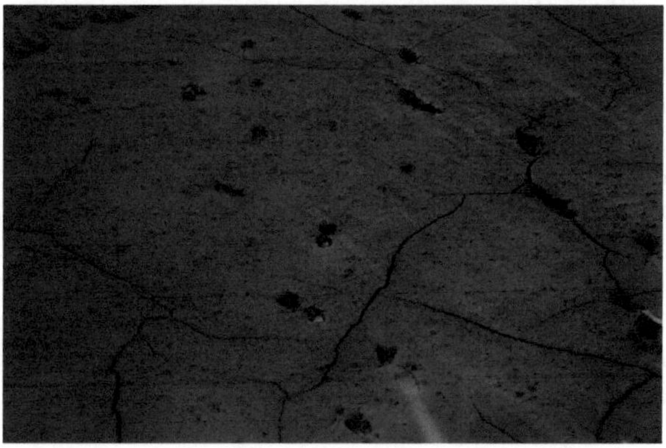

Fig. 6.99 Footprint of Hydropotes inermis found in the Imjin River estuary

Paju Chopyeong Island

Chopyeong island of Munsan-eup, Paju city si, Gyeonggi province is an island located within Imjin River. In the past, a village developed here and got the name Chopyeong-dong. Because of security issues, the access of civilians is prohibited. The region currently belongs to CCZ and is designated as a military facility protection zone where no development is allowed. However, recently it is reviewed to be designated as an ecological park and a wetland protection area for its high ecological value.

As all access has been denied for a long period of time, the habitat is found to be naturally well developed and maintained. The island harbors a river sedimentary wetland where vegetations such as *P. australis*, *S. gracilistyla* Miq., and *Populus nigra* are observed at the edge of the islands. Endangered species including the natural monument *G. japonensis* and *Grus vipio* are observed in shallow waters. Detailed field investigation in the inner part of the islands is not possible due to safety issues. But as observed at Imjin River, the community of *S. koreensis* Andersson is found to be widespread throughout the island. As for the natural species of trees in the island, there is a need for *M. alba* and *P. nigra* to be restored and repropagated. However, as it lies in the middle of Imjin River and is restricted for a long time, the island is expected to revitalize itself in due time into a well-developed habitat with wetland vegetations that can sustain various kinds of species. Figure 6.100 shows *P. nigra* in Chopyeong Island that proves the existence of a village in the past.

Table 6.5 shows the number of species observed in Chopyeong Island as revealed by the investigation performed in 2008. As the target area is a minefield and access is strictly restricted, the investigation on vegetation, inland insects, and aquatic insects were not carried out.

With regard to plants, the community of *S. gracilistyla* Miq., *Salix subfragilis*, *S. koreensis* Andersson, *A. tataricum* subsp. ginnala (Maxim.) Wesm., *Morus bombycis Koidzumi*, *Rhus javanica* L, and *A. altissima* are readily observed in Chopyeong Island including Populus deltoides and *R. pseudoacacia* L., and other exotic species introduced in the area.

For mammals, *H. inermis*, *N. procyonoides*, *Mogera wogura*, *S. scrofa*, and the *P. bengalensis*, a level 2 endangered species is observed at the nearby target area.

A total of 18 bird species are found in this island including *Egretta alba modesta*, *E. garzetta*, *P. colchicus*, *S. orientalis*, *D. kizuki*, *Picus canus*, *Hirundo rustica*, *Hirundo daurica*, *Hypsiperes amaurotis*, *P. auroreus*, *P. webbianus*, *Aegithalos caudatus*, *P. palustris*, *V. oxysepalum Turcz.*, *Parus varius varius*, *P. montanus*, *Oriolus chinensis* and *P. pica*.

Regarding reptiles and amphibians, *Bufo gargarizans*, *H. japonica*, *G. rugosa*, *Rana nigromaculata* and *Rana plancyi* are detected in the island. The level 2 endangered species, *Kaloula borealis*, is confirmed inhabiting the Island.

Lastly for fish, no direct investigation could be made due to the restricted access. But referring to the past literature and materials (1984, Nature of Gyeonggi), 31 species including *A. Japonica* Temminck and *Schlegel*, *Coilia nasus*, *Z. platypus*, *Opsariichthys bidens*, *A. chinensis* Cirinher, *Trichiurus lepturus*, *Cyprnus carpio*, *C. auratus*, *Pseudorasboro Parva*, *S. gracilis majimae*, *Hemibarbus labeo*,

Fig. 6.100 The southern part of Chopyeong island observed at Imjin River

Table 6.5 The number of species observed in Chopyeong Island

	Vegetation and plants	Mammals	Birds	Reptiles and amphibians	Fish	Inland insects	Aquatic species	Total
No. of species observed	–	5	18	6	31	–	–	60

Fig. 6.101 Egretta Alba modesta observed in Chopyeong Island

Pseudogobio esocinus, *Saurogobio dobruyi Bleaker*, *M. anguillicaudatus*, *Misgurnus mizolepis*, *Lefua costata*, *P. fulvidraco*, *Pseudobagrus koreanus*, *Leiocassis nitidus*, *Pseudobagrus emarginatus*, *S. asotus*, *Mugil cephalus*, *Trachidermus fasciatus*, *Siniperca scherzeri*, *Odontobutis platycephala*, *A. hasta* (Temminck et Schlegel), *Tridentiger obscurus*, *Tridentiger trigonocephalus*, *M. chinensis*, and *Takifugu obscurus* are recorded to be existing in the island.

Chopyeong Island is found in a diverse wetland as well as in the estuary of river. As a consequence, it is regarded to be of high ecological value as it provides a spawning area for various kinds of fish. The present investigation is expected to have observed various kinds of species along the well-developed wetland of these areas had it not been for the restrictions and safety reasons (Fig. 6.101).

Paju Jangdan Peninsula Region

Jangdan peninsula is located in CCZ near Imjin River of Paju city, Gyeonggi province. It covers the long riverside and the rice paddy wetland. The region was originally a community of *P. australis*, but now it is cultivated as a flat farmland. Imjin River is flowing in its southeastern region but it is not connected with the farmland due to the bank. The edge of Imjin River is developed with *P. australis* community but some of the edge areas are also cultivated as farmland. The topsoil of the farmland comprises silt loam while the subsoil is made up of moderately coarse textured soil.

The habitat of Jangdan peninsula generally comprises rice paddy ecosystem and aquatic ecosystem found along Imjin River. The artificial bank between the rice paddy wetland ecosystem and Imjin River disconnects the inland habitat connected to the rice paddy wetland ecosystem which is seriously damaged due to the cultivation of the ginseng field. It is also the region used as the shooting range in the past with many blind shell residues. Therefore currently, assessment is being denied (Figs. 6.102, 6.103, 6.104).

Fig. 6.102 Jangdan peninsula around Imjin River (1)

Fig. 6.103 Jangdan peninsula around Imjin River (2)

Fig. 6.104 Jangdan peninsula around Imjin River (3)

Table 6.6 The number of species observed in Jangdan peninsula

	Vegetation and plants	Mammals	Birds	Reptiles and amphibians	Fish	Inland insects	Aquatic species	Total
No. of species observed	92	4	34	7	4	67	16	224

The field surveys on species of Jangdan peninsula were conducted several times. Yet it is subject to some restrictions in specific areas near Imjin River. Table 6.6 presents the number of species observed in Jangdan peninsula.

Among the plants observed in the area are 92 species of vascular plants which include *Bidens tripartita* L., *Persicaria dissitiflora* (Hemsl.) *H.Gross ex Mori*, *Setraria faberii Herrm*, *P. australis*, *Q. aliena*, *Green Bristlegrass*, *Rorippa indica*, *Artemisia annua* L., *E. annuus* (L.) Pers., *Elymus tsukushiensis*, *Equisetum ramosissimum Desf*, *Potentilla supina* L., *Cinnamomum zeylanicum*, *Crepidiastrum sonchifolium*, *J. effusus* var. *d. Buchenau.*, *Carex neurocarpa Maxim.*, *Eragrostis ferruginea*, *Inula britannica* var. japonica, *Carex bostrychostigma* Maxim.,

Solanum nigrum, *Lysimachia barystachys* Bunge, *Acalypha australis*, *Prunella vulgaris* var. *lilacina Nakai*, L. *apetalum*, *Ambrosia trifida*, *Oenothera biennis* L., *Phragmites japonica* Steud., *Commelina communis* L., *Glycine soja*, *Echinochloa crusgalli*, *A. artemisiifolia*, *Kummerowia stipulacea* (Maxim.) Makino, *Mosla punctulata* (J.F.Gmelin) Nakai, *Polygonum aviculare*, *Erigeron canadensis* L., *Avena fatua* L., *Calystegia sepium* var. japonicum (Choisy) Makino, *Persicaria perfoliata* (L.) H.Gross, *Chenopodium album* var. centrorubrum, *P. latifolius* (Steud.) Ohwi, *Artemisia selengensis*, *M. sacchariflorus* Benth., *Veronica undulata* Wall., *Bidens frondosa* L.., *D. sanguinalis*, *Metaplexis japonica*, *Meliosma myriantha* Siebold & Zucc., *C. microiria*, *Duchesnea indica*, *S. koreensis* Andersson, *Stellaria media* (L.) Vill, *Erechtites hieracifolia*, *Lespedeza cuneata* G. Don, *A. subulatus* Michx., *M. alba*, *Torilis japonica*, *Artemisia capillaris*, *Serratula coronata* var. insularis (Iljin) Kitam. for. insularis, *Q. acutissima* Carruth., *Mollugo pentaphylla* L., *Rumex crispus*, Equisetum arvense L., Portulaca oleracea L., L. *virgata* (Thunb.) DC., Cuscuta australis R.Br, *A. princeps*, *R. pseudoacacia* L., *Calystegia hederacea* Wall., *Abutilon theophrasti*, *Persicaria hydropiper*, *Lactuca indica* L., *Eleusine indica*, *Persicaria filiformis* (Thunb.) Nakai ex Mori, *Leonurus japonicus*, *Aeschynomene indica* L., *Melilotus suaveolens* Ledeb., *Arthraxon hispidus* (Thunb.) Makino, *Lespedeza maximowiczii* C.K.Schneid., *A. fruticosa* L., *Chenopodium ficifolium*, *Mazus pumilus* (Burm.f.) Steenis, *P. asiatica* L., *Chamaecrista nomame* (Siebold) H.Ohashi, *Pueraria lobata* (Willd.) Ohwi, *Persicaria modosa* Opiz, *Clinopodium gracile* var. multicaule (Maxim.) Ohwi, *Trifolium repens*, *Eclipta prostrata* and *H. japonicus* Sieboid & Zucc. Aside from the vascular plants, 19 species of exotic and introduced plants were observed such as; *E. annuus* (L.) Pers., *P. supina* L., L. *apetalum*, *A. trifida*, *O. biennis* L., *A. artemisiifolia*, *C. canadensis* (L.) Cronquist, *A. fatua* L., *B. tripartita* L., *A. pilosus* Willd., *E. hieracifolia*, *A. subulatus* Michx., *R. crispus*, *R. pseudoacacia* L., *A. theophrasti*, *M. suaveolens* Ledeb., *A. fruticosa* L., *C. ficifolium*, and *T. repens* (Fig. 6.105).

In the region, 16 species of large size benthic invertebrate are found which include rice paddy *C. chinensis* malleata, *Austropeplea ollula*, *Gyraulus*

Fig. 6.105 Fringe wetland of Imjin River

hiemantium, *Oxyloma hirasei*, *Palaemon paucidens* De Haan, *E. sinensis* H. MILNE EDWARDS, *Cloeon dipterum Linnaeus*, *Potamanthus formosus Eaton*, *Platycnemis phillopoda*, *Micronecta* (*Basilionecta*) *sedula Horvath*, *Hyphydrus japonicus*, *Potamonectes hostilis*, *Potamonectes hostilis* species (larva), Rhantus (Rhantus) pulverosus (Stephens), *Helochares striatus*, and *Anopheles* sp.

Since the region mostly comprises farmland and artificial waterway, it is far from being natural and native. Moreover, due to the dry stream that cuts the continuity of water flow, it is believed that the region is inadequate as a habitat for large benthic invertebrate.

For insects, a total of nine orders 41 family 67 species are found and they include *Sympecma paedisca*, *Eurygaster testudinaria* 'Geoffroy', *Dolycoris baC-CZrum*, *Eurydema gebleri*, *Philaenus spumarius* family sp.1, *Cicadella viridis*, *Orthopagus lunulifer* Uhler, *Dictyophara patruelis* (Stal) genus sp.1, *Pentastiridius apicalis*(Uhler)sp.1, *Harmonia axyridis*, *Pieris melete*, and *Araschnia burejana*. The region is evaluated to be having a lot of limited factors, especially the fact that Diptera is dominating the area.

Regarding fish, a total of seven species are found in the region including *O. bidens* GuNTHER., *A. chinensis*, *Hemiculter leucisculus*, *Cyprinus carpio* LINNAEUS, *C. auratus*, *H. labeo* and Oryzias latipes. There are also four species of reptiles and amphibians including *B. gargarizans*, *H. japonica*, *R. coreana* and *R. nigromaculata* found in the peninsula. *H. japonica* are the only amphibians found in Jangdan peninsula near Imjin River (Figs. 6.106, 6.107).

And as for the birds, 34 species are observed in the peninsula. Among them is the *Accipiter soloensis*, Korea's natural monument No. 323-2. This discovery is highly significant as it is a very rare species (Figs. 6.108, 6.109).

A total of four mammal species are found in the peninsula including; *H. inermis*, *N. procyonoides*, *M. wogura*, Crocidura, and *Suaveolens shantungensis* (Fig. 6.110).

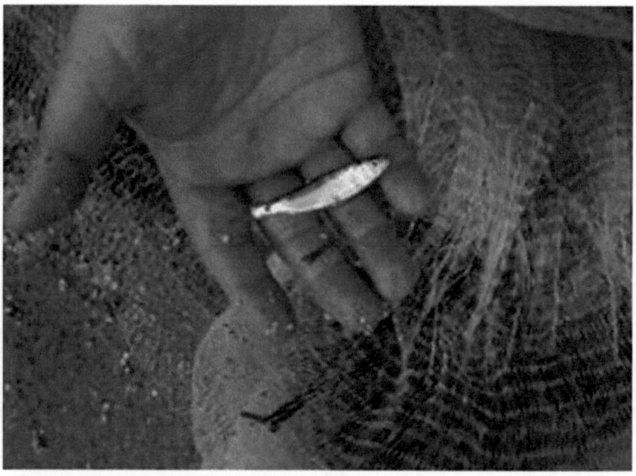

Fig. 6.106 Hemibarbus labeo observed in Jangdan peninsula

Fig. 6.107 Rhantus (Rhantus) pulverosus (Stephens) observed in Jangdan peninsula

Fig. 6.108 Hyla japonica observed in Jangdan peninsula

Yeoncheon Pilseung Bridge Area

The Yeoncheon Pilseung Bridge is located at the upper stream of Imjin river and is a place with wide highwater bed and arenaceous soil. The wetland developed

Fig. 6.109 Picus canus observed in Jangdan peninsula

Fig. 6.110 Nyctereutes procyonoides footprint observed in Jangdan peninsula

around limjin River forms the wetland ecosystem where the water edge ecosystem of the south and north are interconnected. Moreover, the grassland and forest around limjin River connect the inland ecosystem (Fig. 6.111).

The part on the right side of Pilseung Bridge is the upper mainstream area of Imjin River where the sand bar is well developed. Traces of the *P. bengalensis* eating fish were detected as well as various kinds of fish which proved that the place was properly functioning as habitat for fish. The vegetation of the forest area is in poor condition and the community of *M. sacchariflorus* var. gracilis Y.N.Lee is observed in the high water bed area.

Since the region has arenaceious soil, no *P. australis* is found. Instead, *M. sacchariflorus* Benth dominates the area and the floodplain is widely developed.

Because the access of civilians has been denied for a long time, no rice paddies were developed and neither farm fields nor villages are found.

Fig. 6.111 Aquatic ecosystem connecting North and South along Imjin River

The annually performed bushfire is regarded as a major reason for the increase in the density of *M. sacchariflorus* var. gracilis Y.N.Lee community. It is assumed that the estuary wetland formed at the right part of Imjin River was a village in the past since it is named as 'Dangchon' (Figs. 6.112, 6.113, 6.114)

The investigation on the number of species found in the Yeoncheon Pilseung Bridge was conducted in 2008. In this region, legal protection of 10 species is observed including endangered species designated by the Ministry of Environment

Fig. 6.112 Imjin River flowing around Yeoncheon Pilseung Bridge and surrounding wetland

Fig. 6.113 Forest around Yeoncheon Pilseung Bridge

Fig. 6.114 Grassland around Yeoncheon Pilseung Bridge

Table 6.7 The number of species observed in Yeoncheon Pilseung Bridge

	Vegetation and plants	Mammals	Birds	Reptiles and amphibians	Fish	Inland insects	Aquatic species	Total
No. of species observed	13	8	27	1	13	13	–	75

and natural monuments of cultural heritage administration. This fact heightens the need for the extensive preservation of the region's habitat. Table 6.7 shows the species observed in Yeoncheon Pilseung bridge. A thorough field investigation on aquatic insects could not be performed due to the inevitable condition.

S. koreensis Andersson forest is found to be the major species in the riverside wetland. The height of the tree reached 10–16 m. Other trees found in the area are the *R. pseudoacacia* L. forest, and *Q. aliena –Q. acutissima* Carruth. forest, but they are relatively young with a height of less than 10 m.

In the forest of Dangchon village, the height of *S. koreensis* Andersson, *Q. variabilis*, *Phellodendron amurense* Rupr., *Ulmus davidiana* var. japonica (Rehder) Nakai, *Q. aliena* and *Q. acutissima* Carruthhas reached 15–18 m. Found as well in the area is the *U. davidiana* var. japonica (Rehder) Nakai with a diameter at breast height of 60–80 cm.

Further in the mid-lower layer of the forest area is *Q. acutissima* Carruth., Staphylea bumalda DC, and Parthenocissus tricuspidata are detected while bog rush dominates the lowest herbaceous layer. In the vicinity of private houses, 5–8 m tall young *Fraxinus rhynchophylla* and *A. altissima* are found besides the trees smoked by forest fire. Generally, the forest area is dominated by *Q. acutissima* Carruth. In the wetland around Imjin River, *S. koreensis* Andersson forest and *R. pseudoacacia* L. forest trees dominate, while in the village areas, trees comprise *U. davidiana* var. japonica (Rehder) Nakai, *Q. aliena*, *Q. variabilis*, and *P. amurense* Rupr.

Yet, the perennial communities of *P. australis*, and *A. japonica* Steud dominate in stream lowland, representing the wetland vegetation of the region which is maintained in relatively good condition.

Albeit seemingly, the forest vegetation is found to be badly damaged by manmade cultivation, yet it is still regarded as having high landscape ecological value (Figs. 6.115, 6.116, 6.117, 6.118, 6.119, 6.120, 6.121, 6.122).

Fig. 6.115 Phragmites australis and Robinia pseudoacacia L. spread along the eastern part of Imjin River around Yeoncheon Pilseung Bridge

Fig. 6.116 Fresh marsh landscape and aquatic vegetation area around Imjin River of Yeoncheon Pilseung Bridge

Fig. 6.117 Imjin River around Yeoncheon Pilseung Bridge and mountain across the river

Fig. 6.118 Tributary and forest connected to Imjin River around Yeoncheon Pilseung Bridge

Fig. 6.119 Forest formed at Dangchon around Yeoncheon Pilseung Bridge

Fig. 6.120 Ulmus davidiana var. japonica (Rehder) Nakai observed in Dangchon forest around Yeoncheon Pilseung Bridge

Due to time constraints, the investigation failed to observe as many insects as expected. The insects identified include *Sympetrum depressiusculum* (Selys), *M. japonicus*, *N. triguttata*, *Aquarius paludum* (Fabricius), *Ducetia japonica*

Fig. 6.121 The trace of forest fire observed around Yeoncheon Pilseung Bridge

Fig. 6.122 Shell mound of small rocks observed at the private house of Dangchon around Yeoncheon Pilseung Bridge

(Thunberg), *Ognevia longifennis* Shiraki, *Acrida cinerea* cinerea (Thunberg), *Aiolocaria hexaspilota* (Hope), *H. axyridis*, *Meloe proscarabaeus*, *Vespula vulgaris*, *Polygonia c-aureum*, *C. erate*, and *Vanessa cardui*. The region enjoys shore protection from the ecosystem dominated by *M. sacchariflorus* Benth. It is expected that there is physical change in the inland ecosystem and habitat when flood occurs. Therefore, this area can be used as a reference site of natural shore ecosystem.

However, the damage caused by bushfires is the primary concern at the front of the DMZ fence. Environmentalists and specialists are worried about losing the areas' function as a habitat for insects. Figure 6.123 shows *S. asotus* found during the survey.

The Imjin River flowing around Pilseung Bridge produces a large volume of water with gravel at the riverbed and sand bar on the river side. The small stream flowing through *P. australis* community is regarded as an optimum habitat for both fish and aquatic species. Figures 6.124 and 6.125 show *Zacco koreanus* and *Rhinogobius brunneus* found in the region. The result of the research revealed that the area is dominated by *Z. koreanus* and other species observed in the area including

Fig. 6.123 Anapodisma miramae, a type of inland insects, observed around Yeoncheon Pilseung Bridge

Fig. 6.124 Zacco koreanus observed around Yeoncheon Pilseung Bridge

Fig. 6.125 Rhinogobius brunneus observed around Yeoncheon Pilseung Bridge

Hemibarbus mylodon, *R. brunneus*, *Gobiobotia macrocephala*, *Acheilognathus signifer*, *Iksookimia koreensis*, *Hemibarbus longirostris* (Regan), *M. yaluensis*, *Tanakia lanceolata*, *Pungtungia herzi*, *S. asotus*, *Z. platypus* and *Rhynchocypris oxycephalus*.

Again, aggressive investigation could not be made on fish and aquatic species due to security issues. It is assumed though that the areas surrounding the bridge sustain a variety of species as the region is untamed and has an optimal quality of habitat for wild animals.

H. japonica is the only reptile and amphibian species observed in the region because of its seasonal nature.

As for the birds, *V. oxysepalum Turcz.* species, *Cyanopica cyana*, crow, *P. colchicus*, *U. sibiricus*, *P. webbianus*, meadow bunting bird, *P. capillatus*, *Larus argentatus*, and *M. merganser* were found in the Yeoncheon Pilseung Bridge area. Grade I endangered species such as the *G. japonensis* (8) and *A. chrysaetos* japonica (1) and, grade II species such as *B. buteo* (1), *Accipiter gentilis* (1), *A. monachus* (10), and *Circus aeruginosus* (1) were also discovered to be existing in the area.

The natural region on the right side of Pilseung Bridge has the best condition as a habitat for birds. The riverside is observed with rare species of rapacious birds despite its simple vegetation in lower layer.

Despite the fact that the Imjin River is known to be a fast current water system that runs under Pilseung Bridge, *M. merganser* is observed here in large numbers while *A. monachus* are found wheeling overhead.

Meanwhile, in the *P. australis* of the valley, *C. aeruginosus*, *B. buteo* and *A. gentilis* are found inhabiting the place. Ducks drop their feathers on the sand dunes of Imjun River providing clues that they too stop over in these areas.

Mammals such as the *H. inermis*, *N. procyonoides*, *M. sibirica*, *M. wogura*, *M. minutus*, *A. agrarius*, Crocidura *suaveolens shantungensis*, and *Mustela nivalis mosanensis* (Mori) were found at the surrounding areas of the bridge. The low water level stream and surrounding *P. australis* is normally a perfect habitat for mammals such as *Lutra lutra* or *H. inermis* and *M. sibirica*. However, they are nowhere to be found in the region. The absence of the L. *lutra* can be associated with the absence of rocks in the area where they normally enjoy sitting while eating their captured prey. Another reason could be the low density of choiced fish in the area. A further investigation on species of fish and stock biomass is necessary to assess the potentials of the existing habitat. It is further recommended that the follow-up investigation will be preferably done during nighttime in order to observe mammals that left traces of their prey such as dead fish (Figs. 6.126, 6.127).

Sewolcheon Basin

Upper stream region of Paju Sewolcheon

The region includes the whole area of Seogok-ri and Janghwa-dong, Jindong-myeon, Paju city. Before the installation of MDL, it is believed to have had a

Fig. 6.126 Footprint of Hydropotes inermis near Yeoncheon Pilseung Bridge

Fig. 6.127 Droppings of Nyctereutes procyonoides (Parasite) observed near Yeoncheon Pilseung Bridge

village with a farmland along Sewolcheon. The widely developed idle land between the low hills that has an altitude of 50 m is well harmonized with the surrounding natural stream and wetland.

Sewoncheon is flowing in the area where two water branches are merged. Around it, both the slope and the stream wetland are developed. To the right of Sewolcheon is where the Seogok bridge is situated. On the upstream left side of Sewolcheon is where the *H. mylodon*, a legally protected species can be found. The observance of various kinds of wild animals proves that the region is well developed as a habitat for diverse species.

Not many shrub trees are found in this region; however, the forest area is found to be dominated by oak population such as *Q. acutissima* Carruth. with the pine trees as minor species.

The wet grassland is mostly occupied by *P. australis* and *M. sacchariflorus* var. gracilis Y.N.Lee which enhance its function as a wetland habitat. The region maintained a forest that is in stable condition which is used as ecological corridor. Moreover, the development of low layer vegetation and fresh marsh wetland is observed to be in good condition as well (Figs. 6.128, 6.129, 6.130).

Fig. 6.128 A depression wetland observed around Sewolcheon basin

Fig. 6.129 Forest observed around Sewolcheon

Fig. 6.130 Phragmites australis and Miscanthus sacchariflorus var. gracilis Y.N.Lee community of wet grassland observed around Sewolcheon

Table 6.8 The number of species observed in the upper stream of Sewolcheon

	Vegetation and plants	Mammals	Birds	Reptiles and amphibians	Fish	Inland insects	Aquatic species	Total
No. of species observed	102	9	14	1	5	6	6	143

The investigation on the species of Sewolcheon region was carried out in 2008. Table 6.8 presents the number of species observed.

Along the stream, a large scale of *M. sacchariflorus* var. gracilis Y.N.Lee community is observed, spreading vastly in fields and mountains. Vegetations such as L. *maximowiczii* C.K.Schneid., *A. japonica* Steud., *Q. aliena*, *Q. acutissima* Carruth., and pine trees are found in these areas. The tree layer of the forest areas is composed of deciduous broad-leaved trees, including *Q. acutissima* Carruth., *R. pseudoacacia* L., and *Q. dentata* var. fallax Nakai. *Q. acutissima* Carruth is again found dominating the area. Among various target areas, the forest region showed the best condition of vegetation. The pine trees are found at the ridge while *Q. acutissima* Carruth and *Q. aliena* spread out at the slope side.

R. pseudoacacia L. forested the region that meets with abandoned paddy field wetland with an average of 14–16 m in height and average DBH of 18–22 cm. The pine trees have an average height of 12–15 m and an average DBH of 36 cm.

With regard to the plants, *M. sacchariflorus* Benth., *A. tataricum* subsp. ginnala (Maxim.) Wesm., *A. fruticosa* L., *S. koreensis* Andersson, *A. japonica* Steud., *Q. aliena*, *Q. variabilis*, *R. pseudoacacia* L., *M. dianthera* Maxim, *Carex humilis* Leyss. var. nana, Elsholtzia splendens. Nakai, bird, Isodon inflexus (Thunb.) Kudo, *A. princeps*, *Q. acutissima* Carruth, *A. gmelini* Weber ex Stechm., *A. keiskeana*, *L. japonica*, *Salix hallaisanensis* Leveille, *Spodiopogon cotulifer*, *Forsythia koreana* (Rehder) Nakai, *Clematis apiifolia* DC., *A. tataricum* subsp. ginnala (Maxim.) Wesm., *A. subulatus* Michx., *S. virga-aurea* var. asiatica, *S. gracilistyla* Miq., *R. parvifolius* L. for. parvifolius, *C. humilis* Leyss. var. nana, *Potentilla fragarioides* var. major, *S. cotulifer*, *C. epigeios*, *S. wichurae* BOCKLR., *A. contorta*, *F. koreana* (Rehder) Nakai, *Lespedeza cyrtobotrya* Miq., *Lindera glauca* (Siebold & Zucc.) Blume, *G. Bristlegrass*, *O. cynarocephalum* Boiss & Blanche, *A. japonica* Steud., *P. australis*, *Stephanandra incisa* (Thunb.) Zabel var. incisa., *R. trichocarpa*, *Cyperus microiria*, *R. mucronulatum* Turcz. var. mucronulatum, *Woodsia polystichoides*, *F. rhynchophylla*, Lindera obtusiloba, Carex dispalta Boott, pine tree, *L. apetalum*, *Sigesbeckia glabrescens* (Makino) Makino, Viburnum sargentii, *S. prunifolia* for. simpliciflora. Nakai, *S. gracilistyla* Miq are found to have covered this area (Figs. 6.131, 6.132).

In the lowland wetland near the stream, the community of *S. koreensis* Andersson and *A. japonica* Steud. is highly developed around the stream and the water basin. However, the community of *M. sacchariflorus* Benth. clearly dominates the area. The region is mostly a low wetland that transformed from ancient farmlands. The natural wetland formed around the water basin is regarded as an important ecological spot. *A. japonica* Steud., *S. koreensis* Andersson and

Fig. 6.131 The community of pine tree in ridge vegetation and Quercus acutissima Carruth. at the slope of nearby Sewolcheon

Fig. 6.132 The community of Alnus japonica Steud. and Salix koreensis Andersson observed around the lowland water basin of nearby Sewolcheon

A. tataricum subsp. ginnala (Maxim.) Wesm. are observed in lowland tree layers, while *Ligustrum obtuslfolium*, *Z. schinifolium* Siebold & Zucc., *M. bombycis Koidzumi*, *A. tataricum* subsp. ginnala (Maxim.) Wesm., and *L. japonica* are observed in the lowland shrub layers. Based on the *A. japonica* Steud found in lowland having a height of 14–16 m and a DBH of 40 cm, it is assumed that most of the trees developed at the same time. In addition, *A. japonica* Steud., *S. koreensis* Andersson, *M. bombycis Koidzumi*, *L. obtuslfolium*, *A. tataricum* subsp. ginnala (Maxim.) Wesm., *L. japonica*, *S. prunifolia* for. Simpliciflora. Nakai, *A. fruticosa* L., *Miscanthus sinensis*, *M. sacchariflorus* Benth., *Carex siderosticta*, *S. virga-aurea* var. asiatica, *Apus pacificus* pacificus, *J. effusus* var. *d. Buchenau.*, S. maritimus L., *L. maackii* (Rupr.) Maxim., *Typha orientalis* C.Presl, *Z. latifolia* and *A. calamus* L. are believed to be developed at the same places well (Figs. 6.133, 6.134, 6.135, 6.136).

Areas around Sewolcheon maintain good vegetation of *P. australis* and *A. japonica* Steud. that developed in various sizes. At the basin, diverse vegetations are detected and rare plants and protected species can be observed through detailed seasonal field investigation.

Fig. 6.133 The community of Quercus acutissima Carruth. observed in the Sewolcheon region

Fig. 6.134 The community of pine tree observed in the Sewolcheon region

Fig. 6.135 The community of Salix koreensis Andersson observed in the Sewolcheon region

The wetland has a very high landscape ecological value for having various ecologically important patches and corridors. Although it is a wetland surrounded by hill type mountain areas, it also has a fitting habitat for inland insects with the presence of palustrine wetlands and high *P. australis* and *A. japonica* Steud. trees (Figs. 6.137, 6.138, 6.139).

Fig. 6.136 Alnus japonica Steud. observed in the Sewolcheon region

Fig. 6.137 The community of Robinia pseudoacacia L. observed in the concave area of Sewolcheon region

Considering the time of the investigation, the observance of *Sympetrum infuscatum*, *Sympetrum eroticum*, *Oxya chinensis sinuosa* Mistshenko, *H. axyridis*, cast skin of snake, Polistinae and *P. c-aureum* is a huge accomplishment.

Fig. 6.138 Pinus rigida Mill. observed in the Sewolcheon region

Fig. 6.139 Forsythia koreana (Rehder) Nakai

The investigation on fish species was conducted in three spots within the waterways of Sewolcheon, a tributary of Imjin River; (1) Lake located under an old bridge, (2) natural stream which is covered with rocks and gravel, and (3) lake area. Fish such as *Z. koreanus* (dominant), *H. mylodon*, *T. lanceolata*, *O. platycephala* and *R. oxycephalus* are observed. Aquatic species such as; *M. japonicus*, *N. triguttata* Motschulsky, *R. chinensis* Mayr, *R. chinensis* Mayr, *N. hoffmanni*, and *N. denticulata* are also found. The research further revealed that the upper stream of Sewolcheon is the closest home for *H. mylodon*, a natural monument. Even in the survey at CCZ, no *H. mylodon* was observed around Sewolcheon. But in this research only four entities of 20–30 cm long adult *H. mylodon* are observed at the 5 m radius of water regime. And from this result, it is assumed that this region is a stable habitat for the *H. mylodon* (Figs. 6.140, 6.141, 6.142).

Not many reptiles and amphibians are observed since at this time of the year, they mostly went into hibernation. But still, a group of *R. coreana* was detected. The wetland with diverse water level is presumed to be the optimum spawning place for frog (Fig. 6.143).

In the Sewolcheon region, a total of 14 species and 120 entities are found including *L. bucephalus*, crow, *Troglodytes troglodytes* dauricus, *P. canus*,

Fig. 6.140 Odontobutis platycephala observed in the Sewolcheon region

Fig. 6.141 Zacco koreanus observed in the Sewolcheon region

Fig. 6.142 Muljarus japonicus observed in the Sewolcheon region

E. rustica, *P. auroreus*, *P. colchicus*, *V. oxysepalum Turcz.*, *P. webbianus*, *Turdus naumanni* eunomus, and *P. palustris*. The most significant discovery is the presence of a grade I endangered species-3 entities of *G. japonensis*, and a grade II

Fig. 6.143 Rana coreana observed in the Sewolcheon region

endangered species—one entity of *B. buteo* and one entity of *A. gentilis*. Wherefore, this region is also evaluated as an optimum habitat for birds. Moreover, the mountain in this area which is formed through ecological succession is a suitable habitat for birds. The various kinds of bird food and bird preys found in these ecosystems is the major reason for diverse birds found in the region.

The balanced ecosystem of streams and grasslands with high density of frog and fish which is the choice food for many species is an excellent habitat for mammals. The observed mammals in the region include the rare *L. lutra* (retrimentum), the *N. procyonoides*, *S. scrofa*, *M. sibirica*, *M. wogura*, *A. agrarius*, and *M. minutus*. Two entities of *H. inermis* and one entity of *P. bengalensis* were observed as well (Fig. 6.144).

Samicheon Basin

Yeoncheon Samicheon region

The name given to the place is Hakgok-ri, Baehak-myeon, which is derived from its regional characteristics where cranes were detected in groups. It is believed that

Droppings of Prionailurus bengalensis

Droppings of Nyctereutes procyonoides

Footprint of Nyctereutes procyonoides

Fig. 6.144 Mammals observed in the Sewolcheon region

this village of the past can be traced from the rice paddy fields and communities of *Q. acutissima* Carruth. found in the wetland. A wide area of wetland is formed along Samicheon that is flowing through the Yeoncheon plain. The wetland provides a habitat for diverse species as it functions as the buffer zone of the water's edge that connects the ecosystem of both the inland and the waterways. The investigation was carried out in the western region of Yeoncheon plain including Imjin River, Samicheon, and Angaecheon basin. An artificial stream is formed around CCZ of Samicheon basin while a natural stream is found in DMZ. It is believed that in the past, there were residents in these areas near the stream since many graveyards were found in the low hill area.

The water edge ecosystem formed around Samicheon connects the ecosystem of the north and south. It is also connected to the Angaecheon wetland. It is regarded to be having a high connectivity with inland ecosystem linking the nearby grassland and the forest that are found to be in good condition. As an important ecological corridor connecting the ecosystem of north and south, it has an excellent status of forest vegetation that stands in comparison with the natural heritage.

The habitat of the natural monument, *H. mylodon* is found at the high streams of the Angaecheon. Such discovery proves the high value of its ecosystem, and therefore worth to be preserved (Figs. 6.145, 6.146).

A large-scale wetland is found in the western part of the Yeoncheon plain where Samicheon is flowing. It is assessed as maintaining an excellent condition of natural ecosystem which is considered as a typical model of wetland just like the natural type of its stream. The wetland area is observed with various kinds of landscape factors that can only be seen in untamed nature. Specifically, it is

Fig. 6.145 Freshwater marshes observed at Samichoen region (1)

Fig. 6.146 Freshwater marshes observed at Samichoen region (2)

providing an environment that can satisfy diversified species, so it is often regarded as a nature's secret.

The marsh is valued as the standard wetland. The forest zone is characterized as a low hill area with oak dominating the region. Although grassland and shrub forest are not observed within the region, it is regarded to have natural wetland ecosystem of excellent condition that transformed from the farmlands of the past. The Yeoncheon plain that stretched out from Samicheon went through an ecological succession from rice paddy to a natural wetland at present. It has a meander with sandbar at mid-point. The water basins are found all over the plain. The density of *M. sacchariflorus* var. gracilis Y.N.Lee is also found to be high as the result of the bushfire done yearly (Figs. 6.147, 6.148, 6.149, 6.150, 6.151).

Fig. 6.147 Samicheon and nearby wetland: the river continuum

Fig. 6.148 Angae stream at the tributary of Samicheon

Fig. 6.149 Wetland observed at Samicheon region

Fig. 6.150 Wet grassland observed at Samicheon region

The investigation on the species of Samicheon region was conducted in 2008 and the number of species observed is presented in Table 6.9. Samicheon is crossing the Yeoncheon plain, and the survey carried out was on both right and left sides of the Samicheon.

The leading vegetation of the forest area is broad-leaved deciduous trees mixed with *Q. acutissima* Carruth., *Q. mongolica*, and *Q. dentata* var. fallax Nakai. In some parts of the forest, *B. davurica* Pall. community is observed while *R. pseudoacacia* L. community is specifically observed in the lowland. Along the graveyards located at the ridge, pine tree forest is observed.

In the forest area, *A. tataricum* subsp. ginnala (Maxim.)Wesm., *L. cyrtobotrya* Miq., *L. maximowiczii* C.K.Schneid., and *L. maackii* (Rupr.) Maxim. are found as a shrub layer and, *Q. acutissima* Carruth., *Q. aliena*, *Q. mongolica*, *Q. dentata* var. fallax Nakai, *R. pseudoacacia* L., *B. davurica* Pall., *R. multiflora*, *A. tataricum*

Fig. 6.151 Forest observed at Samicheon region

Table 6.9 The number of species observed in Yeoncheon Samicheon region

		Vegetation and plants	Mammals	Birds	Reptiles and amphibians	Fish	Inland insects	Aquatic species	Total
Right side of Samicheon	No. of species observed	68	6	25	1	14	10	5	129
Left side of Samicheon		41	8	25	–	2	10	3	89

subsp. ginnala (Maxim.) Wesm., *M. sacchariflorus* var. gracilis Y.N.Lee, *Potentilla freyniana*, *Melampyrum setaceum* (Maxim.) Nakai, *A. ovata* (Thunb.) DC., *T. triandra* var. japonica (Willd.) Makino, *O. cynarocephalum* Boiss & Blanche, *Adenophora remotiflora* (Siebold & Zucc.) Miq., *Clematis terniflora* var. mandshurica (Rupr.) Ohwi, *C. apiifolia* DC., *S. virga-aurea* var. asiatica, *A. princeps*, *A. keiskeana*, *C. siderosticta*, and *Veratrum maackii* var. japonicum are found as plant species.

As for herbal species, Sagina japonica (Swartz) Ohwi, *T. triandra* var. japonica (Willd.) Makino, *A. pilosus* Willd., *A. trifida*, *G. Bristlegrass*, *A. princeps*, *O. biennis* L., *Spodiopogon sibiricus*, *Calamagrostis arundinacea* (L.) Roth, *C. humilis* Leyss.var. nana, *E. annuus* (L.) Pers., *Panicum bisulcatum* Thunb., *Aconitum pseudolaeve* Nakai, *A. remotiflora* (Siebold & Zucc.) Miq., *A. gmelini* Weber ex Stechm., *Hypericum ascyron*, *P. bisulcatum* Thunb., and *A. contorta* are observed along the checkroads. In some parts of the mountain areas, *Q. acutissima* Carruth.-

Q. mongolica forest are observed to have a height of more than 15 m, with more than medium density of growing stock. However, in the mountain areas near the Samicheon, the *Q. acutissima* Carruth. Forest is observed relatively shorter in height and with low density of growing stock. The place is also dominated by shrub class and *P. australis*. In Paju region, rare *Q. mongolica* is observed but no Quercus serratais found.

The *A. japonica* Steud. forest formed in the wetland around Sewolcheon of Panbu-dong is regarded as a riverside forest with high natural features and superior landscape. Even in the investigation on the wetland within DMZ, the wetland has the most excellent condition with vegetation that has the highest landscape ecological value. However, restoration work is still required for the surrounding forest vegetation that was largely damaged by artificial and human disturbances.

Along the roads in the lowlands, pine trees are seen including *L. maximowiczii* C.K.Schneid., *R. mucronulatum* Turcz. var. mucronulatum, *A. subulatus* Michx., *Kalimeris yomena* Kitam, *S. koreensis* Andersson, *L. barystachys* Bunge, *Betula costata* Trautv., *B. davurica* Pall., *Salix caprea* L., *C. terniflora* var. mandshurica (Rupr.) Ohwi, *R. oldhamii* var., *A. gmelini* Weber ex Stechm., *Adenophora stricta*, *Platycodon grandiflorus*, *A. remotiflora* (Siebold & Zucc.) Miq., *M. sacchariflorus* Benth., *P. australis*, *A. fruticosa* L., Carex dispalata Boott, *A. tataricum* subsp. ginnala (Maxim.) Wesm., *A. japonica* Steud., *A. keiskeana*, *Dioscorea opposita*, *A. pacificus* pacificus, *S. wichurae* var. asiaticus (Beetle) T.Koyama, *A. contorta*, *L. apetalum*, and *R. parvifolius* L. for. parvifolius and *G. Bristlegrass.*

While in the wetland along Angaecheon, *S. koreensis* Andersson and *A. tataricum* subsp. ginnala (Maxim.) Wesm. and *A. fruticosa* L. are observed within the community of *A. japonica* Steud.(DBH 20–30 cm, Height 14–18 m) while *M. sinensis*, *P. australis*, *Typha angustifolia* L., rice and herbaceous communities are observed as herbal species.

In the low wetland around the stream, *S. koreensis* Andersson, *A. japonica* Steud., and *R. pseudoacacia* L. are observed in patchy forms with *M. sacchariflorus* Benth. community dominating the area.

In the tree layers of lowlands, *A. japonica* Steud., *S. koreensis* Andersson, *Celtis edulis* Nakai, *A. tataricum* subsp. ginnala(Maxim.) Wesm. and *M. myriantha* Siebold & Zucc. are found. While the *L. obtuslfolium*, *M. bombycis Koidzumi*, *A. tataricum* subsp. ginnala (Maxim.) Wesm. and *L. japonica* are observed comprising the shrub layer. The average height and DBH of *A. japonica* Steud. and *S. koreensis* Andersson developed in low wetland are found to be 10–15 m and 22–32 cm respectively.

Although the observed species are quite limited due to the restrictions made on the investigation, at least in the wetland around streams, the following species are confirmed appearing: *P. australis*, Carex dispalata Boott, *A. japonica* Steud., *S. koreensis* Andersson, *Q. acutissima* Carruth., *M. myriantha* Siebold & Zucc., *Q. dentata* var. fallax Nakai, *M. bombycis Koidzumi*, *L. obtuslfolium*, *A. tataricum* subsp. ginnala (Maxim.) Wesm., *L. japonica*, *S. prunifolia* for. simpliciflora. Nakai, *A. fruticosa* L., *M. sacchariflorus* var. gracilis Y.N.Lee, *M. sacchariflorus* Benth., *C. siderosticta*, *V. maackii* var. japonicum, *S. virga-aurea* var. asiatica,

A. pacificus pacificus, *J. effusus* var. *d. Buchenau.*, and S. maritimus L (Figs. 6.152, 6.153, 6.154, 6.155, 6.156, 6.157, 6.158, 6.159, 6.160, 6.161, 6.162, 6.163, 6.164, 6.165, 6.166, 6.167, 6.168, 6.169).

Fig. 6.152 Wetland observed in Samicheon-Yeoncheon plain

Fig. 6.153 One of Sewolcheon tributary where stream starting from Panbudong on the *right* and depression wetland on the *left* are observed

Fig. 6.154 Palustrine and depression wetland observed at nearby Panbudong stream (Angaecheon)

Fig. 6.155 The community of Quercus acutissima Carruth.-Quercus mongolica forest

Fig. 6.156 Pine tree observed around the graveyard of the mountain ridge

Fig. 6.157 Soil transect and cave of Halcyon pileata

Regarding inland insects, the habitat of *Sympetrum parvulum* Bartenef, *Sympetrum depressiusculum*, *O. longifennis* Shiraki, *Vespa mandarinia*, *Vespa koreensis* koreensis Radoszkowski and *C. erate* is confirmed. However, as the investigation was carried out in the winter, not many species were detected. But more species are expected to be observed when the research is conducted again

Fig. 6.158 Alnus japonica Steud. and Salix koreensis Andersson observed at the partial wetland of plain of the nearby Samicheon

Fig. 6.159 The appearance of mountainous area right before the joint point of Angaecheon and Samicheon

Fig. 6.160 The community of Alnus japonica Steud. formed around Panbudong stream (Sewolcheon)

either in spring or summer time. There are reports received from residents living near the subjected areas that Luciola lateralis, Lychnuris rufa (Olivier), Allomyrina dichotoma, and Nannophya Pigmaea Ramber can be seen at a distance from their houses during spring and summer time. Based upon its geographical condition

Fig. 6.161 Aster subulatus Michx. observed around the west natural area of checkroad in Samicheon-Yeoncheon plain

Fig. 6.162 Aristolochia contorta (seed) observed around the west natural area of checkroad in Samicheon-Yeoncheon plain

Fig. 6.163 Typha angustata observed in Samicheon area

where natural stream Samicheon flows through the middle- and small-scale wetland developed within the plain, these insects are likely to be seen in these areas.

Investigation on fish was conducted in the Angaecheon area and a total of 14 species are found. Among them 9 were Korean endemic species with appearance having native frequency of 64.3 %, highest compared to other bodies of water.

Fig. 6.164 The panoramic view of lowland in Samicheon-Yeoncheon plain

Fig. 6.165 The community of Quercus acutissima Carruth. observed in the hill area of east natural area of Samicheon-Yeoncheon plain

Fig. 6.166 The community of Salix koreensis Andersson observed in the lowland of east natural area of Samicheon-Yeoncheon plain

Moreover, they are accounted to more than 80 % of the total number of species found in the region.

Besides, from these endemic species, the appearance of natural monuments like the *H. mylodon* and *S. scherzeri* and, the appearance of endangered species like the

Fig. 6.167 The community of Quercus acutissima Carruth. observed in the hill area of east natural area of Samicheon-Yeoncheon plain

Fig. 6.168 The community of Betula davurica Pall. observed in the hill area of east natural area of Samicheon-Yeoncheon plain

Fig. 6.169 The rings of Quercus acutissima Carruth. in Samicheon region

G. macrocephala and *Pseudopungtungia tenuicorpa*, proved its high value as a habitat for fish.

Angaecheon, a tributary of Samicheon, embarks from GP and joins the mainstream right after passing the wetland. It is found to be a good habitat for various

Fig. 6.170 Fish and aquatic species observed in the Angaecheon stream

kinds of vegetations and fish including *P. australis*, *M. sacchariflorus* Benth., *S. gracilis majimae* (Jordan et Hubbs) (dominant), *Z. koreanus*, *M. yaluensis*, *Acheilognathus yamatsutae*, *H. longirostris*, *P. esocinus*, *G. macrocephala*, *H. mylodon*, *I. koreensis*, *P. herzi*, *Coreoleuciscus splendidus*, *P. tenuicorpa*, *R. brunneus*, and *S. scherzeri*.

For aquatic species, N. denticulata, *Semisulcospira libertina*, *N. triguttata* Motschulsky, and *M. japonicus* are found in the region, as well as *Hirondellea gigas*, and Trichoptera in the nearby wetland. Although no detailed survey could be made in this region due to its security issue related to mine residues, it is believed that various aquatic species could be found in these areas. A supplementary investigation is necessary to completely enumerate and describe the species.

No reptiles and amphibians are directly observed since these species are again, seasonal in nature. But it is a region with wide swampy wetlands which have appropriate conditions for the habitant of these species. Based on these environmental conditions, it is presumed that *K. borealis* and *R. nigromaculata* are definitely populating in the region (Fig. 6.170).

The entire area of theYeoncheon plain has Samicheon and the wetlands. It is regarded as a natural gifted habitat for birds, specifically for wildlife birds such as *G. japonensis* and *A. monachus* that naturally avoid the human populace. In the region, a total of 24 species with 731 entities of birds are observed including *D. leucotos* quelpartensis, *C. coccothraustes*, Halcyon pileata, *P. canus*, *P. montanus*, *H. amaurotis*, Alauda arvensis, *A. cinerea*, *E. rustica*, *P. palustris*., *A. albifrons*, *V. oxysepalum Turcz*., *P. colchicus*, *P. pica* serica, crow, *T. troglodytes* dauricus, and *T. naumanni* eunomus, as well as the grade I endangered species *G. japonensis* (44) and grade II species *A. monachus* (47), *A. gentilis* (1), white-naped crane (4), Anser fabalisserrirostris (250) and *B. buteo* (1). The natural monument Accipiter nisus (1) is also found in the investigation. Specifically, 48 entities of *G. japonensis* and white-naped cranes, 47 entities of *A. gentilis*, *B. buteo* and Anser fabalisserrirostris were observed several times during the investigation.

Unlike *G. japonensis*s detected in the farmland rice paddy of Dongsong region at Cheorwon, *G. japonensis*s found in this region were preying in Samicheon and in the surrounding small basins. It is probably because of the availability of fish, reptiles, and amphibians which are sources of proteins that the birds needed to strengthen their muscles for flight. Some *G. japonensis*s are also detected in the rice paddy of nearby CCZ where intake of carbohydrates is also taken from the grains of rice left on the rice paddy (Figs. 6.171, 6.172, 6.173, 6.174).

As for mammals, it is confirmed that *H. inermis*, *N. procyonoides*, *M. sibirica*, *P. bengalensis*, *A. agrarius*, and *M. wogura*inhabit the area. The marsh wetland provides a perfect activity area for *H. inermis* and *N. procyonoides*. In the region, the droppings of *N. procyonoides* were observed. The plain is regarded as a major habitat for *N. procyonoides* (Figs. 6.175, 6.176).

Fig. 6.171 Grus japonensis observed in the Yeoncheon plain

Fig. 6.172 Aegypius monachus observed in the Yeoncheon plain

Fig. 6.173 The habitat of Grus japonensis

Fig. 6.174 White-naped crane and its habitat

Fig. 6.175 Droppings of Prionailurus bengalensis

Fig. 6.176 Droppings of Hydropotes inermis

Hantan River Basin

The vicinity of Cheorwon Woljeong-ri station

Woljeong-ri station is located in Cheorwon county. It is basically a flat wetland and waterways coming from the water of North Korea that flows to Yeokgokcheon stream. The soil of this wetland is mostly composed of clay.

But most of the area was developed as farmlands, except for the areas surrounding the Samyeongho and its tributary. It is not a conducive habitat for wild animals because of possible soil and water contamination from agricultural pesticides and chemical fertilizers used in the farmland (Fig. 6.177).

The region has *P. australis*, *S. koreensis* Andersson, *P. hydropiper*, as well as the exotic plants; *B. frondosa* L.. and Amorpha fruticosa as its main marshy vegetation.

There are three mammal species found in the marshes namely *M. wogura*, *N. procyonoides*, and *H. inermis*.

Fig. 6.177 Hantan river tributary and its wetland of nearby Woljeong-ri station

In addition, 16 bird species are also found here including *A. alba*, *E. garzetta*, *A. cinerea*, *P. webbianus*, Oriolus chinensis, *P. capillatus* as well as the endangered species; the *G. japonensis*, and white-naped cranes.

As for fish, a total of 11 species including; *R. oxycephalus*, *A. yamatsutae*, *H. longirostris* and *R. giurinus* are observed, as well as the endemic species; *A. yamatsutae*, *I. koreensis* and *Coreoperca herzi*. Especially, the stream flow fish such as *Z. platypus* were also found to be dominating the area.

For reptiles and amphibians, 3 order and 5 species including Rana dybowskii and *R. tigrinus* are observed. For insects, 8 order, 18 family, and 23 species including *Calopteryx atrata*, *Cercion hieroglyphicum*, *S. infuscatum*, *Ectmetopterus micantulus*, *Adelphocoris suturalis*, *Lygocris spinolae* Meyer-Dur, *Polymerus cognatus* sp. 1, *Rubiconia intermedia*, *Diostrombus politus*, *A. japonica* Steud., *Chrysomela populi*, chironomid family sp. 1, *Allograpta balteata* (de Geer), Sphaerophoria menthastri (Linnaeus), Campiglossa hirayamae (Matsumura) were detected.

Lastly, for fresh water invertebrate species, a total of 33 species including Baetiella tuberculata (Kazlauskas), Caenis KUa, Protohermes grandis, and Cheumatopsyche brevilineata are regarded as major species.

Cardina denticulata denticulate was observed eating rotten prey. Predacious species such as the dragonflies and Hemiptera are found in the upper streams of Sanmyeongho Lake. In downstream, various kinds of species are observed around riffles including dayfly and caddis fly, representative species of stream ecosystem. The large number of benthic invertebrates living in the fast currents is considered as a typical feature distinguishing the area from the upper stream.

Cheorwon Plain Basin

The region comprises residential areas and farmlands with an artificial waterway. The waterway is restricting the growth of certain types of vegetations. It covers the Cheorwon plain rice paddy wetland that was formed around Hantan River of Galmal-eup, Cheorwon county, Gangwon province which also comprised a reclaimed plain. Most of the area is flat and a small-sized tributary flows through

Fig. 6.178 Hantan river tributary and its adjacent farmland around Cheorwon plain

Fig. 6.179 Hantan river tributary around Cheorwon plain

the residences and farmlands. The organic soil covering the area is just right for cultivation.

The region is completed with stream improvements and the rice farmland is developed along the streams. Annual herbaceous plant is found to be reseeding around the bank of the stream while the vicinity of stream is occupied by Persicaria thunbergii and *H. japonicus* Sieboid & Zucc (Figs. 6.178, 6.179).

In the rice paddy wetland near Hantan river, following species are found: Persicaria thunbergii, *H. japonicus* Sieboid & Zucc., *P. hydropiper*, Corydalis speciosa, *C. communis* L. and Crypsinus hastatus, including various kinds of exotic plants such as *B. frondosa* L.

A total of 3 family and 3 other mammal species such as *M. wogura*, *N. procyonoides*, and *H. inermis* are observed.

Regarding birds, 8 species including endangered wild animals such as Bubo bubo, as well as Korea's natural monument No. 323—Falco tinnunculus and *A. gentilis*—are observed.

But for fish, only 1 species, *M. mizolepis*, is detected in tributary water system. But it is expected that more species will be detected in the mainstream water system.

The reptiles and amphibians found in this area, including *Bombina orientalis* and *B. gargarizans*, with a total of 3 family and 4 other species, with 12 entities observed.

For insects, a total of 7 order, 19 family, and 26 species are found and they include *S. infuscatum*, Phaneroptera nigroantennata, Oecanthus indicus, Gryllidae sp 1, Oedaleus infernalis, Arcyptera fusca albogeniculata Ikonnikov, Orthocephalus funestus Jakovlev, Nysius plebejus Distant, Eoscatopsis assimilis, *A. balteata* (de Geer), Eristalomyia tenax, and Celastrina argiolus (Linnaeus),

Eleven species of freshwater invertebrate species are found including *Physa acuta*, *Uracanthella rufa* (Imanishi), *Davidius lunatus*, Simuliidae sp., and Tanypodinaesp. 1.

Lastly, benthic large invertebrates are observed in high population density within the stream receiving human disturbance, especially that flowing through the farmland.

Anhocheon Basin

The region is a wetland near Anhocheon of Hyunnae-myeon, Goseong county, Gangwon province forming a natural type of mountain area. Although it is surrounded by high mountains, the inner part of the target area is either flat or formed with a water basin and valley which provides a good geographical condition for wetland development. Moreover, with its low altitude of 0–100 m and with a gentle slope on its sides, it qualifies the condition where water gate can be joined together.

The flow of Anhocheon is connected from mountain to ocean and its hydrosphere flows along the valley of the mountain side. Ponds are found right along it and as an artificial ecological waterway, it influences the formation of wetland.

The water source of Anho reservoir is assumed to be either underground water or rainfall, as no single water flow is detected to be emanating from the nearby streams though it has an inlet.

Regarding the type of soil, sandy soil is found in sand dunes connected to the ocean, while organic soil is found in valleys and alluvial wetlands in which *P. australis* and *S. koreensis* Andersson are found as dominance species.

The target area mostly comprises wetland and with the widespread mountain area around the wetland, a perfect habitat for wild animals is provided. Moreover, facilities like the unification observatory, parking lots, railroad, and roads are observed. Donghae railroad and road promote the transformation of wetlands into land by disturbing its existing hydraulic and hydrologic system. As a result, the expansion of the habitat for *A. fruticosa* L. is confirmed. The Inter-Korean Cooperation Economic development, as well as the railroad and road connecting between two Koreas is increasing the pressure of developments in this area. They should consider appropriate countermeasures to preserve the natural environments (Fig. 6.180).

The following species are observed in the areas around Anhocheon:

First, for wetland vegetation, *T. japonica*, *P. australis*, and *S. maritimus* L. are found in the area. As for the dominance in each area, the following vegetation and its corresponding areas of dominance are designated: *P. australis* for alluvial wetland, *A. japonica* Steud. community for mountain areas and grass and sedge family for lower layers.

Fig. 6.180 The panoramic view of Goseong Anhocheon stream and its floodplain: surrounding area of Goseong Unification Observatory

Regarding mammals, a total of 11 species are found in the Anhocheon basin which include *M. wogura*, *N. procyonoides*, *M. sibirica*, *P. bengalensis*, *S. scrofa*, *H. inermis* and Sciurus vulgaris coreae. Among them, *P. bengalensis* is a grade II endangered species which is worth preserving.

A total of 39 species of birds are detected in the areas such as wetland, mountain areas, and regions connected to the seaside. With the grade I endangered wild animals such as Falco peregrinus japonensis and Haliaeetus albicilla, as well as the grade II species *A. monachus*, Korea's natural monument No. 323 Falco tinnunculus and No. 327 *A. galericulata* are found in the Anhocheon basin areas, and also *Cettia diphone*, *Motacilla alba*, *E. elegans* and *Anas platyrhynchos*.

As for the fish population, a total of 5 species are observed including grade II endangered species Lethenteron camtschaticum and Pungitius sinensis kaiborga, as well as *Cobitis granoei*, *Choenogobius urotaenia* and *Phoxinus logowskii*.

Lastly, regarding reptiles and amphibians, a total of 6 species including *B. gargarizans*, *B. orientalis* and *R. tigrinus* are found.

Regarding insects, various kinds of inland and aquatic insects are found in the area where the wetland and mountains are connected, as well as in the coastal sand dune. Especially in the wetland, species such as Brahmaea certhia, Atractomorpha lata, and Aquarius paludumare are found, while 178 species including *C. viridis*, *Brahmaea certhia*, and *Apatura ilia* are observed in the sand dunes.

As for the freshwater vertebrate species, a total of 14 species including Molanna moestaand *P. paucidens* De Haan are observed. The appearance of macrura proves the excellent quality of the water.

Chapter 7
Wetland-Type Classification

Reviews of wetland-type classification are provided by Kim (2006) and Cowardin et al. (1979). A new type of classification called the hybrid approach was developed by Kim (2006). The system is structured in six levels in total.

Level 1 is classified into inland, estuarine, and coastal areas. Level 2, the target area, is classified as an eco-region at a watershed level, and level 3 is classified into depression, riverine, slope, flat, and fringe areas based on a HGM approach. Level 4 is classified into detailed wetland types based on specific characteristics of wetlands. Level 5 is classified into marsh and swamp based on grasslands and shrubs and forest trees. Level 6 indicates the dominant vegetation communities.

The result of the application for the DMZ is described in this chapter.

7.1 The Need for Wetland-Type Classification

Recently, as the functions and values of wetlands, which used to be considered only wastelands, have been disclosed, much attention has been paid to the cautious and wise management of wetlands. In modern years in particular, despite their various functions, wetlands have been destroyed or eliminated significantly by developments or various forms of disturbance. The amount of wetland lost is difficult to quantify since the total area of wetland in the world is uncertain. There are, however, some figures for individual countries that indicate the scale of the problem. For example, the USA has lost some 87 million hectares (54 %) of its original wetlands, primarily to agricultural production (Barbier et al. 1996). Therefore, conservation and management efforts are urgently required. For such an approach, data on wetland types are necessary.

Currently, wetland classification systems used in Korea as well as in Asia mostly follow criteria set by the Ramsar Convention, and, thus, there is a great necessity for wetland-type classification systems and plans tailored to Asia. First of all, wetland-type classification systems need to be identified in Asia.

Four major objectives of a wetland-type classification system are (1) to describe ecological units that have certain similar ecological characteristics, (2) to arrange

K.-G. Kim, *The Demilitarized Zone (DMZ) of Korea*,
DOI: 10.1007/978-3-642-38463-9_7,

these units in a unified framework for the classification and characterization of wetlands that will aid action planning for conservation and management of wetlands, (3) to identify classification units for inventory and mapping, and (4) to provide uniformity on concepts and terminology (Mitsch and Gosselink 2000). In addition, it needs to assess biodiversity, functions, destruction and restoration possibilities of wetlands, wetland classification and mapping, which have to be performed at a watershed level.

In other words, wetlands have different functions and values, and understanding these functions and values is essential when developing measures for management, including conservation and restoration. Therefore, classification and mapping as well as recognizing wetlands are an essential process for conservation and restoration of wetlands and their sustainable use.

7.2 How to Classify Wetland Types

In 1974, the US fish and Wildlife Service began a national wetlands inventory (NWI). This inventory was designed to fulfill several management objectives, and a new wetland-type classification was developed (Cowardin et al. 1979).

A wetland-type classification of the NWI is based primarily on their substrate material, flooding regime, and vegetation life forms (Mitsch and Gosselink 2000). Its detailed classification system is shown in Fig. 7.1a.

The Ramsar system has 32 classes, divided into a marine/costal group and an inland group. This classification is based on geographical features and natural/man-made areas (http://www.ramsar.org/).

All wetlands have common hydrologic, soil, and vegetative characteristics. However, in the process of time, these general similarities of wetlands exhibit wide variation in terms of their size, complexity, and physical, chemical and biological characteristics and process (Mitsch and Gosselink 1993; Cowardin et al. 1979). The hydrogeomorphic method (HGM) classification system was developed specifically to accomplish this task. For example, the HGM classification system in Arkansas, USA, is shown in Fig. 7.1b.

In the 2000s, with the growing importance of the assessment of wetland functions, interest in the HGM rose. The HGM is drawing much attention because it can assess wetland functions, while the NWI classification, which has been used often in the past, cannot assess wetland functions (Richard Hauer et al. 2002). The HGM classifies wetlands by the geomorphology, water source, and hydrodynamics. Its detailed classification is given in Table 7.1.

When the NWI and HGM are compared, they have the following strengths and weaknesses (Marin et al. 1953; US Army Corps of Engineers 1996a; Klimas et al. 2005).

- In the NWI classification, palustrine wetland include many wetlands, and their distinctions are not clear.

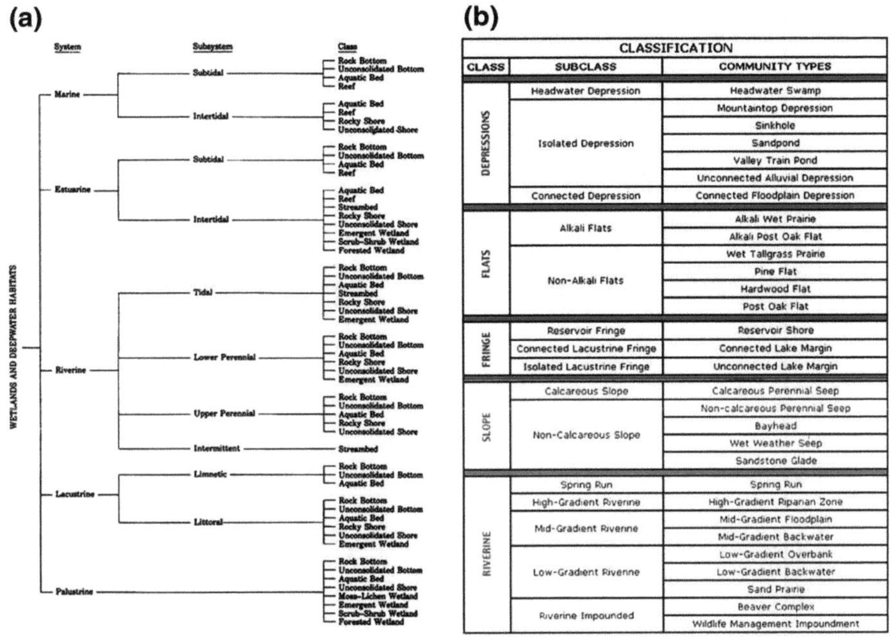

NWI wetland classification HGM wetland classification

Fig. 7.1 Wetland classification systems of the NWI and HGM. **a** NWI wetland classification. **b** HGM wetland classification

- The USA is currently shifting from the NWI classification to HGM classification.
- In the functional aspect of the wetlands, geomorphology and hydrology are more important than vegetation. The water-retaining capacity is determined by the geomorphology (depressions, gradients, etc.).
- Therefore, the HGM is better for a functional assessment of wetlands.

In addition, this study reviewed the wetland classification adopted by the Ramsar Convention, regions and countries [USA (Arkansas), Germany, the Netherlands and Japan]. An integrated analysis of these wetland classifications and some countries and their criteria are given in Table 7.2.

In most countries, vegetation, hydrology, and topography are major criteria in determining and classifying wetlands. While the NWI of the US places stress on vegetation, the HGM classification place relatively less weight on it. Wetland classification in Germany is focused on wetlands located within biotopes. In the Netherlands, hydrology is an important criterion in classifying wetlands.

One thing we need to pay attention to is the wetland classification system in Arkansas, USA. In fact, it classifies territory into eco-regions before the wetland classification system is applied. For the wetland classification and characteristics, Arkansas uses five major eco-regions that are designated to reflect major variations in climate, geology, topography, soils, and vegetation within the state. The five eco-regions are shown on the state map (Fig. 7.2).

Table 7.1 HGM wetland classification

Category	Geomorphic setting	Water source	Hydrodynamics	Cross-section
Depression	Depression	Return flow from groundwater and interflow	Vertical	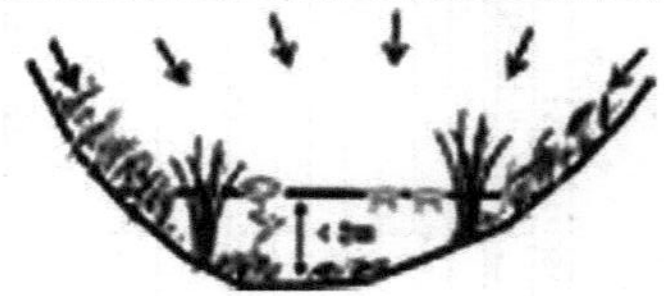
Riverine	Riverine	Overbank flow from channel	Unidirectional and horizontal	
Slope	Slope	Return flow from groundwater	Unidirectional, horizontal	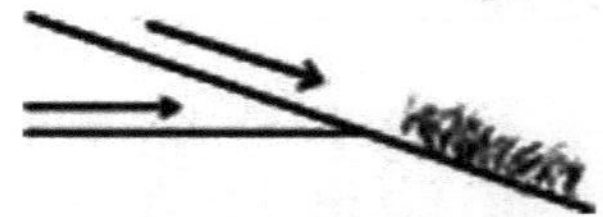
Flat	Flat	Precipitation	Vertical	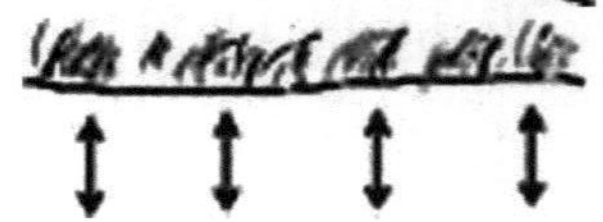
Fringe	Estuarine fringe, lacustrine fringe	Overbank flow from lakes and estuary	Bidirecional, horizontal	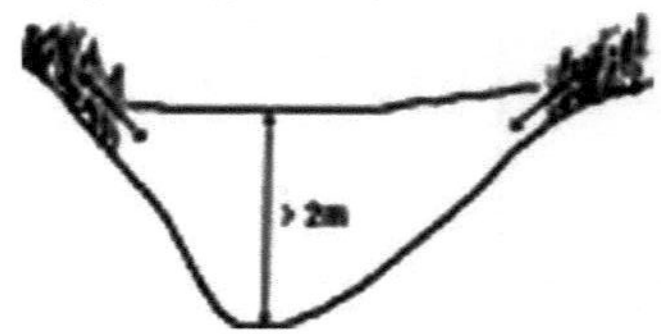

Table 7.2 Wetland classification and criteria by international conventions and countries

Item	Wetland classification	Criteria	Reference	Remarks
Western countries	NWI of US (Cowardin-based wetland classification)	Substrate materials and flooding regime or vegetation life form	Cowardin et al. (1976, 1977a, b)	Consider vegetation
	HGM of US (Arkansas)	Geomorphic setting, water source and hydrodynamics	Richard Hauer et al. (2002), Daniel Smith et al. (1995), Arkansas Multi-Agency Wetland Planning Team (2001)	Relatively low stress on vegetation
	Germany	Chemical properties of water, source of water, water level of groundwater and surface water	Riecken et al. (1994), Bundesamt fuer Naturschutz (1995), Succow and Joosten (2001)	Classify wetlands with a focus on biotope
	The Netherlands	Table and movement of water, nutrient source	Best and Baker (1993)	Consider dynamics of water
International convention	Ramsar	Geographical location, natural/man-made, hydrological conditions, vegetation format	The Ramsar Convention on Wetlands (2006)	Consider location and vegetation
Asian country	Japan	Substrate materials and flooding regime, vegetation life form	Ministry of the Environment (2006)	Consider vegetation

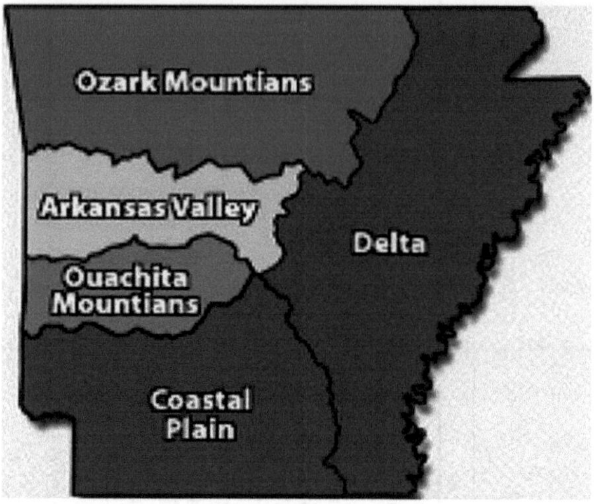

Fig. 7.2 Eco-region division in Arkansas. *Source* http://www.mawpt.org/wetlands/classification/divisions.asp

In Korea, studies on wetlands began in the 1990s when their importance was recognized. In the beginning, efforts mostly focused on ecological surveys to determine the ecological value of wetlands (Korea Wetlands Society 2000; Korean Nature Conservation Institute 1997).

Then, in the 2000s, studies on Korean wetland classifications and mapping began to come into full swing. They considered vegetation and features of the natural environmental in wetland classifications and performed wetland-type mapping as well (Hyohaemi 2000; Kim et al. 2001; Bon-hak 2002; The Korea Forest Service 2004).

7.3 New Classification System: Summary of Wetland Types

Detailed methods on wetland classification and mapping have been approached scientifically as the following based on the literature and case studies reviewed earlier. The wetland classification and mapping method developed in this study may be referred to as the "hybrid approach."

The basic approach of this study is as follows:

- Science- based: focus on functions.
- New data are developed as needed.
- Analysis is comprehensive, inland-wide.
- The process incorporates outreach and review of other methods.
- Products are transparent and user-friendly.

By utilizing the following information resources, theoretical classification of wetland types will be reviewed.

- Publications.
  - A survey using public domains on the Internet.
  - A review of existing reports and reference publications is carried out.
  - A survey of the data is kept by Seoul National University (SNU) environmental ecological planning lab (EEPL).
- Resource persons are involved.
  - Expert contact is maintained.

7.3.1 Overall Study Process

The overall process of the wetland classification and mapping system developed in this study is undertaken as the following from phase 1 through phase 5 (Fig. 7.3).

7.3.2 Detailed Methods of Each Phase

7.3.2.1 Wetland Identification

Wetlands are identified based on the following three parameters.

The first phase is hydrology. As hydrological characteristics are based on flooding or waterlogging duration, wetlands are recognized through inundated traces, sediments running down along water, depositing basins and soil saturation, washed out and exposed areas, wetland drainage patterns, moss on tree branches, formative adaptation of trees, traces of submergence or washout, leaves with fur and field-confirmed hydric soils (US Army Corps of Engineers 1998).

The second phase is hydrophytes. According to the US fish and wildlife Service (2000), survey data (currently available vegetation maps, etc.) are divided as follows:

- Obligate wetland (OBL) is observed almost always (estimated probability >99 %) in wetlands under natural conditions.
- Facultative wetland (FACW) is usually found in wetlands (estimated probability 67–99 %), but occasionally are observed in non-wetlands.
- Facultative areas (FAC) are equally likely to be observed in wetlands or non-wetlands (estimated probability 34–66 %).
- Facultative upland (FACU) is usually found in non-wetlands (estimated probability 67–99 %), but occasionally is observed in wetlands (estimated probability 1–33 %).

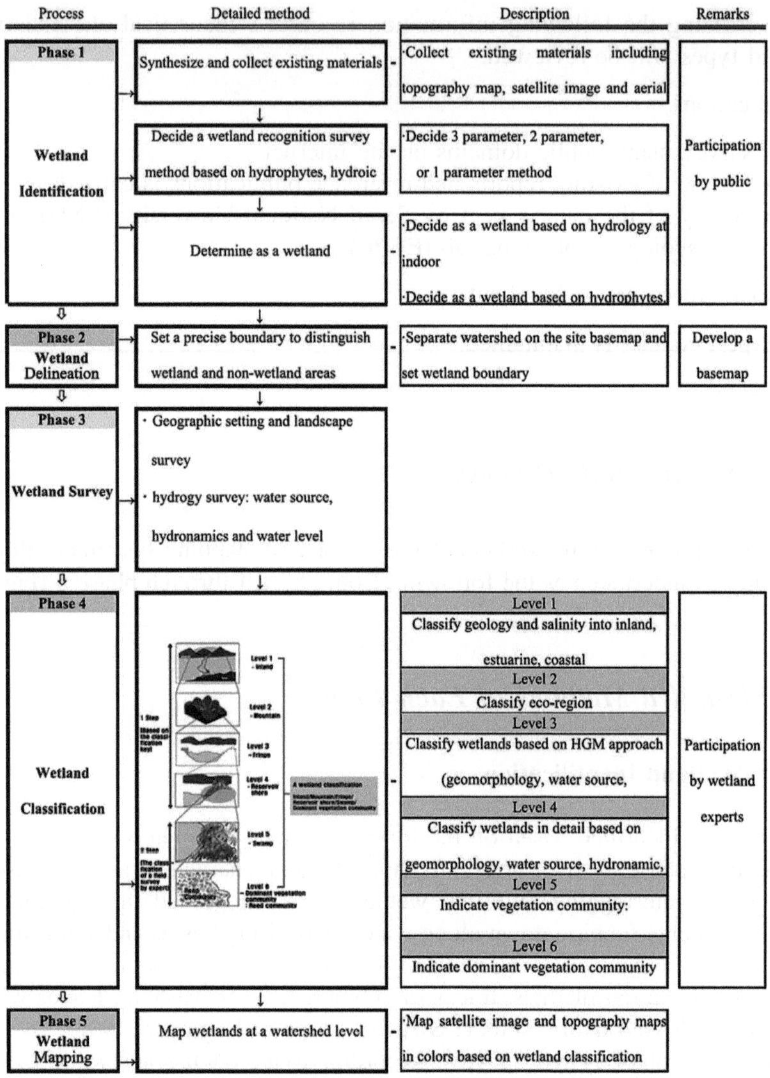

Fig. 7.3 Overall process of wetland classification and mapping system

- Obligate upland (UPL) is observed in wetlands in another region, but is observed almost always (estimated probability >99 %) in non-wetlands under natural conditions in the region specified.

Hydric soil is formed under conditions of saturation flooding or ponding long enough during the period. Unlike inland soil, its drainage is poor and has been submerged for a long period and, thus, maintains an anaerobic environment.

7.3.2.2 Wetland Delineation

Wetland delineation distinguishes wetland areas and non-wetland areas by identifying wetlands. To this end, wetlands are determined based on the above-mentioned parameters required for wetland recognition (hydrology, hydrophytes, and hydric soil). This is a sort of pre-survey, and wetland are delineated indoors by using satellite images, aerial photos and topography maps (US Army Corps of Engineers 1987).

7.3.2.3 Wetland Survey

An ecological survey is conducted in order to classify wetland characteristics and types for a target area. Ecological survey items include topography, landscape, biota, pollutant sources, and land use status. Detailed survey methods are as follows: Dargle et al. (1994), US Army Corps of Engineers (1996b), Richardson and Vepraskas (2001), Kilmas et al. (2005).

- *Geomorphic setting and landscape*. Based on the appearance of wetland geomorphology, it is classified into depression, slope, riverine, flat and fringe, and its wetland landscape is surveyed.
- *Hydrology*. The water source, hydrodynamics and water level of the water environment that affects the wetlands are surveyed.
- *Biota*. Dominant plant communities and fauna (mammals, birds, amphibians, and reptiles and insects) living in the wetlands are surveyed.
- *Soil survey*. Organic and mineral soils are determined.
- *Pollutant sources and land use*. Point sources and non-point sources that affect wetlands within a water and major land use are surveyed.

7.3.2.4 Wetland Classification

The finalized wetland classification is based on the HGM approach, but criteria on natural/man-made areas and vegetation relevant to the Ramsar and NWI have been added. Therefore, a wetland classification in Korea has been developed through a hybrid approach based on the HGM.

The system is structured in six levels in total. Step 1 is a general survey based on wetland classification criteria, and step 2 is conducted through a field survey by experts. In order to classify wetlands accurately, step 1 as well as step 2 need to be conducted (Fig. 7.4).

In the developed wetland classification, the detailed classification of each level is as follows:

- *Level 1*. To classify land into inland, estuarine and coastal areas based on the location of the wetlands and salinity of the water.

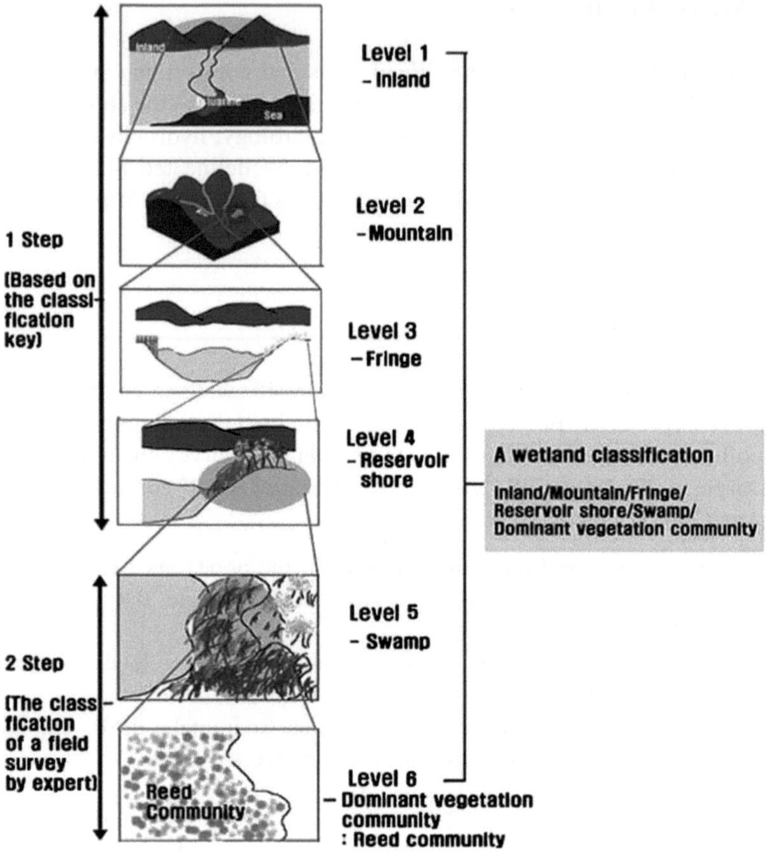

Fig. 7.4 Illustration of the finalized Korean-style wetland classification

- *Level 2*. To classify land into mountain, plain, river, lake, and reservoir in inland and delta areas, into estuary, plain, river, lake, and reservoir in estuarine and coastal areas and into mountain, plain, river, lake, and reservoir in coastal areas based on the eco-region. Eco-regions in this study are classified based on topography and hydrology at a watershed level and are slightly different from eco-regions in Arkansas which are macro-classified considering climate, terrain, soil and vegetation; eco-regions in this study are micro-classified based on topography and hydrology.
- *Level 3*. To classify land into depression, riverine, slope, flat and fringe areas based on the HGM approach that classified according to topographical conditions, water source and hydrological dynamics.
- *Level 4*. To classify land into detailed wetland types based on specific characteristics of wetlands (topography, soil, hydrology, vegetation, etc.)

- *Level 5*. To classify land into marsh and swamp based on the grasslands and shrub and forest trees in the wetland site.
- *Level 6*. To indicate dominant vegetation communities inhabiting the wetland site.

7.3.2.5 Wetland Mapping

In the mapping phase, specific boundaries are drawn based on wetland types, which allow wetland classifications to be understood at a glance. This phase is performed on topographical maps or satellite images that show the entire site relatively accurately and precisely. Correctional actions are taken based on field survey data. The final map is coded in different colors representing various types.

7.4 Application for the DMZ: Overview of the Study Site Wetland Classification

A site has been selected to apply and map the wetland classification developed in this study for validation.

7.4.1 Overview of Study Site: The DMZ Paju Phanmun Area

7.4.1.1 Location

In this study, the developed classification was applied to the DMZ Phanmun area. In terms of administrative district, the southern part belongs to Gunnae-myeon and Jinseo-myeon, Paju city. Gyeonggi province, and the northern part is located at Phanmunjeom-ri, Dongchang-ri, Limhan-ri and jogang-ri in phanmun county, Gaeseong city (Fig. 7.5). The spatial scope of the survey of this study is the southern section of the Phanmun field area, which spans about 6,296,875 m^2. As the Sacheon river runs across the center of the Phanmun field area, it belongs to the Sacheon river watershed. Since the site is limited to the southern part of the Phanmun field, it belongs to the sub-watershed of the Sacheon river (Fig. 7.6).

7.4.1.2 Rationale for the Site Selection

This site was selected for the following reasons.

First, this study is a part of the DMZ, where access by civilians is restricted and, thus, no surveys have been conducted to date.

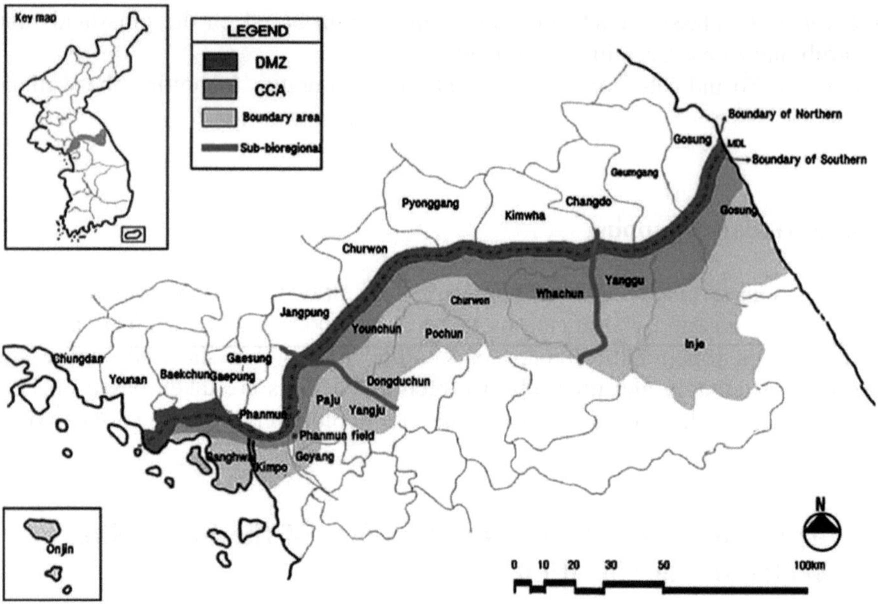

Fig. 7.5 Location of the case study area

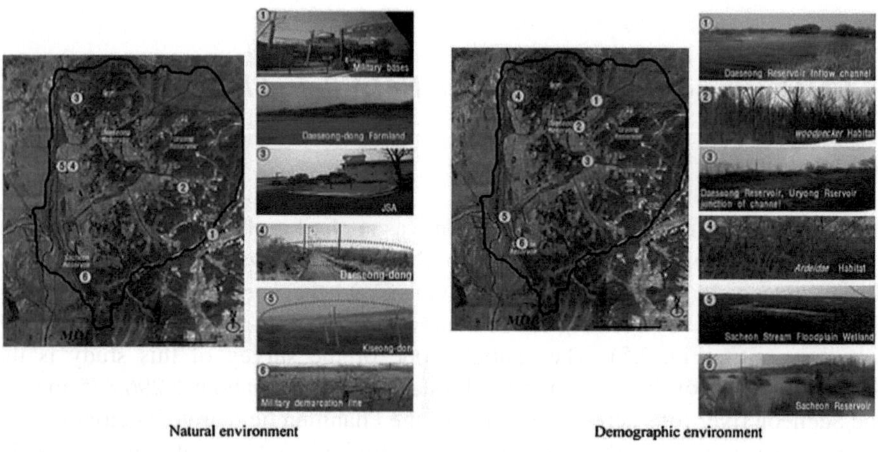

Fig. 7.6 General overview of the study site

Second, the Sacheon river runs near the study site. Reservoirs and waterways are developed, and the wetlands are very well developed.

Third, a literature review confirms that this study site is inhabited by rare birds, including the red-crowned Crane and Bewick's swan. It is highly valuable as an important habitat.

Fourth, this study site potentially has various wetland types. It is most appropriate to verify the wetland classification developed in the study.

Therefore, the Phanmun field area in the southern DMZ area has been selected as a case study area.

7.4.1.3 General Overview

Overall, the area is made up of hills and plains. As the Sacheon river runs across the area, it belongs to the Sacheon river watershed. In addition, three reservoirs—the Daeseong reservoir, Uryong reservoir and Sacheon reservoir—and an agricultural channel have been created, and, thus, wetlands are well developed. In particular, wetlands are well developed in inundated areas near the Sacheon river and a long-deserted area adjacent to Phanmunjeom. The Phanmun field area located within the DMZ is special in terms of periods and space and includes military bases, the joint security area (JSA), the military demarcation line (MDL), Daeseong-dong village in South Korea and Kigeong-dong village in North Korea.

7.4.2 Wetland Identification

Based on the existing data, areas that were affected by water and developed into wetlands and those not developed into wetlands were separated on a satellite image. The outcome of the wetland identification is shown in Fig. 7.7. Through indoor activities, areas that were created into wetlands as the result of hydrological effects were classified. Through field surveys, areas inhabited by wetland plants and having hydric soil were recognized as wetlands. In particular, the Sacheon riverine wetlands were created extensively. Its image is shown in Fig. 7.8.

7.4.3 Wetland Delineation

Based on a field survey on areas identified as wetlands through indoor activities, a precise boundary between wetlands and non-wetlands was clearly defined. To this end, a base map has been developed for this study site to classify watersheds. Classification of watersheds allows the viewing of wetland types in a broader context, and the wetland creation process can be understood. In this study site, watersheds can be divided into five areas (Fig. 7.9). Watershed 1 is the Phanmun field area where the mainstream Sacheon river flows into the watershed. Watershed 2 includes a tributary that runs into the Daeseong reservoir. Watershed 3 is formed by a tributary that flows into the Eoryong reservoir. Watershed 4 is an area where a tributary running from the Daeseong reservoir and a tributary from the Eoryong reservoir meet. Lastly, watershed 5 is the Sacheon reservoir zone that includes the Sacheon reservoir.

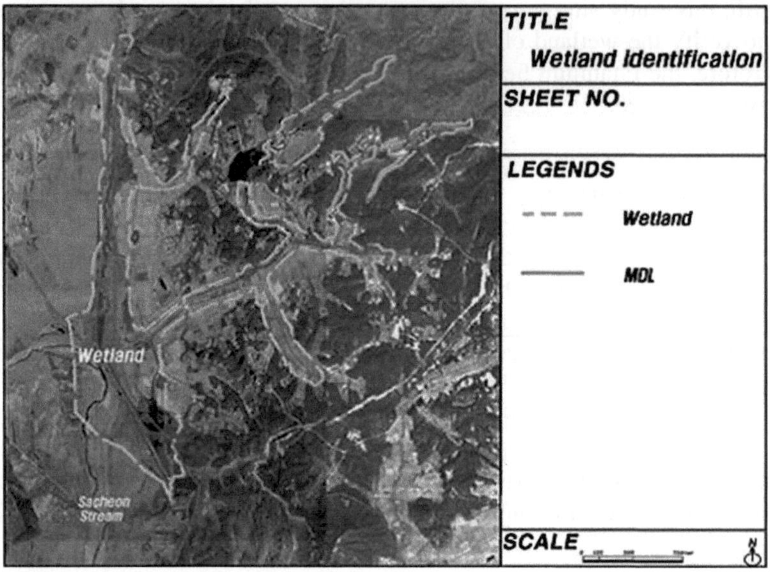

Fig. 7.7 Watershed-based wetland identification map

Fig. 7.8 A view of riverine wetland in Sacheon stream

Areas that were identified as wetlands, but where field surveys were not possible due to the special and uncontrollable nature of the site in this study were identified as wetlands based on nearby wetland areas. In addition, they identified based on satellite images. Areas identified as wetlands during a field visit are shown in Fig. 7.10.

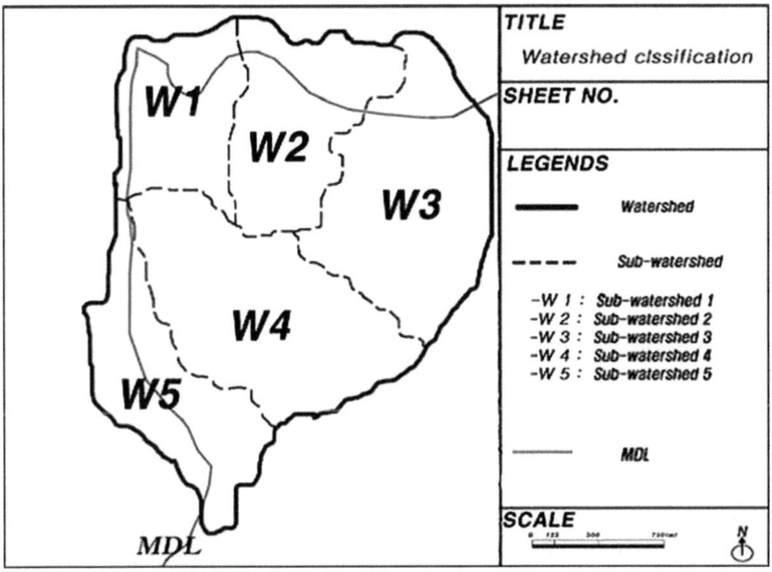

Fig. 7.9 Watershed classification of the study site

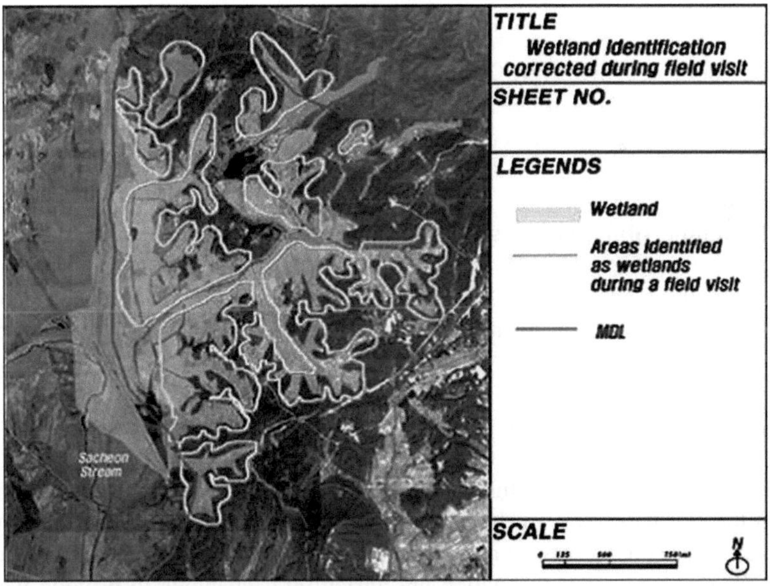

Fig. 7.10 Wetland identification corrected during a field visit

The study site has been used as farmland in the past. Today, forever, many places have been deserted for a long period as land mind controlled areas and have

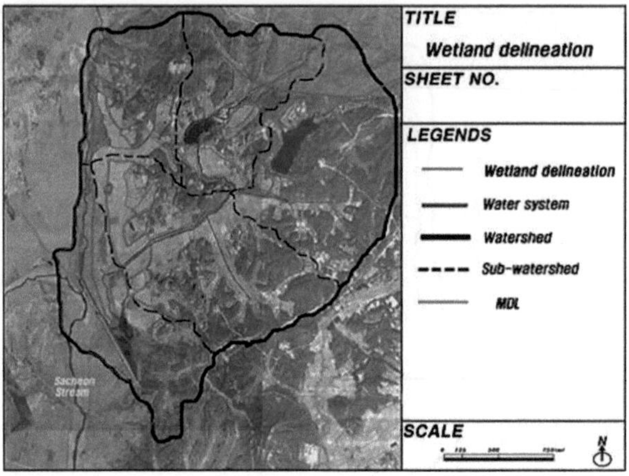

Fig. 7.11 Wetland delineation map

been turned into natural wetlands. Precise delineation of wetlands has been defined by including these areas well. Its details are shown in Fig. 7.11.

7.4.4 Wetland Survey

Wetland field surveys were conducted six times on a monthly basis from November 2005 to November 2007. Surveys of the overall study site and surveys on individual wetlands were conducted in parallel.

7.4.4.1 Geomorphic Setting and Landscape Survey

This study site is an extensive open field with low hilly spots here and there. As shown in the Fig. 7.12, an open field is created extensively up to the northern part.

7.4.4.2 Hydrological Survey

In the watershed flowing in this study site, water flowing in from the south runs toward the Sacheon river in the MDL. It has three large reservoirs: the Eoryong reservoir, Daeseong reservoir and Sacheon reservoir. Several waterways are separated from these reservoirs, which are agricultural channels to provide water to nearby farmland (Fig. 7.12).

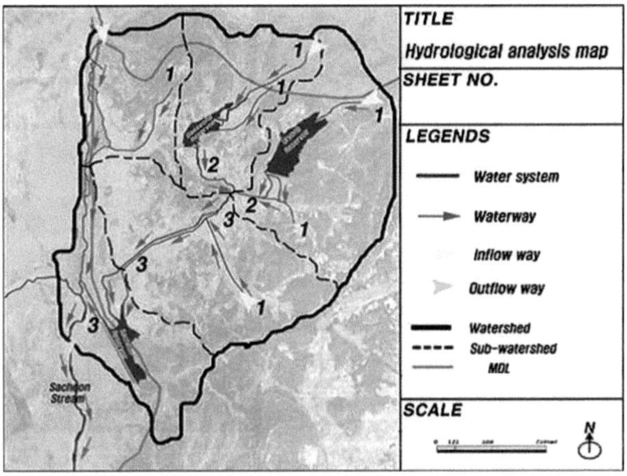

Fig. 7.12 Hydrological analysis map

7.4.4.3 Soil Survey

The soil survey on wetlands in this study site was conducted briefly. At the current status photo on soil, the soil was organic and had conditions favorable to the growth of vegetation.

7.4.4.4 Biota Survey

In this study site, a total of 89 species were found, including nine plant species, 34 bird species, seven mammal species, two amphibian and reptile species, 29 fish species, three insect species, and five crustacean species. These included the small-eared cat and common buzzard, which are protected species designated by the Ministry of Environment, and the red-crowned crane, an internationally protected species and a natural monument. Three reservoirs and waterway located in the site highly used by indigenous fishery.

The distribution of the species listed in the Table 7.3 is shown in Fig. 7.13. Species are highly distributed in the Eoryong reservoir, Daeseong reservoir and Sacheon reservoir.

In particular, when legally protected species, including endangered species designated by the Ministry of Environment, red data, natural monument species and indigenous species, were identified from these species, out of a total of 89 species, 17 species (10 bird species, two mammal species, and five fish species) were identified as specially designated species as given in Table 7.3.

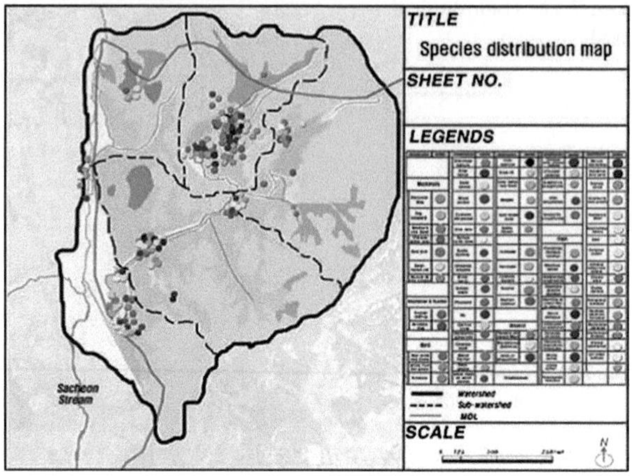

Fig. 7.13 Species distribution map

7.4.5 Pollutant Sources and Land Use Status

Land in the study site was mostly made up of forest and agricultural areas. Its details are shown in Fig. 7.14. Land was mostly farmland in agricultural areas, while it was mixed forest areas. The images of forest ecosystems are shown in Fig. 7.15.

7.4.6 Threatening Factors

Although the study site is located within the DMZ where human activities are controlled, there still are a number of threatening factors to the wetlands. Wetlands have disappeared to a large extent as local residents have reclaimed wetlands as farmlands. Wetlands are also threatened by reclamation of mountain areas into ginseng fields and the use of fertilizers. In addition, fishing nets have been set up in the three reservoirs and waterways to catch fish. However, the nets have been abandoned for a long period and are disturbing the activities of fishery, and many fish were found dead within the nets. Sacheon reservoir was inhabited by bass, a foreign species, which is threatening fish species native to Korea.

7.4.7 Wetland Classification

Classification of inland, estuarine, and coastal areas. This study site is located within Paju inland, and the Sacheon river is fresh water and belongs to the inland.

Table 7.3 The current status of specially protected species in the Phanmun field

Classification	Endangered species designated by the ministry of environment	Red data	Natural monument species	Indigenous species	Total species
Vegetation	–	–	–	–	0
Mammals	Small-eared cat	Chinese water deer, small-eared cat		–	2
Birds	Golden eagle, red-crowned crane, white-naped crane, marsh harrier, swan goose, cinereous vulture, common buzzard, goshawk, bean goose	Red-crowned crane, white-naped crane, hooded crane, cinereous vulture, swan goose	Red-crowned crane, white-naped crane, golden eagle, goshawk, marsh harrier, kestrel, swan goose	–	10
Amphibians and reptiles	–	–	–	–	0
Insects	–	–	–	–	0
Fish	Common Korean bitterling	–	–	Korean rose bitterling, Squalidus gracilis majimae, Abbottina springeri Banarescu et Nalbant, Korean spotted sleeper	5
Total	–	–	–	–	17

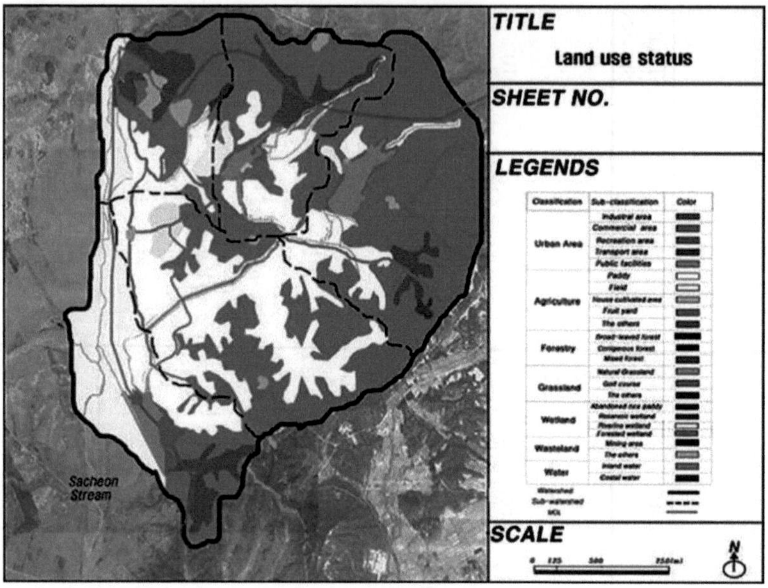

Fig. 7.14 Land use status

Fig. 7.15 Images of forest ecosystems (*top* a forest on the way to Eoryong reservoir, *bottom* forest area near Daeseong reservoir)

7.4.7.1 Classification of Eco-region

When classification comes to an eco-region level, this study site has mountains, plains, rivers, lakes and reservoirs and, thus, has topographical conditions to create various types of wetlands (Fig. 7.16).

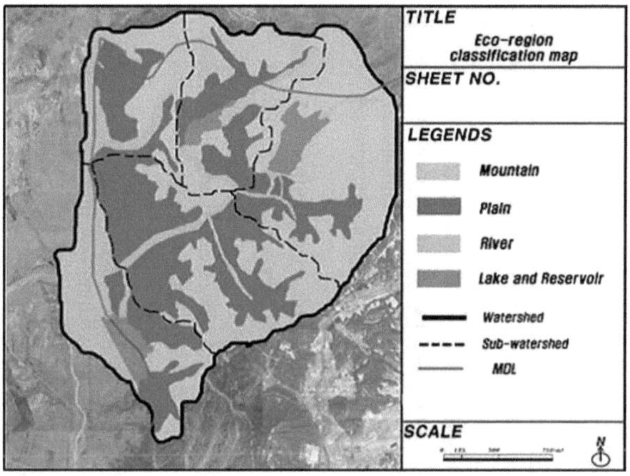

Fig. 7.16 Eco-region classification map of the study site

7.4.7.2 Detailed Wetland Classification

Wetlands found in this study site have been classified through levels 1–6 as developed in the previous chapter, and they are given in Table 7.4. A total of 19 wetland types have been classified, indicating that wetland types are diverse.

7.4.7.3 Wetland Mapping

Through wetland classification, wetland types were mapped on the base map of this study site (Fig. 7.14). The result shows the rice paddy depression wetlands are connected through agricultural channels and river watersheds.

The areas of these wetland types are calculated in Table 7.5, which shows the areas of inland/plain/depression/agricultural abandoned rice paddy depression/swamp/alder community and other wetland types.

7.5 Summary of Wetland Types

For reference, wetland types in Korea classified by the study team led by the authors, with the support of the UNDP/GEF KOREAN Wetland Project, are summarized in this section.

In the wetland classification system newly developed in this study, the total number of wetland types in Korea was 125: 74 inland wetland types, 36 estuarine wetland types, and 15 coastal wetland types. This result is from classifications down to level 4. Levels 5 and 6 are generated through field surveys. Therefore, when these are included, a greater number of types will be generated. However, this study classified the wetland types down to level 4 only.

Table 7.4 Wetland types

Level 1	Level 2	Level 3	Level 4	Level 5	Level 6	Wetland type
Inland	Plain	Depression	Rice paddy depression	Marsh	*Oryza sativa*	Inland/plain/rice paddy depression/marsh/*Oryza sativa*
Inland	Plain	Depression	Agricultural abandoned rice paddy depression	Unknown	Unknown	Inland/plain/depression/agricultural abandoned rice paddy depression/unknown
Inland	Plain	Depression	Agricultural abandoned rice paddy depression	Swamp	Alder community	Inland/plain/depression/agricultural abandoned rice paddy depression/swamp/alder community
Inland	Plain	Depression	Agricultural abandoned rice paddy depression	Swamp	Willow community	Inland/plain/depression/agricultural abandoned rice paddy depression/swamp/willow community
Inland	Plain	Depression	Agricultural abandoned rice paddy depression	Marsh	Reed community	Inland/plain/depression/agricultural abandoned rice paddy depression/marsh/reed community
Inland	Plain	Depression	Agricultural abandoned rice paddy depression	Marsh	*Miscanthus sacchariflorus community*	Inland/plain/depression/agricultural abandoned rice paddy depression/marsh/*Miscanthus sacchariflorus* community
Inland	River	Riverine	Low gradient overbank	Marsh	Reed community	Inland/river/riverine/low gradient overbank/marsh/reed community
Inland	River	Riverine	Low gradient overbank	Swamp	*Virgularia gustaviana* community	Inland/river/riverine/low gradient overbank/swamp/*Virgularia gustaviana* community
Inland	Plain	Depression	Farm pond	Marsh	Reed community	Inland/plain/depression/farm pond/marsh/reed community
Inland	River	Riverine	Low gradient floodplain	Swamp	Willow community	Inland/river/riverine/low gradient floodplain/swamp/willow community

(continued)

Table 7.4 (continued)

Level 1	Level 2	Level 3	Level 4	Level 5	Level 6	Wetland type
Inland	River	Riverine	Low gradient floodplain	Marsh	Reed community	Inland/river/riverine/low gradient floodplain/marsh/reed community
Inland	River	Riverine	Low gradient floodplain	Unknown	Unknown	Inland/river/riverine/low gradient backbank/unknown
Inland	River	Riverine	Artificial water channel	Marsh	Reed community	Inland/river/riverine/artificial water channel/marsh/reed community
Inland	River	Riverine	Artificial water channel	Swamp	*Virgularia gustaviana* community	Inland/river/riverine/ artificial water channel/swamp/ *Virgularia gustaviana* community
Inland	River	Slope	Moderate gradient agricultural levee	Swamp	*Virgularia gustaviana* community	Inland/river/slope/moderate gradient agricultural levee/swamp/*Virgularia gustaviana* community
Inland	Lake and reservoir	Fringe	Reservoir shore	Swamp	Willow community	Inland/lake and reservoir/fringe/reservoir shore/willow community
Inland	Lake and reservoir	Fringe	Sand island	Unknown	Unknown	Inland/lake and reservoir/fringe/ sand island/unknown
Inland	Lake and reservoir	Fringe	Reservoir shore	Marsh	Reed community	Inland/lake and reservoir/ fringe/reservoir shore/marsh/reed community
Aquatic ecosystem						Aquatic ecosystem

The finalized model was applied to 84 wetland sites commonly found in Korea and 10 wetland protection areas designated by the Ministry of Environment to classify wetland types. As a result, the following 126 wetland types were identified. Ten wetland protection areas designated by Ministry of Environment showed 11 wetland types in total. Their detailed types are shown below.

7.5.1 General Wetlands

7.5.1.1 Inland Wetland

Inland/mountain/depression/headwater depression, inland/mountain/depression/high forested depression, inland/mountain/depression/limestone sinkhole, inland/mountain/depression/volcanic forested wetland, inland/mountain/depression/connected alluvial valley depression, inland/mountain/depression/unconnected alluvial depression, inland/mountain/depression/agricultural abandoned rice paddy depression, inland/mountain/depression/gravel pits depression, inland/mountain/depression/farm pond, inland/mountain/depression/rice paddy depression, inland/mountain/depression/artificial pond, inland/mountain/depression/linear depression, inland/plain/depression/agricultural abandoned rice paddy depression, inland/plain/depression/artificial pond, inland/plain/depression/linear depression, inland/plain/depression/floodplain depression, inland/plain/depression/underflow depression, inland/river/depression/sand island, inland/river/depression/agricultural abandoned rice paddy depression, inland/river/depression/rice paddy depression, inland/river/depression/artificial pond, inland/river/depression, inland/lake and reservoir/depression/sand island, inland/lake and reservoir/depression/agricultural abandoned rice paddy depression, inland/lake and reservoir/depression/rice paddy depression, inland/lake and reservoir/depression/artificial pond, inland/lake and reservoir/depression/linear depression, inland/mountain riverine/high gradient spring run, inland/mountain/riverine/mid gradient riparian, inland/mountain/riverine/mid gradient backbank, inland/mountain/riverine/mid gradient overbank, inland/mountain/riverine/artificial water channel, inland/plain/riverine/artificial water channel, inland/river/riverine/low gradient floodplain, inland/river/riverine/low gradient backbank, inland/river/riverine/low gradient overbank, inland/river/riverine/artificial water channel, inland/lake and reservoir/riverine/artificial water channel, inland/mountain/flat/alkali fen, inland/mountain/flat/acid hardwood flat, inland/mountain/flat/acid wet grass, inland/mountain/flat/acid bog, inland/plain/flat/alkali fen, inland/plain/flat/acid hardwood flat, inland/plain/flat/acid wet grass, inland/plain/flat/acid bog, inland/mountain/slope/steep gradient waterfall seep, inland/mountain/slope/steep(moderate) gradient non calcareous perennial seep, inland/mountain/slope/steep(moderate) gradient calcareous perennial seep, inland/mountain/slope/steep(moderate) gradient wet weather seep, inland/mountain/slope/low gradient sandstone glade, inland/mountain/slope/steep(moderate, low) gradient agricultural levee, inland/plain/slope/steep(moderate, low) gradient

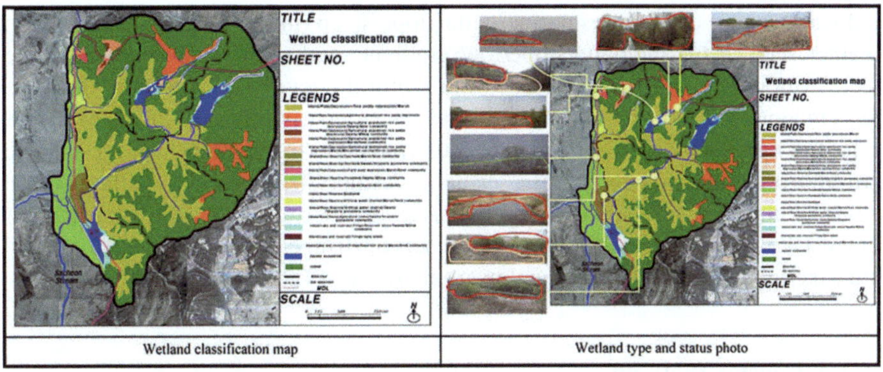

Fig. 7.17 Wetland classification map

agricultural levee, inland/plain/slope/low gradient sandstone glade, inland/river/slope/low gradient non calcareous perennial seep, inland/river/slope/steep(moderate) gradient wet weather seep, inland/river/slope/steep(moderate, low) gradient agricultural levee, inland/river/slope/steep(moderate, low) gradient abandoned rice paddy levee, inland/lake(reservoir)/slope/steep(moderate) gradient wet seep, inland/lake and reservoir/slope/steep(moderate, low) gradient agricultural levee, inland/mountain/fringe/connected lake margin, inland/mountain/fringe/unconnected lake margin, inland/mountain/fringe/reservoir shore, inland/plain/fringe/reservoir shore, inland/plain/fringe/connected lake margin, inland/plain/fringe/unconnected lake margin, inland/lake and reservoir/fringe/connected lake margin, inland/lake and reservoir/fringe/unconnected lake margin, inland/lake and reservoir/fringe/reservoir shore (Fig. 7.17).

7.5.1.2 Estuarine Wetland

Estuarine/delta/depression/sand pond, estuarine/delta/depression/sand island, estuarine/delta/depression/alluvial valley depression, estuarine/delta/depression/floodplain depression, estuarine/delta/depression/unconnected alluvial depression, estuarine/delta/depression/rice paddy depression, estuarine/delta/depression/abandoned rice paddy depression, estuarine/delta/depression/farm pond, estuarine/delta/depression/farm pond, estuarine/delta/depression/artificial pond depression, estuarine/delta/depression/underflow depression, estuarine/delta depression/linear depression, estuarine/plain/depression/rice paddy depression, estuarine/plain/depression/abandoned rice paddy depression, estuarine/plain/depression/farm pond, estuarine/plain/depression/artificial pond depression, depression, estuarine/plain/depression/linear depression, estuarine/river/depression/sand island, estuarine/lake and reservoir/depression/lagoon, estuarine/delta/riverine/artificial water channel, estuarine/farmland/riverine/artificial water channel, estuarine/river/riverine/mid (low) gradient floodplain, estuarine/river/riverine/mid (low) gradient backwater, water channel, estuarine/delta/slope/steep (moderate, low) gradient wet weather seep, estuarine/delta/slope/low gradient sand seep, estuarine/delta/slope/steep (moderate, low) gradient agricultural levee, estuarine/plain/slope/steep(moderate, low) gradient agricultural levee, estuarine/river/slope/steep(moderate, low) gradient wet weather seep, estuarine/delta/fringe/delta shore, estuarine/estuary/fringe/tidal marsh margin, estuarine/estuary/fringe/tidal mudflat margin, estuarine/plain/fringe/reservoir shore, estuarine/lake and reservoir/fringe/connected lake margin, estuarine/lake and reservoir/fringe/unconnected lake margin, estuarine/lake and reservoir/fringe/reservoir shores (Table 7.5).

7.5.1.3 Coastal Wetland

Coastal/Coastal/depression/sand pond, Coastal/Coastal/depression/rocky shores, Coastal/plain/depression/abandoned salt farm, Coastal/plain/depression/active salt farm, Coastal/plain/depression/aquaculture pond, Coastal/plain/depression/exploitative land, Coastal/river/riverine/low gradient tidal channel, Coastal/Coastal/slope/sleep gradient cliff, Coastal/Coastal/slope/low gradient sand seep, Coastal/plain/slope/steep(moderate, low) gradient agricultural levee, Coastal/river/slope/low gradient wet weather seep, Coastal/Coastal/fringe/beach shore, Coastal/Coastal/fringe/salt flat marsh

- Wetland protection areas by Ministry of Environment

Estuarine/river/depression/sand island, estuarine/delta/fringe delta shore, inland/mountain/flat/acid bog, inland/lakefringe/connected lake margin, inland/mountain/depression/volcanic forested wetland, inland/mountain/flat/acid wet grass, coastal/coastal/depression/sand pond, inland/mountain/depression/high forested depression, inland/river/riverine/low gradient floodplain, inland/mountain/depression/high forested.

Based on the study results reviewed so far, the following may be summarized. First, the hybrid-approach classification system is a new wetland classification system that is based on the HGM and considers the geomorphology, water source, hydrodynamic, vegetation and soil.

Second, the hybrid approach is classified into six levels in total. Levels 5 and 6 are classified by experts through field surveys.

Third, wetland-type mapping is performed in five steps in total, and each step has been defined in detail.

Fourth, when the new classification system was actually applied to the study site, it not only represented the characteristics of the wetlands well, but also highlights the values and functions of wetlands in the wetland survey and classification mapping stages. Accordingly, it may be used in wetland conservation and management in the future.

Table 7.5 Wetland area by wetland types

Wetland type	Area (m^2)
Inland/plain/depression/rice paddy depression/marsh/*Oryza sativa*	1,279,183
Inland/plain/depression/agricultural abandoned rice paddy depression/unknown	5,371
Inland/plain/depression/agricultural abandoned rice paddy depression/swamp/alder community	151,361
Inland/plain/depression/agricultural abandoned rice paddy depression/swamp/willow community	4,944
Inland/plain/depression/agricultural abandoned rice paddy depression/marsh/reed community	27,814
Inland/plain/depression/agricultural abandoned rice paddy depression/marsh/*Miscanthus sacchariflorus* community	4,969
Inland/river/riverine/low gradient overbank/marsh/reed community	14,926
Inland/river/riverine/low gradient overbank/swamp/ *Virgularia gustaviana* community	28,095
Inland/plain/depression/farm pond/marsh/reed community	151
Inland/river/riverine/low gradient floodplain/swamp/willow community	23,848
Inland/river/riverine/low gradient floodplain/marsh/reed community	6,051
Inland/river/riverine/low gradient backbank/unknown	29,876
Inland/river/riverine/artificial water channel/marsh/reed community	21,847
Inland/river/riverine/artificial water channel/swamp/*Virgularia gustaviana* community	4,916
Inland/river/slope/moderate gradient agricultural levee/ swamp/*Virgularia gustaviana* community	2,941
Inland/lake and reservoir/fringe/reservoir shore/swamp/willow community	13,000
Inland/lake and reservoir/fringe/sand island/unknown	3,019
Inland/lake and reservoir/fringe/reservoir shore/marsh/reed community	17,628
Aquatic ecosystem	91,554
Area total	1,731,494

Fifth, the study site in the Phanmun field is a part of the DMZ area, and access to the area is still restricted. As a result, it has not been studied before. Therefore, the area has a high value as a study site. In fact, agricultural wetlands and idle abandoned rice field wetlands have developed in many spots, and a variety of wildlife can be observed. This supports that this area is a very important habitat in terms of biodiversity.

Sixth, the study so far shows that inland wetlands account for 58 % of total the wetland types, which indicates that inland wetlands are well developed in Korea. One unique feature is that many wetlands have originated from agricultural land and idle rice fields.

In order to establish the national wetland policy, the extent of wetlands should be estimated first and then classified later. In other words, the delineation between each wetland should first be identified and then the diverse indicators, criteria and system of classification have to be defined based on these identified wetlands. The extent of wetlands in a country can be estimated differently depending on the methods employed for wetland identification, wetness and topography or position in the landscape.

From the perspective of 'wetness', the area where the land surface is under water all year round is easily identified because an understanding of the hydrologic process in the watershed is required.

Wetlands occur in areas where water naturally flows or accumulates, such as in the floodplain of streams and rivers, along smaller creeks and washes, in low spots and depressions, in flat or rolling landscapes, around seeps and springs, along the fringes of reservoir and lakes, and in coastal areas affected by tides. Therefore, the geomorphology helps a lot in recognizing wetlands.

In the USA, the wetland classification has mostly focused on the appearances of the wetland types rather than their functions. Recently, however, attention has shifted to emphasize the latter as wetland functions need to be evaluated in order to assess the impacts of proposed projects subject to government regulations (Tiner 1999).

The Ramsar and the NWI methods have been traditionally used. The HGM is the most recently developed method, but there are certain pros and cons in regards to its introduction on the national level in the USA. Consequently, the so-called hybrid method is being developed. As the Ramsar Convention has 148 contracting parties, and these countries use the methods based on Cowardian, the most common method used at the global level can be said to be the Ramsar method.

However, since the eighteen Ramsar congress was held in Uruguay, there has been a movement to classify the farming land in the areas under the monsoon climate influence as wetlands, and also the problem with the Ramsar classification has been identified: this system is centered too much on vegetation and soils.

As a result, the argument that the Ramsar method has to be improved and complemented through analysis of the strengths and weaknesses of the HGM method is persuasive. In the USA, for financial reasons, a one-parameter approach has been carried out for wetland identification and classification by using hydrology or topography. A four-parameter approach, on the other hand, can be carried out by using hydrology, topography, vegetation and soil, if there is enough financial support. This has a profound effect on wetland identification and classification, which can be applicable to Korea.

The wetland-type classification system described in this book introduces not only the HGM elements to evaluate the functions and values of wetlands, but also the vegetation sector of the NWI wetland classification system to develop a new hybrid-approach wetland classification system. This allows the understanding of wetland characteristics easily through wetland classification and evaluates the functions and values of wetlands. Therefore, it may be used in conserving and managing wetlands in the future.

The Ramsar Convention suggests surveying the biota (vegetation and animal group) in order to understand the bio-physical features when surveying wetlands. Wetland surveys by the NWI of the USA focus on vegetation, soil and hydrology. The hybrid approach addressed in this book conducts surveys of wetlands by focusing on topography, hydrology, soil, and vegetation. Additionally, animal groups are surveyed. When animal groups are surveyed, the habitats of animals living in wetlands may be understood. Then, this can be used in conserving and managing wetland animals. Therefore, further discussions are required on the necessary of animal group surveys in wetland surveys.

Asia as well as Korea natural and cultural environment difference compared to the Western countries, and, therefore, wetland types are also different. In particular, in the case of Asia, its culture is based on farming, and many wetlands originate from farmlands. However, Asia currently does not have an established wetland classification system or mapping methods and uses Ramsar wetland classification system. Therefore, the study described in this book may be a starting point for the development and evolution of wetland classification systems and mapping methods in Asia.

Through the Ramsar Convention, a total of 152 countries use the Ramsar wetland classification. Its basic framework is similar to the NWI (Cowardian) wetland classification system, and, naturally, the two systems share the same problems. In addition, evaluating the functions and values of wetlands based on simplified absolute criteria poses problems to wetland conservation and management. This is because the Ramsar wetland criteria for evaluating wetland functions and values basically consider wetlands while focusing on migratory birds. Therefore, an advanced wetland classification system should be developed by introducing the HGM elements.

Chapter 8
Biological Integrity: Assessing the Health of Wetland Ecosystems

Environmental agencies in Korea have searched for adequate methods to assess anthropogenic impacts on the environment. In this chapter, the conceptual paradigm of bio-assessment is addressed together with the wetland integrity assessment process. The results of the DMZ case study for assessing the health of wetland ecosystem are presented for a illustration purpose.

8.1 Concept and Objectives of Bio-assessment

Biological integrity is defined as a balanced, integrated, and adaptive community of organism having a species composition, diversity, and functional organization comparable to that of the natural habitat of the region (Karr and Dudley 1981). Several advanced countries including the USA and Australia have already adopted 'Integrity' as the assessment method for wetlands. As a healthy looking person can possibly be diagnosed with a serious disease, the same is true with wetlands. Although they might seem fine on the outside, it is only by thorough evaluation that the ecological and structural damage of the wetland can be specifically sorted out.

In structural aspects, the diversity of species can be the yardstick for assessing its integrity but it does not necessarily represent the health of the wetland. The accurate assessment lies on the ratio of annual and perennial plants, i.e., herbs and trees, exotic plants included. Biological integrity is determined upon whether the structure of ecosystem and its features are well balanced and applied appropriately.

In the functional point of view, the health of the wetland can be assessed upon the consideration of productivity and circulation rate of nutrients (influx/outflow of the nutrients).

The wetland integrity can be used as a core value in assessing the quality and value of a wetland ecosystem. By adopting such evaluation concept, the degree of damage made to the wetland could be grasped and on then, the most appropriate method for protection and management could be established.

Many investigators have tested the reliability of the indices of biological integrity that are available in the existing literature and abstracts to assess the

K.-G. Kim, *The Demilitarized Zone (DMZ) of Korea*,
DOI: 10.1007/978-3-642-38463-9_8, © Springer-Verlag Berlin Heidelberg 2013

extent of environmental degradation in wetlands including rivers and lakes. Nevertheless, scientists and scholars continue to discover new indicators befitting the specific environments they investigate. This means that nature will always find a way to adapt though the outcomes or contents of the ecosystem may appear more or less the same. For example, landscapes are shaped by nature and the contents of the ecosystems are sometimes peculiar to the other though they are exposed to the same conditions. Though learning from surveys of the abstracts may guide researchers in their quest to assess biological integrity, a continuous development of qualifiers and indicators for specific targeted environment is preferable.

The ultimate objective of the assessment is to diagnose the health status of the wetland ecosystem as a basis for planning appropriate strategies and the methodologies in preserving and restoring the ecosystem.

Finally, the scores will be ranked. The ordinal scale of the ranking will be classified into Very Healthy, Healthy, Fair, and Poor. Very Healthy and Healthy will be applied to conservative zones, Fair to restorable zones while Poor will be applied to wetlands with no sustainability in its ecosystem.

The output of this evaluation shall provide data-based decision making in selecting preservation methods and strategies for the wetlands that are still in good conditions and restoration or quality improvement methods for the wetlands with poor conditions.

8.2 Wetland Integrity Assessment Process

A research procedure was agreed among the team of specialists led by Dr. Kwi-Gon Kim specifying the series of activities to be conducted in the target areas within their expected timeframe. In each specific activity, the responsibility assignment is respectively clarified. This process is done in order to standardize the procedures and synchronize activities of the teams. The process implemented for the wetland integrity assessment is then itemized in a matrix presented in Table 8.1.

8.3 Methodology and Design for Assessing the Health of DMZ

The stages described in Table 8.1 are discussed in detail.

Stage 1. Standardization of assessment criteria

The research team has developed the major ecological factors, assessment indicators, and the criteria of wetland integrity based on relevant literatures, case studies, and research abstracts. In this assessment, scores from 1 to 3 were assigned for each assessment indicator and the detailed steps of stage 2 are presented as follows.

Table 8.1 The process of applying index and criteria of wetland integrity assessment to case study areas

Assessment process			
Stage 1	Developing the criteria for evaluation index and major ecological factor of wetland integrity assessment		
Stage 2	Application of developing index and criteria of wetland integrity assessment	Site selection	Select the site that is believed to be having the characteristics of a wetland
		Survey and analysis on the present condition	Survey and analysis on the present condition of ecological environment, species, wetland habitat, and disturbing factors
		Application of criteria per index on each site	Implementation of the GIS analysis and overlay of the established items upon the application of the criteria
		Calculation of evaluated scores by summing up the scores of each index	Preparation of the general chart via superimposition of the suitability index of each item
		Ranking upon the score	Ranking upon the score. The scale of the ranking will be classified into Very Healthy, Healthy, Fair, and Poor. Very healthy and Healthy will be applied to conservative zones, while Fair to restorable zones and Poor to wetland with no sustainability in its ecosystem will be applied

Stage 2. Site selection, survey and analysis

- Site selection

Environmental scanning of the target area was conducted to identify specific sites for the research locale. The sites identified as having the value of a wetland are then demarcated for scrutiny. In this study, a total of 12 sites are selected and illustrated in Fig. 8.1.

- Site survey and analysis

The survey and analysis on the present conditions of the identified sites are carried out specifically scrutinizing for the ecological infra features, biota. In this study, ecological features refer to the topography, hydrology, soil, and watershed. Biota, on the other hand, refers to the mammals, birds, reptiles, and amphibians, insects, fish, and invertebrates in freshwater. Further, respective specialists are assigned to develop into evaluating more complicated conditions of the wetland which includes; the wetland habitat, the existing land use and land holdings, and

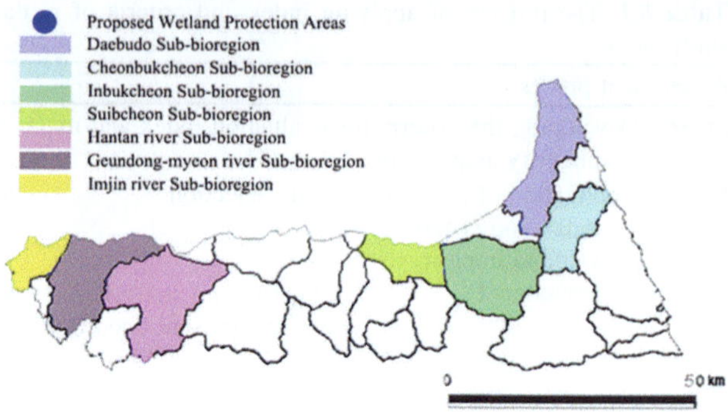

Fig. 8.1 Location map of the target areas for wetland integrity assessment

equally important is the assessment of the disturbing factors. The items of survey and analysis and their methods used for the DMZ case study are presented in Table 8.2.

Stage 3. Application of criteria for each indicator

The Grid Cell Analysis of Arc View 3.2 is the standard utilized in analyzing the specific indicators attributed to the integrity of the wetland.

Stage 4. Scoring and nominal scaling of values

In calculating the scores as a result of evaluation, the summation of all the nominal values per indicator is computed at a micro level. To get the overall score for the specific site, the computed values per indicator are again summed up. Then the general chart was drawn to view the overall status of the wetland integrity at a macro level.

In order to nominally rank the summation of scores, the standard conversion of the summated scores is then scaled according to a range of 0.0–1.0. The sum of the evaluated scores per criteria is divided by the sum of the maximum value per standard. The process is illustrated in a map shown in Fig. 8.2.

Table 8.2 Item of survey and analysis and their methods

Item of survey and analysis	Method of survey
Survey on currently presenting vegetation in wetland. Survey on distribution of exotic plants	Survey of vegetation
Analysis on hydrology	Using the existing digital map
Survey on species inhabiting	On-site survey per species by group
Survey on the present condition of legally protected species per group	On-site survey per species by group
Analysis on the landscape pattern of wetland habitat	Survey on landscape pattern via site survey
Analysis on the damaged area of wetland habitat	Survey on existing land use via site survey

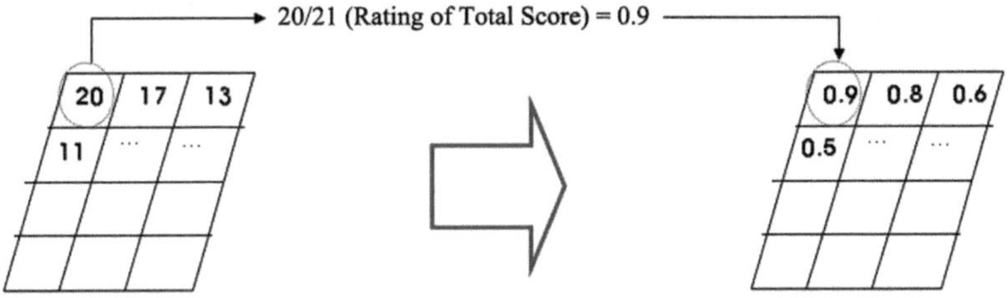

Fig. 8.2 Calculating the evaluation score

Score	Ranking	Legend (color)
1–0.8	Very Health	
0.8–0.6	Health	
0.6–0.3	Fair	
0.3–0	Poor	

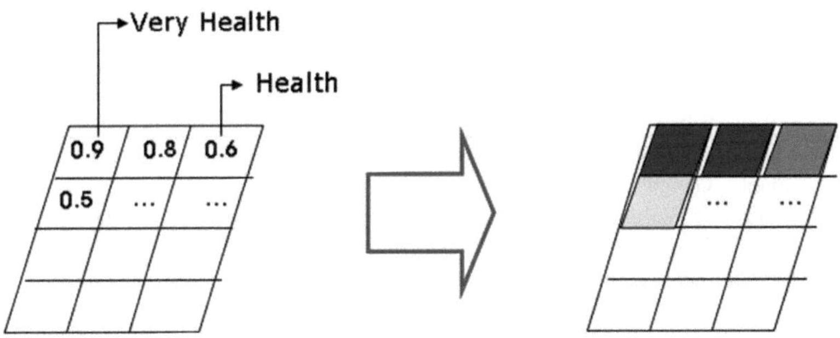

Fig. 8.3 Process for ranking

Stage 5. Ranking

An ordinal scale is a qualitative measurement of the indicators ranked accordingly from highest to lowest quality or vice versa. The computed scores in stage 4 are interpreted accordingly to the qualitative ordinal scale indicated here. In this case study, the ordinal scaling and its equivalent nominal scale of values are presented and color coded in Fig. 8.3.

8.4 Case Presentation

The methods reviewed earlier regarding the integrity of the wetlands was used for the DMZ study. The results of the application for the DMZ are described in this section.

8.4.1 Site Selection

There are 12 wetland sites designated as protected areas by the Ministry of Environment located near Gangwon province and, accordingly were selected as the target areas for the research. These areas are within the CCZ which are regarded to have biological diversity.

In this study, the location, size, characteristics, wetland type, and ecosystem of the sites were investigated. The location and the size of the researched sites are shown in Table 8.3.

Then the habitat and species were analyzed and the threats per target area were identified. This is for the purpose of identifying the most preferred three wetlands

Table 8.3 Location and size area of researched sites

Site No.	Location	Area (km^2)	Remarks
1	Wetland near unification observatory of Hyunnae-myeon, Goseong county, Gangwon province	0.12	CCZ
2	Wetland near Woljeong-ri station of Cheorwon-eup, Cheorwon county, Gangwon province	2.01	CCZ
7	Cheorwon plain wetland near Hantan river of Galmal-eup, Cheorwon county, Gangwon province	1.87	CCZ
8	Baekma Highland forest valley wetland of Galmal-eup/Dongsong-eup, Cheorwon county, Gangwon province	0.81	CCZ
9	Namdaecheon wetland of Kimhwa-eup/Galmal-eup, Cheorwon county, Gangwon province	1.47	CCZ
10	Mt. Gyewoong forest valley wetland of Geundong-myeon, Kimhwa-eup, Cheorwon county, Gangwon province	1.55	CCZ
11	Wetland near peace dam of Hwacheon-eup, Hwacheon county, Gangwon province	2.93	CCZ
12	Wetland near punch ball of Haean-myeon, Yanggu county, Gangwon province	1.16	CCZ
13	Eulji observatory forest valley wetland near punch ball of Haean-myeon, Yanggu county, Gangwon province	0.03	CCZ
14	Stream valley wetland of Sihwa-myeon, Inje county, Gangwon province	0.37	CCZ
15	Hyangnobong forest valley wetland of Sudong-myeon, Goseong county, Gangwon province	0.47	CCZ
16	Bukcheon downstream wetland of Gansung-eup, Goseong county, Gangwon province	1.32	CCZ

with the highest ecological value. With the end in mind, that the management scope and the boundaries of three sites were established, as well as the methods to be utilized were identified.

8.4.2 The Process

To establish the scope and boundary, and to identify the appropriate method to be used in management of the preferred three target sites, the first was to identify ecological factors for zoning. Then thorough survey and analysis on the major ecological factors was the second step. The third step was the establishment of zones within the site based on analysis. The zones are categorized according to the following; core zone, buffer zone, transition zone, and restoration zone, which were drawn from the results of the investigation. The fourth step was the establishment of the concrete boundaries and scope of management, including the identification of appropriate methods per specific type of wetland. Figure 8.4 shows the overall process of the research.

8.4.3 Results

The results of applying the process to the selected sites shown in Fig. 8.5 are described in this section.

A series of maps were developed to illustrate the present status of the targeted wetlands by factor. The thematic maps drawn are based on the assessment of each

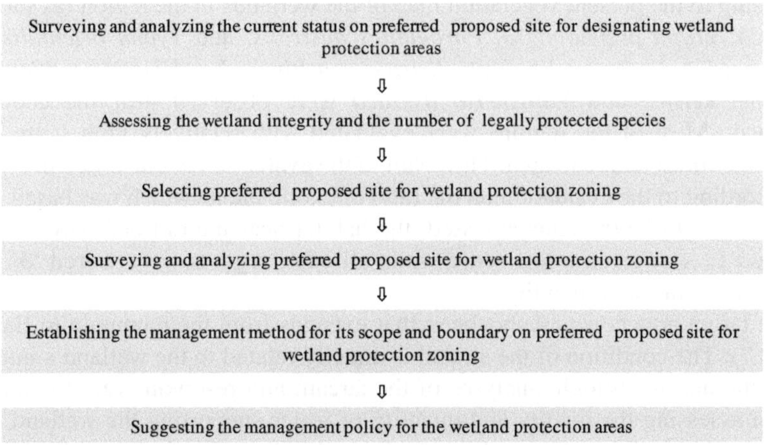

Fig. 8.4 Overall process of the case study

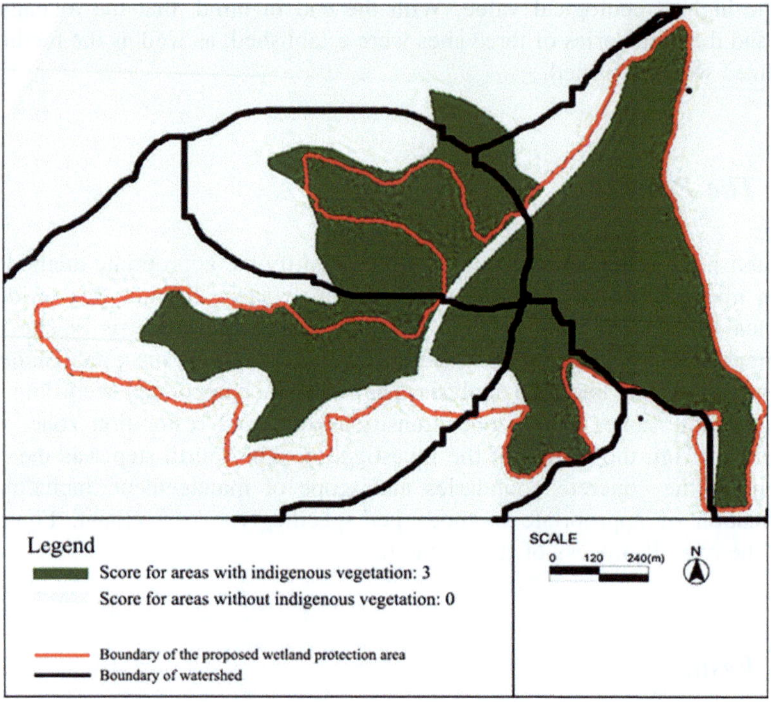

Fig. 8.5 Evaluation on abundance of indigenous species

factor. The assessment was conducted according to the criteria on the diversity of species and its value by watershed within the DMZ.

Vegetation

According to the present vegetation map of the wetlands in the region (as shown in Fig. 8.5), *Trapa pseudoincisa*, *Phragmites australis,* and *Typha orientalis* were found near Anho reservoir while *Scirpus maritimus* L., *Alisma canaliculatum*, *Aneilema keisak,* and *Vallisneria asiatica* were observed near the ecological waterway. Most of the regions were evaluated with relatively high score of '3' except the forest and road area. The result of the evaluation is illustrated in Fig. 8.6.

According to the evaluation on the ratio of exotic plants which was based on the distribution of exotic plants assessed, the habitat near the railroad with *Amorpha fruticosa* L. was given a low score while all other regions were scored '3'.

Water quality and quantity

The target area covers Anhocheon that expands from the mountain to East Sea (Fig. 8.7). The condition of the stream is directly related to the wetland's integrity. Hydraulic and hydrologic analyses of the stream and reservoir were the methods used in assessing the quality and quantity of water supporting the wetland.

The region is located near the seaside so that it shows features of a brackish ecosystem unlike the streams flowing in the inland areas. Both the midstream and

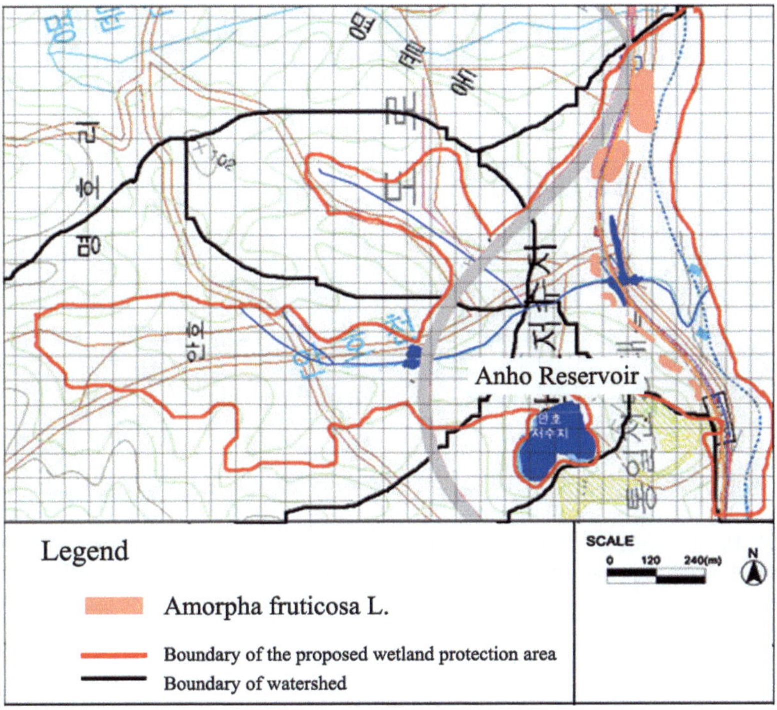

Fig. 8.6 Distribution of exotic plants

downstream of Anhocheon and Anho reservoir received the score of '3' and regions located farther than these two received lower scores. Distance, therefore form the water sources, affects the status of the wetlands.

Biodiversity

There were 19 species of plants, 11 species of mammals, 39 kinds of birds, 5 types of fish, 6 species of reptiles and amphibians, 178 species of terrestrial insects, and 14 species of invertebrate in fresh water that were observed in Anho reservoir. The type and number of species are significant indicators for the diversity and density within a habitat. Density of species is found to be directly related to the value of the region as habitat. Most of the inhabiting species are found in and around of the wetland. They are the major indicators for assessing the health of the wetland. The score of '3' achieved by the region was primarily, based on the integrity of the 272 species observed in the region.

Extent of damage

The wetland habitat of the target region including Anho reservoir is comprised of a buffer zone and atypical matrix. The damaged land in the region is accounted to

Fig. 8.7 Habitat corridor of the Anho reservoir area; the landscape consists of mosaic of reservoir, waterway, pond, alluvial valley wetland, sand dune, estuarine and coastal wetland, and forests

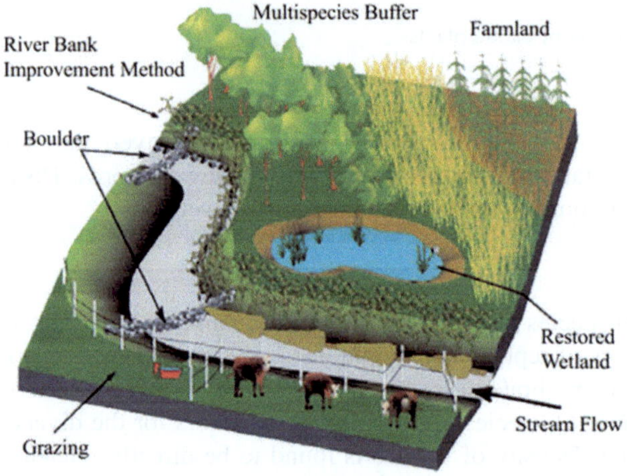

Fig. 8.8 Mimic of buffer vegetation zone at the edge of water: an illustration

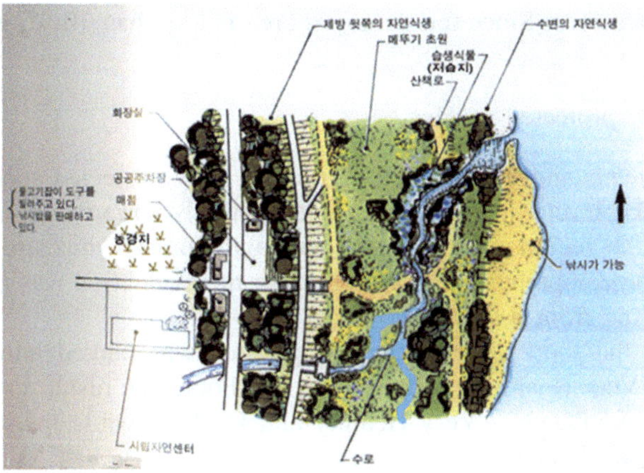

Fig. 8.9 Establishment of river water edge vegetation zone for the enhancement of biological diversity

Fig. 8.10 Mimmic of multiplex buffer zone

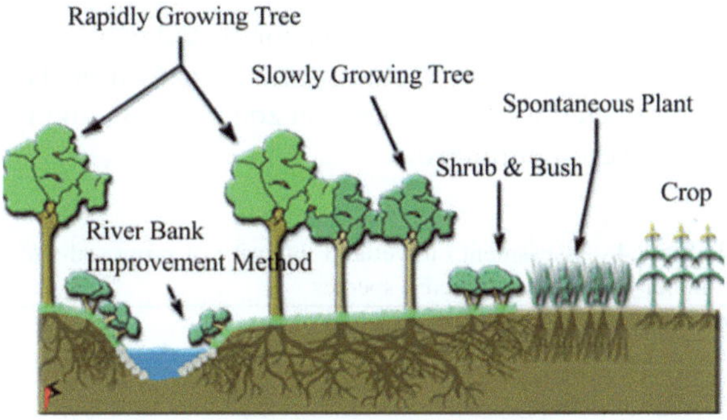

Fig. 8.11 Stabilization of stream applying the biological method of shore protection

4.9 % of the total area. Since the damaged area is less than 10 %, it received the score of '3'.

Status of legally protected species

Aegypius monachus and Eurasian buzzard are legally protected species observed near Anho reservoir and they are covering a relatively large radius in this specific habitat. Thus, the region with *A. monachus* and Eurasian buzzard received the score of '3'. The comprehensive score chart for the evaluation was produced with the scores ranging from 0 to 21.

Based upon the scores, the wetland ecosystem was ranked into four ordinal scales namely; Very Healthy, Healthy, Fair, and Poor. The result revealed that the wetlands were evaluated as Very Healthy and Healthy, while the inland regions and roads were evaluated Fair.

8.4.3.1 Wetland Health Assessment and Assessment by Number of Protected Species

The ranking resultant from the conducted evaluation on wetland integrity and the number of legally protected species is shown on Table 8.4. For the ranking of the final selected wetland protection zone, the ranking made in the previous valuations above were rearranged.

Table 8.4 Assessment on wetland integrity of proposed site of wetland protection zone and the number of legally protected species

Proposed site of wetland protected zone	Area ratio of wetland in good condition (%)	Number of legally protected species
Site 1. Anho reservoir near unification observatory in Goseong county	77.9	8
Site 2. Wetland near Woljeong-ri station of Cheorwon county	11.1	5
Site 7. Cheorwon plain wetland near Hantan river in Cheorwon county	6.2	1
Site 8. Wetland near Baekma Highland of Cheorwon county	85.3	5
Site 9. Wetland near Namdaecheon of Cheorwon county	60.3	6
Site 10. Wetland near Mt. Gyewoong of Cheorwon county	88.8	13
Site 11. Wetland near peace dam in Hwacheon-county	95.6	10
Site 12. Wetland near punch ball in Yanggu county	6.3	2
Site 13. Wetland near Eulji observatory forest valley in Yanggu county	2.5	2
Site 14. Wetland near Inbukcheon valley in Inje county	39.7	6
Site 15. Wetland near Hyangnobong forest valley in Goseong county	11.4	2
Site 16. Wetland near Bukcheon downstream in Goseong county	40.4	4

Table 8.5 The preferred proposed site of wetland protection area

Site 11. Around the wetland of peace dam in Hwacheon county
Site 10. Around the wetland of Mt. Gyewoong in Cheorwon county
Site 1. Around the Anho reservoir of unification observatory in Goseong-county

8.4.3.2 Proposed Site of Wetland Protection Zone

Based on the results of the assessment and evaluation carried out by the research team on wetland integrity as well as the results of the survey on the number of legally protected species, this study came up with three wetland protection areas as shown in Table 8.5.

8.4.3.3 Detailed Survey and Analysis of Proposed Site of Wetland Protection Area

Site 1, out of three proposed sites has a very small area of farmland since most of the region is mountainous. Only a small part of the south is occupied by civilians. The north of CCZ is a military controlled zone and the west is a mountainous area, and to the east is the eastern sea. With such geographic features, several types of wetlands are developed, namely, inland, estuarine, and coastal wetlands.

Among the endangered species found in these wetlands are *Peregrine Falcon*, Haliaeetus *A. monachus. A. monachus*, *Lethenteron camtschaticum,* and *Pungitius sinensis*. Among the species designated as natural monuments and heritage are *Aix galericulata*, *Falco tinnunculus* and the *L. camtschaticum* of the IUCN Red Data. The species which should be approved by the Ministry of Environment for taking out of the country is the Apatura ilia are also observed with high density in this site.

The target area is surrounded by high mountains but inside, it has low altitude of 100 m with plain, basin, and valley. It is often regarded as having perfect conditions for the formation of wetlands. The Anhocheon is flowing out of the mountains is connected to the ocean. Ponds are formed along its waterways. Moreover, the ecological waterway is artificially created and all these features are having a significant influence on the formation of the wetland. The sand dune land connected to the ocean is sandy loam soil, while the valley and alluvial wetland has an organic soil in which *P. australis* and Salix koreensis Andersson have grown pretty well.

The habitat of the region is mostly comprised of wetlands and further appears in various other formations such as; water way, reservoir, pond, alluvial valley, sand dune, and beach wetland. The shrub and bush in this area provide an advantageous habitat for various kinds of species as well. Furthermore, habitats and species of this region can be organized according to the following:

In the inner circle of the railroad, the community of *P. australis* and Alder tree is found while the number of *A. fruticosa* L. increases approaching the railroads. Near the Anho reservoir, *T. pseudoincisa*, *P. australis,* and *T. orientalis* are observed, while *S. maritimus* L., *A. canaliculatum*, *A. keisak,* and *V. asiatica* are found near the ecological waterway. For mammals, traces of 11 species are detected including *Mogera wogura*, *Nyctereutes procyonoides*, *Mustela sibirica* manchurica, *Prionailurus bengalensis*, cat, *Sus scrofa*, *Hydropotes inermis*, Korean Hare, Eurasian Red Squirrel, Squirrel, and black-striped field mouse. The habitat harmonized with wetland and mountain areas which provided a perfect place for birds to live. Hence, endangered species Eurasian Buzzard and *A. monachus* and the natural monument *A. galericulata* are observed in the area. A total of six reptiles and amphibian species are found in the region that includes Asian toad, *Hyla japonica*, Black-spotted pond frog, *Glandirana rugosa*, Korean fire-bellied toad, and Red-Sided Water Snake. However, the only fish found in the Anho reservoir during the investigation is the floating goby, and worse, it was small in size. It is assumed that the isolation of the reservoir from other water systems caused the inbreeding and the genetic defect is expressed in the sizes of the fish. Nevertheless, a relatively large quantity of freshwater invertebrates are detected in Anho reservoir unlike streams of other basins and this is due to the imbalance in the food chain where there are numerous prey for the existing primary consumer but the predator group is minimal.

This region was a village in the past but after the division of Korea, the access has been denied so it is currently maintained as a natural wetland. Most of the region is comprised of wetlands and mountains but approximately 86 % of the total area is private lands. Recently, facilities such as the unification observatory and the railroads are filled in and more facilities are planned to be developed. The region is in direct need for ecological management for possible designation of the area as a Ramsar Site in the nearest future.

8.4.3.4 The Result of Establishment of Management Scope, Boundary and Practices on Preferred Proposed Sites of Wetland Protection Area

The establishment of management scope and detailed zoning classification of core, buffer, transition, and restoration zone was made based on the information described in the previous section.

The scope of management is classified into two, namely, the wetland protective zone and the wetland fringe area. In the wetland conservation statute, the following are designated; the core zone, the buffer zone (including restoration zone) as wetland protection zone, and the transition zone (including restoration zone) as wetland fringe area.

8.4.3.5 Wetland Protection Area Management

Establishment of the management practices

The management practices should be established with the consideration of species inhabiting in the region. Moreover, there is a need to develop measures for conservation and management, especially in taking linear ecosystem as a core zone which is connecting the inland ecosystem of Anho stream and coastal ecosystem of Donghae. In addition, the target area must have inland, estuarine, and coastal wetland that can be managed in an integrated form of management for the watershed as a whole. Further, proper management practice should be undertaken for inland plants such as *A. fruticosa* L. as their numbers within the wetlands are increasing ever since the development of railroads and roads.

Management practices for wetland protection zone by type

The types of wetland are divided into riverine, lacustrine, and palustrine wetland. The suggested management practices for each type are presented below.

Riverine wetlands need the management of its water quality and biological diversity.

For water quality protection, the establishment of water purification wetlands will be required in order to remove the nonpoint pollution sources and minimize the influence from surroundings. In addition, the creation of riparian buffer forest is highly recommended.

For the promotion of biological diversity, the damaged stream should be transformed into a natural stream and an integrated watershed-based management approach is encouraged. Also, in order to provide an advantageous habitat for various kinds of species, multi-species buffers should be installed and should consider adopting biological methods for shore protection (Figs. 8.8, 8.9, 8.10 and 8.11).

Lacustrine wetlands should maintain the water cycle and the stability of water levels. Water quality must be monitored regularly in order to control and prevent eutrophication.

Palustrine wetlands need to establish a database of the wetland for systematic management and continuous monitoring. The sincere recognition of the importance of management information system (MIS) and the development of self-evaluation program are recommended. Standardization of wetland usage should be advocated and establishment of policies and institutional strategies is required. These are the core principles of the effective management of the palustrine wetland.

At a national/state level and local levels, guidelines for the management of wetlands should be set up and made accessible to the public. Continuous study on the wetland should be conducted periodically, and the network with research institutions should be strengthened.

8.4.3.6 Considerations for Livelihood for the Residents in the Target Area

Environmental issues are definitely community issues and the effective management of the environment can only be possible with the active participation of the locals and the many publics it served. Therefore, people in the community should be empowered to perform their ethico-legal responsibilities as stewards of the environment.

First, as a non-regulatory approach, economical incentives should be considered by the government such as; promoting tax cut or exemptions, establishment of economical foothold and, convenience businesses as a means of enhancing the cooperative relationship with the local residents.

Second, it is highly recommended to involve the local residents and landowners in the development of restoration projects to promote and enhance their sense of ownership of the program. This is also a perfect venue for sharing community-based strategies, settling differences, and for information dissemination, creating a win–win situation. People who are directly in touch with the environment have a very profound knowledge of the environment, especially, the knowledge handed down by their ancestors. Their knowledge too cannot be underestimated. Implementing community-based technologies had proven to be a reliable indicator for sustainability. It is, wherefore, recommended that the cooperation of the nearby residents and landlords should be relentlessly initiated, in creating various types of wetland habitats, restoration of the connectivity of the wetlands to water edges and forests, and establishment of parks among other tasks.

Third, control on the use of chemical fertilizers and agricultural pesticides should be strictly imposed. Measures should be implemented to prevent soil erosion caused by cultivation. The propagation of the environment friendly farming and management of wetland on the watershed level are recommended as well.

Fourth, to some extent, the government should purchase the habitats around the wetland so that as a public land, the control and management of the habitats would be easier. Implementers of the government can use it as the space for connecting other habitats and the restoration and preservation management of the wetland can be made upon the reference wetland.

Fifth, the adoption of civilian wetland monitoring system. Especially, when authority is given to the local residents, the government can empower them to directly manage the wetland. This approach will boost their ego and implant in them the sense of ownership of the restoration/conservation project. Also, by considering the opinion of the local residents, their participation in the project is induced and they themselves will seek for ways to activate the management and monitoring system of the wetlands.

The output of the evaluation discussed in this chapter shall provide data-based decision making in selecting methods and strategies for conservation and management of the wetlands that are still in good conditions and for restoration or quality improvements for the wetlands with poor conditions.

Chapter 9
Threatening Factors in the DMZ Area

This book is again another attempt to convince all nations to unite, to do whatever they can to save the environment, and save the world they live in. This book speaks of landscapes, ecology, biodiversity and biologic health. All that has been learned from the facts presented through research revelations, the bottom line is, DMZ is "crying for help."

But coming to think of it, is it really the world's environment like DMZ that needs help? For crying out loud, IT IS THE HUMAN SPECIES THAT IS CRYING FOR HELP. For several decades now, the world is threatened by the devastating effects of climate change. People all over the world are aware that this is the result of the irresponsible disregard of humans for the very environment that sustained them. Yet, little has been done to revive and sustain it.

There are only few ecosystems in the world that has been sustained. They function as buffers to global warming and depleted oxygen levels on earth's atmosphere. Nevertheless, it is not enough to prevent climate change. Proof to that are the super typhoons and tsunamis that affects even the countries that never experienced such in several lifetimes. But then again, the worst is yet to come.

This chapter describes some lessons learned from the DMZ experiences, focusing on threatening factors in the DMZ area.

9.1 Homo Sapiens, an Endangered Species

Mother earth has gone through with the ice age and all the eras when no humans could possibly survive. Yet, nature will always find a way to heal itself because it can wait. But definitely, humans cannot, or else end to extinction just like the poor mammoth. Though heavily populated, the lifespan of the human species is decreasing in recent years. At the advent of the emerging incurable and dreadful diseases, not to mention the global calamities, the Homo sapiens are actually an endangered species.

Throughout history, ancestors taught succeeding generations to live in harmony with nature. It fell on deaf ears. Now, humans should learn to find a way to heal

K.-G. Kim, *The Demilitarized Zone (DMZ) of Korea*,
DOI: 10.1007/978-3-642-38463-9_9,

themselves. The only way to do that is to revive and preserve the earth's ecology that sustains biodiversity. Biodiversity therefore, should not be just viewed as the existence of the species of birds and the bees, the flowers and the trees, but come to think of it, it is also the existence of the Homo sapiens. And that is a fact. Then heed to the call of nature.

9.2 SWOT Analysis of the DMZ Ecosystem

Korea is one of the few countries that is blessed with an immediate environment that is still healthy and magnanimous in providing hope for the world. The DMZ is in fact hope for the world, as another buffer system to diffuse global warming and climate change, a wholesome refugeof the regular and endangered species. DMZ is worth preserving. A SWOT analysis is presented below.

9.2.1 Strengths

- Unique evolving process of natural and semi-natural ecosystem
- Strict control of human disturbance
- Increased international interest in the untamed ecosystem
- Magnificent landscapes and wholesome scenic views
- Very spiritual and therapeutic atmosphere
- Government programs providing incentives to local residents who are involved in bird feeding
- Presence of experts, specialists, and scientists
- Active involvement of universities and other educational institutions
- Increased interest of local and international tourists.

9.2.2 Weaknesses

- Invasion of exotic species
- High vulnerability to forest fire
- Fragmentation of habitat by iron fences
- Separate governance: division between North and South Korea
- Presence of private lands
- Human resettlement in some areas
- Limited cooperation and involvement from human settlers
- Modern developments: railroads, roads, coastal business establishments, artificial banks, etc. that initiate human activities.

9.2.3 Opportunities

- Securing a range of environmental enhancement
- Educational and ecotourism programs
- Human and Ecological Security: opportunities for the cooperation of international agencies symbolizing world peace and coexistence of man and nature
- Feelers from the North on their joint cooperation for the conservation and protection initiatives of the South, an opportunity for peace between the two countries
- Increased economic cooperation between North and South Korea
- Opportunity for the furtherance and continuous research/investigation on the status of the ecology, habitats, and the biodiversity of DMZ
- Opportunities for international agencies symbolizing world peace and coexistence of human beings and nature.

9.2.4 Threats

- Multilayered regulation for national security
- Land mines spread over the DMZ
- Stagnation of regional economy due to unplanned land use
- Continued use of plastics and other non-biodegradable materials
- Soil erosion due to cultivation of farmlands
- Devastating effects of climate change.

9.3 Development Pressure

After the Inter-Korean summit held in 15 June 2000, the economic cooperation between North and South Korea increased. As a result, interest on the development of DMZ surged up. From then on, series of projects have been carried out till today including; the construction of Gyeonggi line, Donghae line, North–South connecting roads, CIQ, distribution center, dam for controlling flood, industrial complex, farming lands, and many more as planned.

However, most of the projects are biased to industrialization projects rather than to preservation and restoration projects. The constructions alone demolished some naturally significant areas of DMZ, disturbed the surrounding ecosystems and even introduced unwanted and destructive species. Worse, it is even expected to turn wetlands into the land to pave the way for developments, disturbing natural hydrology, and connectivity of habitats. Take note: "when water is lost, nature will eventually die; a natural death".

The ecosystems of DMZ are diverse and they are worth to be preserved. But recently, they are exposed to lots of threats for development. Indeed, the development in the DMZ is a significant step in bridging cooperation between North and South Korea. In as much as the intension is true enough, the means may not justify the ends. It is also true that the development is causing serious damage to the natural environment.

Recently, the construction plan for Gyeongwon line has been released. The strong apprehension is on the secondary contamination it may cause the environment as it is already being widely used in route to Pyongyang and Mt. Geumgang. The major problem of the construction of the railroad and road is that it is disconnecting ecosystems that had been maintained for 60 years. Constructions like this must seriously consider the impact it may cause to the surrounding environment and must have an accompanying plan for the restoration. What North and South Korea needed is "Responsible Development" that heeds to the global call for the preservation of the environment. Figures 9.1 and 9.2 present the damaged habitat due to the construction of Gyeonggi, Donghae line and North–South connecting road. It is an example of the destruction of areas that are of high value for existing habitats.

Fig. 9.1 Destruction of habitat due to the construction of Gyeonggi line

Fig. 9.2 Destruction of habitat due to the construction of North–South connecting road

On the other hand, the development of the Gaesung industrial complex raised new problems on the main water resources of DMZ including the downstream of Imjin river and Han river causing the water pollution in the area.

9.3.1 Threats by Farming and Fishing

DMZ and CCZ have huge wetlands with significant ecological value to the revival of nature. As surface and underground waters are connected in the wetlands, the development in one area has a direct impact on the ecosystems of the surrounding areas.

Threats have attributed to agricultural and fishing activities. The access to most of DMZ area is denied to the civilians due to military and political conditions. However, there are areas that are opened including the Daesungdong that is used to rice and ginseng farming. Such cultivation activities damage and disturb the ecosystem. It is not only causing the disappearance of habitat but also secondary damages including; diversion of waterways to irrigate farms, water pollution due to the agricultural pesticides, creation of farm roads to market, and noise pollution caused by agricultural machineries. For details, please refer to Figs. 9.3 and 9.4.

Daesungdong area is suitable for farming as reservoirs are located everywhere. However, the cemented dams are causing serious consequences to the survival of creatures living around the Daesungdong area. Moreover, the agricultural pesticides and chemical fertilizer used for farming contaminated the soil and the surrounding rivers. For details, please refer to Fig. 9.5.

Moreover, fish boxes made of nets are permanently left in the streams and reservoirs of Panmunbeol. This is not only disturbing the movement of those fish but also fatal, as evidenced by the rotten fish observed inside the fish boxes. The

Fig. 9.3 Cultivated farm land in Panmunbeol

Fig. 9.4 Ginseng field in Panmunbeol

Fig. 9.5 Waste bottles of agricultural pesticides and chemical fertilizers found in riversides

decaying fish add to the pollution of the waters. For details, please refer to Figs. 9.6 and 9.7.

9.3.2 Threats to Military Activities

DMZ is a special zone where North and South Korea are vigilantly on guard with each other. Military facilities are being constructed (Figs. 9.8 and 9.9) and various kinds of military activities are performed in the place. The military regularly clear the area by burning the surrounding vegetation for the purpose of easy visibility of anticipated invasion. This exercise aggravated the already devastating bush fires that beset the area. When fire is started, it can spread wherever the wind blows and

Fig. 9.6 Breakwater of Daesungdong reservoir in concrete

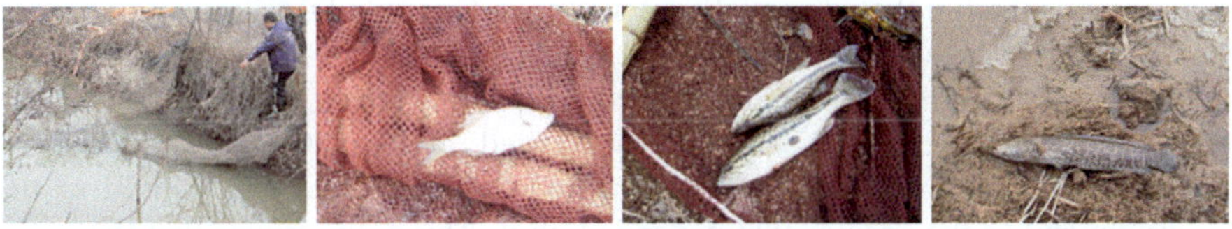

(a) (b) (c) (d)

Fig. 9.7 Damage on fish due to the fishing activity (**a**) Fishing net installed in stream. (**b**) Dead Acheilognathus yamatsutae in fishing net. (**c**) Bass: Introduced species. (**d**) Dead Northern snake head due to the inflow of concrete toxic substances to the stream

Fig. 9.8 Forest wetland damaged by the installation of military facilities

Fig. 9.9 Forest damaged by the installation of military facilities

the extent of damage is beyond the intended area to be cleared. Along with the trail of the fires, all living things die. But the greatest problem is the eventual disconnection of the species that are interdependent within the established natural habitat of the DMZ.

9.3.3 Forest Fire and Natural Disaster

DMZ is also not free from natural disasters including bush and forest fires. Worth mentioning is the fire that is annually occurring in Goseung DMZ causing so much damage to large areas of forestlands and wetlands Fig. 9.10.

It was 23 Feb 2009 when fire enveloped Chopyeong island caused by one of the regular military exercises. The fire swallowed almost 20 % of total about 1,600,000 m^2 of area. One can only imagine the damage it had caused to the ecosystem and to the species that depended on it. This fully proves that the range of damage of such military activities can be wider than expected. Figures 9.11 and 9.12 present the actual fire that occurred on that day.

Fig. 9.10 The burning ecosystem of Chopyeong island wetland

Fig. 9.11 The burning ecosystem of Chopyeong island wetland

Fig. 9.12 Damaged habitat due to the forest fire

While most of the forest fires are man-made, the damage caused by flood and hail is regarded as serious natural threats. Therefore, urgent restoration work is required in this area.

Access to Chopyeong island is denied since 1970s because of the presence of land mines. This island is located in the inner part of the civilian controlled zone of Jangsan-ri, Munsan-eup, and Paju city. It harbors a river sediment wetland that has been preserved and its natural features remained untouched. Having such high values of ecological preservation, it was designated as the proposed site for the creation of the Peace ecology park.

A map of the habitat at Chopyeong island is presented in Figs. 9.13 and in 9.14, showing the cross-sectional diagram of Chopyeong island's edge that was damaged by the fire. In the pictures, wetland plants such as *Miscanthus sacchariflorus* Benth.,*Phragmites australis*, *Salix gracilistyla* Miq., Populus deltoides MARSH, Alder tree, and Acer ginnala are observed (Figs. 9.15 and 9.16).

Mud and shallow waters covered the edges of the island where gray heron, Northern Lapwing, duck, and wild goose were found, and Haliaeetus albicilla, and Eurasian Buzzard were noticed hovering in the skies. The natural monument 202-Grus japonensis and 203-white-naped crane is detected by the local residents besides from cormorant, owl, Cygnus columbianus jankowskii ALPHERAKY, Eurasian tawny owl, Aix galericulata, black-crowned night heron, oystercatcher, and owl. The island depicts a true paradise of birds. Moreover, various kinds of rare species are also found in this island including; cormorant, Falco tinnunculus, and Aegypius monachus. In the streamlets of the island, both snakes and rat snake are observed. It is also a perfect habitat for pelophylax chosenicus. Cries of narrow-mouth frog were heard by the researchers last summer.

Upon site survey conducted right after the fire occurred, scattered droppings of Hydropotes inermis were detected. Actually they were observed on the spot during the survey. Proofs of the inhabitation of the Sus scrofa were seen during the hearing survey.

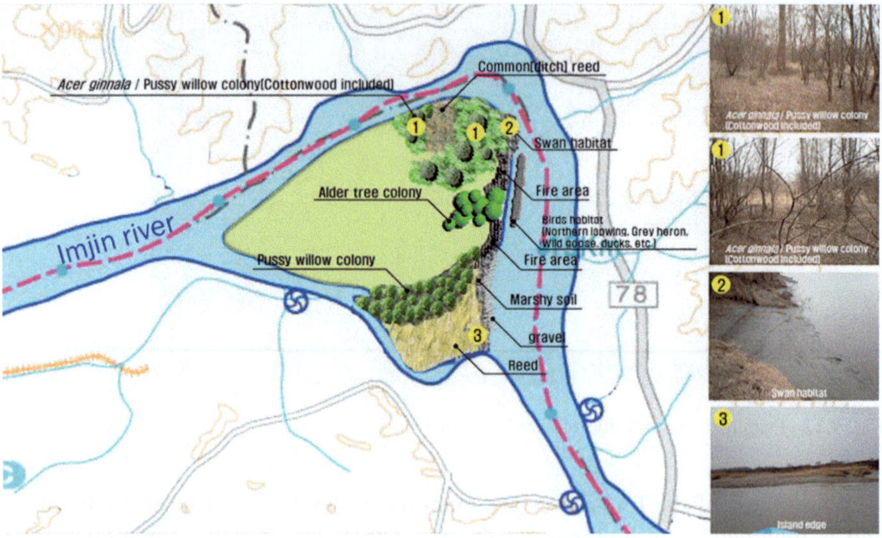

Fig. 9.13 Habitat map of Chopyeong island

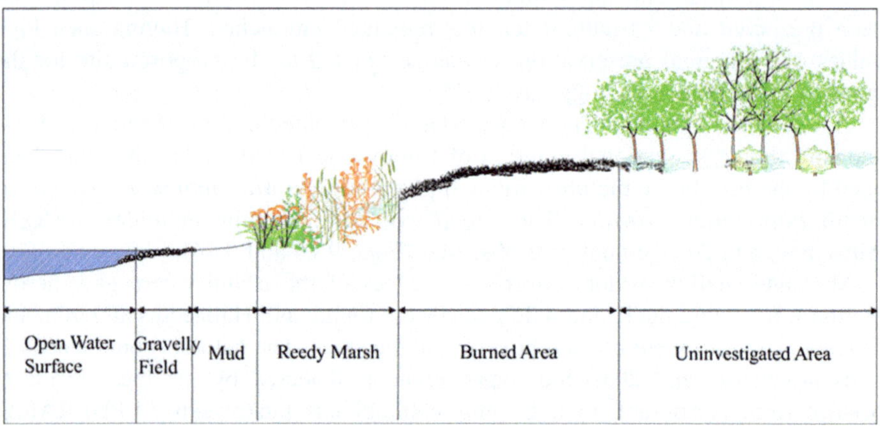

Fig. 9.14 Cross-sectional diagram of ecosystem at the edge of Chopyeong island

Fig. 9.15 The northern part of Chopyeong island before the fire (July 25 2008)

Fig. 9.16 The northern part of Chopyeong island after the fire (24 February 2009)

The negative impact of the fire to Chopyeong island's ecosystem is organized into three aspects: First, the rise of water temperature; Second, the dryness of soil, depletion of nutrients and increase in pH and; Third, the destruction of habitat. The moisture of the soil disappeared and exotic plants which are relatively stable in damaged environments are introduced, changing the entire structure and feature of the ecosystem in Chopyeong island. Chances are, before the desirable vegetation could even attempt to sprout, their growth would be deterred by the already fully grown exotic plants. Therefore, measures should be implemented in controlling the growth of exotic plants that hinder the growth of the natural vegetation. Close monitoring is necessary, especially, when nature tries to revive itself on the aftermath of destruction. Figures 9.15 and 9.16 show features of Chopyeong island before and after the fire.

9.3.4 Use of river levees

The river levee focused in this discussion is specifically situated in Haean-myeon Yanggu county, Gangwon province. Along the banks of the river is natural wetland vegetation. Primarily, the natural vegetation found here is functioning as a stable barrier from flood and at the same time preventing soil erosion and silt from inland. But this buffer vegetation zone at the water's edge also plays a significant pivotal role in preserving the ecosystem as it also acts as the ecological corridorfor wild animals shown in Fig. 9.17.

Perhaps river levees are regarded harmful in Korea as most of these levees are used as roads and farms rather than an ecological corridor for wildlife. This sad transformation disrupts the movement of wildlife causing eventual dislocation of the species, destroys habitats and introduces the dreaded exotic plants.

Fig. 9.17 River levee comprised of natural wetland vegetation

Fig. 9.18 River levee before the removal of buffer forest

The actual river levee bank of Mandaeri, Inje county is shown in Figs. 9.18, 9.19 and 9.20. A lot of damage had been made to this area as well. The buffer forest before the removal is shown in Fig. 9.18 which displays a thick forest of willow. While Fig. 9.19 presents the aftermath upon the removal of the buffer forest using weed killer and farming machineries. No willow could be found around, as well as other wetland herbaceous vegetation. In Fig. 9.20, the buffer forest was transformed into a corn field.

The river levee is also transformed into roads as shown in Fig. 9.22. The natural bank vegetation was totally removed and replaced with exotic plants. This development posed a negative impact to the surrounding ecosystems and habitats. In the following pictures; Figs. 9.21, 9.22, 9.23 and 9.24, various uses of the river levee at Andae-ri are illustrated.

Perhaps realizing that the inland lost its protective buffer zone against floods, some damaged areas are attempted to be restored using vegetation mats made of straw (Fig. 9.25). The damage in vegetation of the river levees caused not only the destruction of habitat but the flooding. The threat is definitely not just on wildlife but also to humans. Therefore this restoration initiative should be shared, duplicated, and applied to wider areas for restoration.

Fig. 9.19 River levee after the removal of buffer forest

Fig. 9.20 River levee being used as a corn field

Fig. 9.21 Case where river levee is used as a farm land

Fig. 9.22 Case where bank is used as the road

Fig. 9.23 Case where bank vegetation are removed from river levee via weeding

9.3.5 Streams Transformed into Muddy Water

The river basin and tributary of Soyang river are turned into muddy water, especially when flooded, as shown in Fig. 9.26. It was observed that this phenomenon started when development projects are built within the vicinity of the river basin. The muddy water made the river an inadequate habitat for aquatic

Fig. 9.24 Case where exotic plants are introduced to river levee

Fig. 9.25 Restoration work for river levee using vegetation mat

wildlife including fish and invertebrates that thrive in fresh water. Due to turbidity, the water is no longer potable and in poor quality for any creature to survive.

The same phenomenon is happening in connecting streams of Mandae-ri Punch ball, Inje county. The flow of muddy water after a heavy rain fall in the basin is shown in Fig. 9.27. Moreover, since the wetland vegetation of the banks around

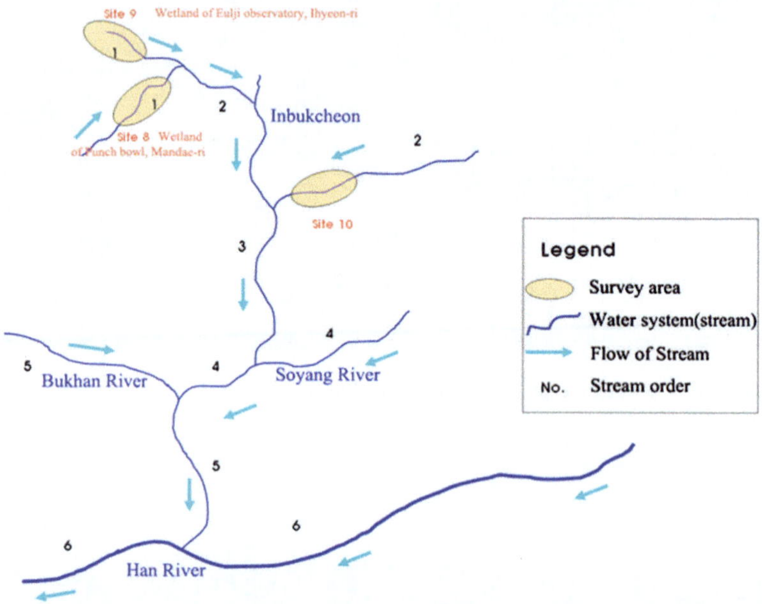

Fig. 9.26 Basin of Soyang river

Fig. 9.27 The flow of muddy water in Mandae-ri punch ball, Inje county

the streams is all removed, the expanse of damage caused by the muddy waters became wider and even bigger. The muddy water is the greatest threat to both river ecosystem and humans in the area. Therefore, an urgent measure should be untaken in order to prevent further damage.

Fig. 9.28 Reclamation of forest valley wetland into wild plant cultivation field, Haean-myeon, Yanggu county, Gangwon province

Fig. 9.29 Former alluvial wetland cultivated into farmland

9.3.6 Conversion of Wetlands into Farmlands

Recently, many of the water's edge and forested wetlands near the streams of DMZ and CCZ are cultivated into farmlands. Based on site inspection of the research team, the forest valley wetland area of Haean-myeon, Inje county, Gangwon province was once used as the wild plant cultivating field shown in Fig. 9.28. Before that it was once grassland filled with *Erigeron canadensis, Oenothera biennis* L., and *Aster yomena*. The region is a seaside basin with an alluvial fertile land (Fig. 9.29). It is now cultivated into a farmland and researchers could not find any traces that it was once a wetland (Figs. 9.30 and 9.31). There were even forest areas located in Haean-myeon punch ball, but now they are all cultivated into farmlands and fields of radish.

Fig. 9.30 Radish field created under the mountain of Haean-myeon punch ball

9.4 Disturbance of the Ecosystem

9.4.1 Invasion of Exotic Species

9.4.1.1 Invasion Ecological Approach

Non-native species
Public and academic recognition of the problems associated with biological invasions have grown exponentially over the past decade. The reasons for this growth are three-folds: First, the biological invaders on native species are increasing. Thus, increasing numbers of scientists are studying and managing non-native species in an effort to minimize the effects of biological invaders on native species and human economies. Second, the number of species being moved out of their native ranges is increasing. Therefore, not only the problems caused by non-native species becoming blatantly obvious, but also the overall number of problems appears to be growing. Third, with so many invasive species, it is very hard to do ecological field research without encountering invaders and eventually including them in investigations even if those investigations are for basic research. Invaders offer some new insights, and it is very difficult for curious scientists to pass up the opportunity to explore these new avenues (Julie et al. 2007).

Invasion Process Model

In order for non-native species to have impact in either the ecological or economic aspects to its surroundings, it must go through at least the following 3 steps: (as shown in Fig. 9.31).

First, all those exotic species must go through a ‘Transport’ stage in which they are transferred from their own territory to a new one. After being moved, some dies but some will survive as a self-sufficient group. The surviving group will reach the next stage called ‘Establishment’. In this stage, some of the survivors are still expected to be possibly wiped out but there will be successful ones that can adapt and settle into the new environment. Among the species that succeeded in adaptation, some will eventually form small-scale and large-scale groups and

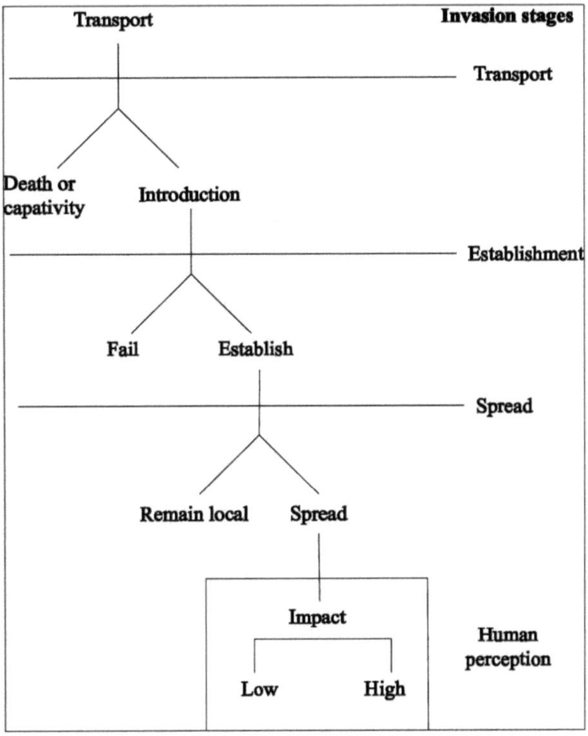

Fig. 9.31 Invasion process model depicting the discrete stages an invasive species passes through as well as alternative outcomes of each stage avenues (Julie et al. 2007)

eventually reach the stage called 'Spread' where they expand their geographical territory, and finally attain the name 'invasive species'.(exotic/non-native species).

This section will use the "Invasive Process Model (Fig. 9.31)" developed by Julie L. Lockwood, et al., as a framework used in describing the non-native species of DMZ.

9.4.1.2 Exotic Species that Invaded DMZ

The research team found various kinds of invasive species within DMZ and CCZ. The invasion is found to be more active among vegetation. False *acacia Amorpha fruticosa L.*, Daisy Fleabane and white heath aster are found in large quantities and in some areas, they were dominated the areas in their numbers. In addition, pokeberry, *Ambrosia artemisiifolia var. elatior*, *Humulus japonicus* Siebold and Zucc., Cocklebur, Diodia teres, and buffalo-weed are also detected in the region.

Invasion of exotic plants

In some disturbed or newly developed areas, the number of introduced species is found to be unmatched in number compared to other plants. This phenomenon is frequently observed near wetlands. These species are often found in groups in disturbed flat areas at the edge of the water.

Fig. 9.32 The cross-sectional diagram and vegetations of Hyangnobong forest valley wetland

In the US, introduced species is managed by classifying them either into volunteer or invasive groups. In this book, the introduced species is classified based on the period they were introduced:

- Introduced species before history: Plants that are uncertain of their introduced period. We assume they were introduced before the prehistorical period ex) *Humulus japonicus Siebold* and *Zucc.*, *Foxtail*, *americus*, *wild millet*, *copperleaf*, *Polygonum aviculare L.*
- Former introduced species: Also unclear of its introduced period but those assumed to be introduced at historical period (ex) asian plantain, *Cerastium holosteoides var. hallaisanense*, *Stellaria alsine var. undulata*, *Lamium amplexicaule L.*, *Rumex acetosa L.*
- New introduced species: Plants introduced after Chosun dynasty (ex) *Erigeron canadensis*, *Daisy Fleabane*, *Oenothera biennis* L., *Ambrosia artemisiifolia var. elatior*

Ecosystem disturbance is frequently detected within DMZ and CCZ which is usually caused by cultivating the wetland into the farmland or by setting up military facilities. The result of the inspection on species in Hyangnobong, Inje county, Gangwon province is shown in Fig. 9.32. It was originally a buffer forest zone but after it has been developed as shooting range of the Ministry of National Defense, it turned into grassland. As a consequence, various kinds of introduced species invaded the area.

Erigeron canadensis, *Beggar-Ticks,* and *Oenothera biennis* L. are species commonly detected within DMZ and CCZ. The spots where these species were found are shown in Fig. 9.33 and in Table 9.1.

Erigeron canadensis

Eriger-on cana-densis is native to North America and commonly found in fields of disturbed regions. The ecological research was conducted at the upper river of Soyang lake in Inje county Fig. 9.34. The community of *Erigeron canadensis* is

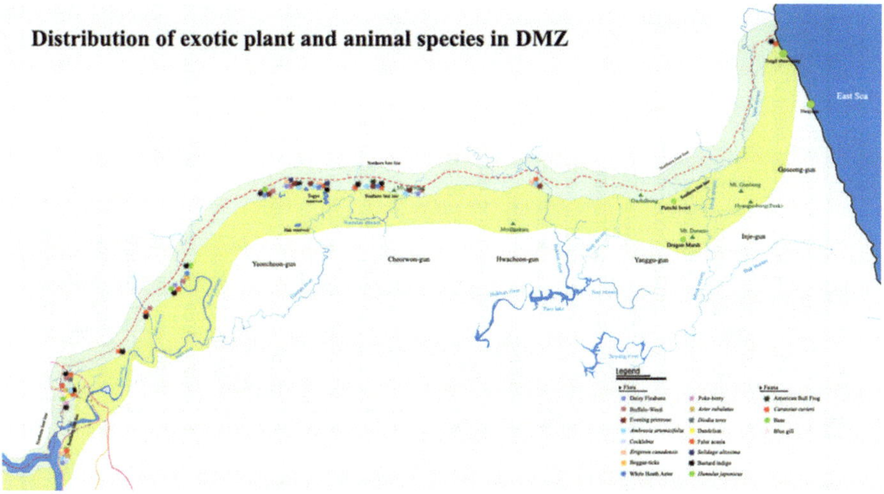

Fig. 9.33 Distribution of exotic plant and animal species in DMZ

found in large quantities at the edge of the water canals dug up for irrigation purposes.

Erigeron canadensis is a sun plant that grows in wet areas as well. Figure 9.35 shows the *Erigeron canadensis* observed in the upper river of Soyang lake Inje county and the community of willow dominating the area in Fig. 9.36.

Beggar-Ticks

Beggar-Ticks is native to North America and detected in wet and disturbed areas especially in the forest valley wetland near Eulji observatory of Haean-myeon, Yanggu county, Gangwon province. The area was cultivated as bellflower field at first, but now left as grassland, and *Alkekengi*, and *Beggar-Ticks Humulus japonicus Siebold,* and *Zucc.* are observed in the region. Among them are Beggar-Ticks, a plant that is easily found in wet regions. According to the survey team, *Beggar-Ticks* were found not only in wetlands but also in riparian buffer zones, specifically in water edges shown in Fig. 9.37.

Figures 9.38 and 9.39 show *Beggar-Ticks* found in the forest wetland near Eulji observatory and Soyang lake. Based on reports of site inspections in the DMZ and CCZ, *Beggar-Ticks* was propagating itself within the wetland in competition with indigenous plants. It was described as growing faster in the wetland with plenty of sunshine. Therefore in order to prevent their further spreading, trees should be planted in the area to block the sunlight.

Oenothera biennis L.

Oenothera biennis L. is native to South America and often found in disturbed areas in Korea. Seohwacheon region is the area where military facilities are installed. *Oenothera biennis* L. is commonly observed in the area as shown in Fig. 9.40. *Oenothera biennis* L. is also detected in wetland areas together with willow and in areas where the community of *Erigeron canadensis* is found.

Table 9.1 Exotic plant and animal species in DMZ

Exotic plants	Spots discovered	Exotic plants	Spots discovered
Daisy Fleabane	• Gowangsan Choryang valley natural area	*Buffalo-Weed*	• Natural area on right bank of Pilseung bridge
	• Gokneungcheon natural area		• Samicheon Yeoncheon plain natural area
			• Daeseong reservoir natural area
	• Yeokgokcheon natural area		• Yeokgokcheon natural area
	• Kimhwa Namdaecheon natural area		• Deokeungol natural area
			• Kimhwa Namdaecheon natural area
Evening primrose	• Yeokgokcheon natural area	*Ambrosia artemisiifolia var. elatior*	• Gokneungcheon natural area
	• Deokeungol natural area		• Dorasan station natural area
			• Yeokgokcheon natural area
	• Naepo Gangsan natural area		• Naepo Gangsan natural area
	• Yugok plain natural area		• Yugok plain natural area
	• Mando plain natural area		• Kimhwa Namdaecheon natural area
	• Kimhwa Namdaecheon natural area		• Hajinhyun valley natural area
	• Hajinhyun valley natural area		
Cocklebur	• Gokryucheon natural area	*Beggar-Ticks*	• Natural area on right bank of Pilseung bridge
Poke-berry	• Geogok-ri natural area	*Diodia teres*	• Daeseong reservoir natural area
Dandelion	• Dorasan station natural area	*Aster subulatus Michx.*	• Gokneungcheon natural area
			• Dorasan station natural area
Solidago altissima L.	• Yeokgokcheon natural area	*Humulus japonicus Siebold and Zucc.*	• Natural area of Choryang valley in Mt. Gowang
			• Deokeungol natural area
	• Naepo Gangsan natural area		• Natural area on the right bank of Imjin river
Erigeron canadensis	• Gokneungcheon natural area	*White Heath Aster*	• Natural area on right bank of Pilseung bridge
	• Yeokgokcheon natural area		• Deokeungol natural area
			• Naepo Gangsan natural area
	• Naepo Gangsan natural area		• Yugok plain natural area
	• Hajinhyun valley natural area		• Mando plain natural area

(continued)

Table 9.1 (continued)

Exotic plants	Spots discovered	Exotic plants	Spots discovered
False acasia	• Natural area of Choryang valley in Mt. Gowang • Samicheon Yeoncheon plain natural area • Gokneungcheon natural area • Daeseong reservoir natural area • Sewolcheon natural area • Dorasan station natural area • Naepo Gangsan natural area • Hantan River dandelion field natural area • Yugok plain natural area • Mando plain natural area • Hajinhyun valley natural area • Donghae Line natural area	Amorpha fruticosa L.	• Natural area of Choryang valley in Mt. Gowang • natural area on the right bank of Imjin river • Samicheon Yeoncheon plain natural area • Daeseong reservoir natural area • Sewolcheon natural area • Geogok-ri natural area • Dorasan station natural area • Yeokgokcheon natural area • Deokeungol natural area • Naepo Gangsan natural area • Hantan River dandelion field natural area • Yugok plain natural area • Mando plain natural area • Donghae Line natural area
Rana catesbeiana	• Naepo Gangsan natural area • Yugok plain natural area	*Carassius cuvieri*	• Gyeonggi Line natural area
Blue gill	• Gyeonggi Line natural area	*Bass*	• Sacheon reservoir Line natural area

Fig. 9.34 The community of *Erigeron canadensis* near the edge of water in Soyang river (Upper river of Soyang lake in Inje county)

Fig. 9.35 *Erigeron canadensis*

Fig. 9.36 *Erigeron canadensis* competing with neighbouring willow

Oenothera biennis L. is found to grow well in both wet and dry zones as shown in Fig. 9.41. It is actually detected in various types of habitats such as; near roads and streams, even at the edge of mountains.

Other exotic species

Various kinds of exotic plants are found within DMZ and CCZ. Invasive species like the *Erigeron canadensis*, *Beggar-Ticks,* and *Oenothera biennis* L. show dominance not only in DMZ and CCZ but also in other regions as well. Other exotic species were also discovered such as; the *Amorpha fruticosa L.* found along the Donghae line and the communities of Daisy Fleabane found in Inbukcheon of Gangwon province. For details please refer to Figs. 9.42 and 9.43, respectively. Practical measures are urgently needed because *Amorpha fruticosa L.* is harmful to native plants.

9.4.1.3 Invasion of Exotic Animals

Bass is a typical exotic animal found in both DMZ and CCZ. It is a freshwater fish preying on carps, a native fish regarded as special by Asians especially the Koreans, Chinese, and the Japanese. It is first designated as an exotic species by the Ministry of environment. Since it has a very high fertility rate, it is now regarded as the major offender disturbing the river ecosystem. The research team discovered the expanse

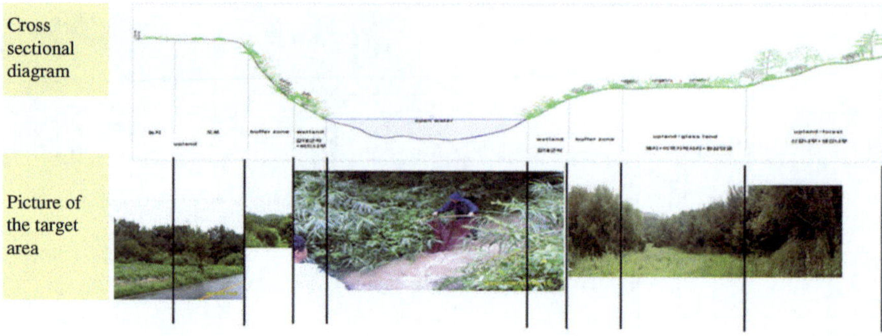

Fig. 9.37 Forest valley wetland near Eulji observatory of Haean-myeon, Yanggu county, Gangwon province

Fig. 9.38 *Beggar-Ticks* in upper stream of Soyang lake of Shinnam-ri, Inje county

Fig. 9.39 *Beggar-Ticks* near the forest wetland of Eulji observatory

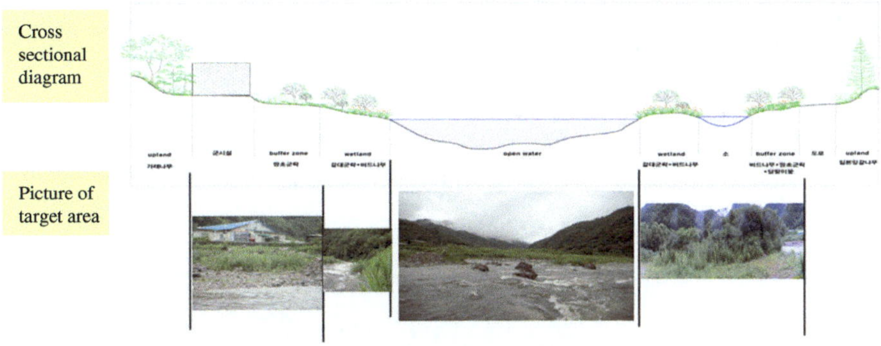

Fig. 9.40 Seohwacheon Seohwa-myeon, Inje county, Gangwon province

Fig. 9.41 Detected in various locations. **a** Oenothera biennis L. near the road of Soyang river in Shinam-ri, Inje county. **b** Evening primrose near Bukcheon of Gansung-eup,Goseong county. **c** Oenothera biennis L. inhabiting at the edge of mountain in the tributary of Nam river, Hyangnobong. **d** Oenothera biennis L. found near Seohwacheon of Inje county

Fig. 9.42 *Amorpha fruticosa L.* invaded into the wetland near Donghae Line, DMZ (The *dark green* plant at the *bottom* of this picture)

Fig. 9.43 The community of Daisy Fleabane found in Inbukcheon of CCZ Gangwon province

of its population from the Sacheon reservoir of Paju extending to the north and south with a ratio of 2:1, and that is doubled compared to native fishes. Sacheon reservoir is considered as one of the cleanest reservoir in the DMZ. Based on the research, Bass is also found in North Korea. It is believed that the reservoir of South Korea is connected to the waterways in North Korea (Fig. 9.44).

9.4.2 Disappearance of Indigenous Species

In order to prevent flood and for the practical use for farmlands, the artificial channel straightening project has been carried out. Consequently, both the population of inhabitants and indigenous species dropped sharply. Fish are often washed out by the powerful current and wetland vegetation at the edge of the waterways disappears as the number of habitat diminishes. Various kinds of

Fig. 9.44 Bass discovered in reservoir connected to Sacheoncheon, DMZ

Fig. 9.45 Wetland at the edge of water in poor condition due to the river levee work

indigenes including lenok were observed in the punch ball area of Haean basin before. But now they are hard to find. Here are some major reasons for their disappearance:

Foremost, the wetland vegetation at the edge of the waters failed to grow and develop (Fig. 9.45) due to the river levee project. Moreover, the influx of sewage and wastewater from near private houses and farms is the major culprit for the water pollution in the streams (Fig. 9.46). In addition, the disappearance of fish habitats including in sand bars is due to the channelization project (Fig. 9.47).

The function of the habitats in the river ecosystems has weakened, if not yet destroyed. The change will definitely have a direct impact on the number of species and its diversity. Figure 9.48 illustrates the removal of wetland vegetation due to the river levee project. Drastically, it ceases to function as habitat for both fish and

Fig. 9.46 Water pollution in stream due to the influx of contaminant

Fig. 9.47 Loss of habitat due to the channel straightening project

wild animals and even threatening the inhabiting indigenes in the overall ecosystem. Lastly, Fig. 9.49 shows the ill-effects of the channel straightening work. As the running speed of the stream increases, fish, and vegetation are washed out resulting to poorly developed wetland and to the eventual loss of habitat.

9.5 Environmental Pollution

9.5.1 Soil Contamination

The major sources of soil pollution are blamed on the manufacturing and storage of petroleum and toxic chemicals, the solid waste landfills, closed mines, factories, industrial facilities, and military facilities. In the case of the DMZ, outputs from

Fig. 9.48 Loss of habitat due to the disappearance of wetland vegetation

Fig. 9.49 Malfunction of habitat due to the poor condition of wetland

the military facilities and the chemical fertilizers and agricultural pesticides of the farmlands are pointed out as the major causes of soil contamination.

Despite the fact that civilian access to the DMZ is still restricted, the various kinds of project plans announced to be built in the CCZ cannot be disregarded as potential risks for soil contamination, among other identified threatening consequences to the ecosystems.

To take a serious look at the oil storage facility, waste landfills, storage for waste oil, machineries, shooting exercises, and maintenance activities within the military camp, all are storing and handling soil contaminants. Most of the wasted oils and chemical substances produced by these facilities are drained and absorbed by the soil without passing through proper purification or sanitation process before it can be safely disposed. The prolonged and direct exposure of the soil to these toxic substances causes serious problems.

Some wetlands and forest regions of DMZ and CCZ including Daeseong-dong, Paju, are converted into farmlands and ginseng fields. Chemical fertilizers and agricultural pesticides used in those lands are found to be polluting both soil and water of the area. Biomagnification of agricultural pesticides, for example, can be induced via the food chain, from primary producer to the end consumers like birds and big mammals, including humans. Eventually, it could bring irreversible damage to the ecosystem of DMZ, to say the least. Especially, the non-biodegradable agricultural pesticides that is very dangerous to human health. It can remain in soil and underground water for a long time, and slowly pile up in the body of living creatures. The sewage, on the other hand, can leak through the middle and deeper layers of the earth and contaminate underground water as well.

Pollution disturbs the growth of helpful microorganisms and earthworms, and even changes the ingredients of the soil. Underground water contamination can cause stunted growth and development of vegetations including herbs and forest woods. Such damage eventually triggers the disappearance of the vegetation strip which is regarded as the foundation of the ecosystem.

Soil and plants are having an interdependent relationship. The soil provides the nutrients and water necessary for the growth of plants, and plants in return increase the fertility level of soil and prevent surface erosion. In addition, both are very important factors of the ecosystem. The soil propagates the plants and the plants provide food and shelter for wildlife animals. This is a symbiotic relationship significant to the development of the habitat and the entire ecosystem as a whole. When this relationship is disturbed, especially by toxic pollutants, a considerable damage will beset the natural environment of the DMZ.

DMZ is thought to be an uncontaminated area and, to some extent, that is still true. But as they are so clean, they are more vulnerable to contaminants. When soil contamination issue comes into view, like cancer, it is often too late to be controlled. Therefore, extensive research should be conducted immediately to fully describe the level of soil contamination so that proper damage control can be designed and implemented.

9.5.2 Water Contamination

Water contamination can be classified into two categories: point pollution source and nonpoint pollution source.

Point pollution source often consists of domestic sewage and factory waste and, generally not much difference found in water-outflow regardless of rainfall as certain amounts are consistently emitted.

However, nonpoint pollution source refers to contaminants that are spilled out to the surface and let down when rain falls heavily. It includes dusts and wastes of metropolitan area, chemical fertilizers, and agricultural pesticides sprayed to farmlands, soil wash, effluence of cattle sheds, residues of natural creatures, and fallout of air pollutants. Nonpoint pollution source has significance difference in its

Table 9.2 Water quality investigation area within DMZ and CCZ

Types of basin	Target area within DMZ	Target area within CCZ
Estuary	• Estuary of the Han river(Gokneungcheon), Manwoori area where Han river and Imjin river meet, Estuary of east coast	• Wolgot, Kimpo
Lagoon	• Gamho	• Hwajinpo, Youngrangho
Stream and waterway	• Samicheon, Yeokgokcheon, Namgang, Sewolcheon, Myeolgongcheon, Sungnaecheon, Suipcheon	• Imjin river, Hantan river, Bukhan River
Reservoir and lake	• Eoryong reservoir, Sacheon resorvior, Daeseong reservoir	• Hak reservoir, Tokyo resorvoir, Sanmyoungho, Anho, Paroho lake

outflow per period and its amount is hard to be predicted nor quantified. This limitation is affected by factors that cannot be controlled artificially such as weather condition, geological features, and geography.

In the investigation, the DMZ is found to be uncontaminated, especially that it is still relatively undisturbed by humans. Excellent water conditions are maintained in most of the streams, lakes, and wetlands within this region. However, unlike DMZ, some contaminations are detected within CCZ and most of the water contamination in this area is found to be caused by nonpoint pollution sources, specifically due to the spraying of chemical fertilizers and agricultural pesticides in farmlands. This needs immediate action from the government and requires strong conviction and political will to stop the use of chemical in food production to prevent the biomagnification of the food chain. Organic farming should be strictly implemented and chemical fertilizers and pesticides be removed from the market.

The areas which need the water quality testing in the DMZ regions are shown in Table 9.2. Based on the results, monitoring and management of water quality should be implemented. It is highly recommended that the focus of management will be on nonpoint pollution sources, specially applying integrated approach at the watershed level.

Chapter 10
Conservation and Management Strategies

This chapter is carefully planned to enhance the biodiversity and ecosystem services of the DMZ and CCZ through the development of conservation and management strategies.

10.1 Land Use Strategies

This section provides a powerful way of managing ecological systems, as it attempts to sustain biodiversity and ecosystem services in the face of global climate change and other development pressures.

10.1.1 Site Selection

Kim (2000) has conducted "A Detailed Study and Proposal for ESSD in Northern Gyoenggi province."

The study area, which is located in the west of DMZ, has been under a lot of development pressure recently and underwent changes in land use including diversion of reed fen and wetland area, particularly in CCZ, as agricultural land. When such changes are left as they are, negative impact is likely to affect natural resources in the study area. Therefore, it was imperative to develop specific land use plan to prevent such negative impact and to manage effectively.

10.1.2 Process

10.1.2.1 Applying Land Use Model

Through literature survey and expert's advice, development process and method of land use strategy adopted in this study have been generated.

K.-G. Kim, *The Demilitarized Zone (DMZ) of Korea*,

DOI: 10.1007/978-3-642-38463-9_10,

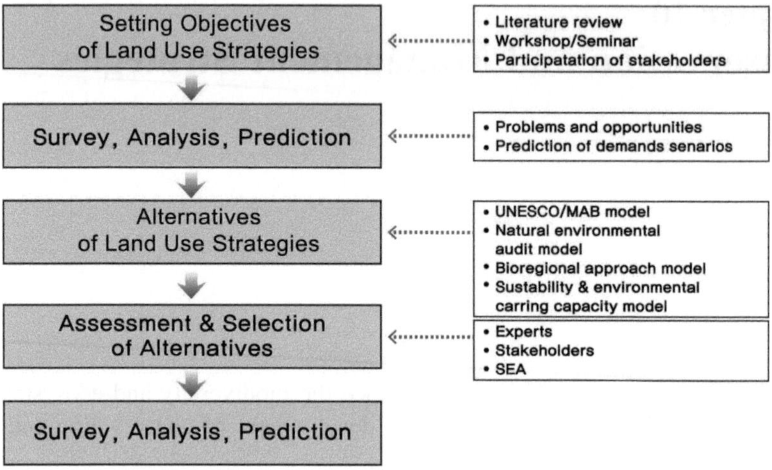

Fig. 10.1 Landuse strategy process

The generated process is to describe process applicable to this study area. It is useful in developing more reasonable and objective land use and management strategy.

Details of each process are shown in Fig. 10.1.

10.1.2.2 Alternatives of Land Use Strategies

An integrated model was developed by defining models developed globally as environment-friendly model in land use and models suitable for Korean domestic situation. Such models include UNESCO/MAB model, nature environment audit model, bioregional approach, and environmental carrying capacity and suitability analysis. These models are most desirable models that may be applied to this study area which is mostly covered by natural environment such as forest, river and stream, and wetland. This was utilized in developing more reasonable and objective land use and strategy.

The application results of each model are as follows.

10.1.3 Application of UNESCO/MAB Model

10.1.3.1 Process

UNESCO/MAB zoning model is shown Fig. 10.2.

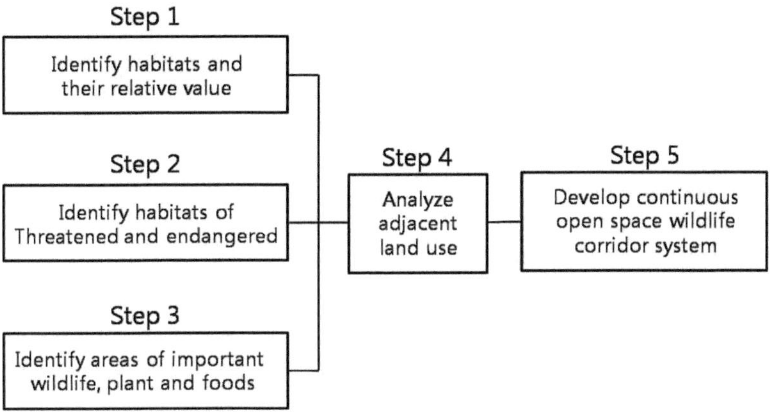

Fig. 10.2 The application process of UNESCO MAB zoning model

10.1.3.2 Application Results

In Fig. 10.3, red color indicates core area and pink color shows buffer zone, while white color indicates flexible transition area or area of cooperation.

10.1.4 Application of Nature Environmental Audit

10.1.4.1 Process

Nature Environment Audit model of this study was applied to Paju through the following procedures:

1st Phase
Identify critical natural capital (CNC), supportive capital (SC), constant natural asset (CNA), and priority enhancement zone (PEZ)
▼
2nd Phase
Collect existing analysis data (literature survey and case study)
Define local rare resources (rare species and protected species by various criteria)
Understand change features through analysis of historical changes
Apply assessment criteria
▼
3rd Phase
Set overall goals
Set area-specific objectives
▼
4th Phase
Set natural area within a target area

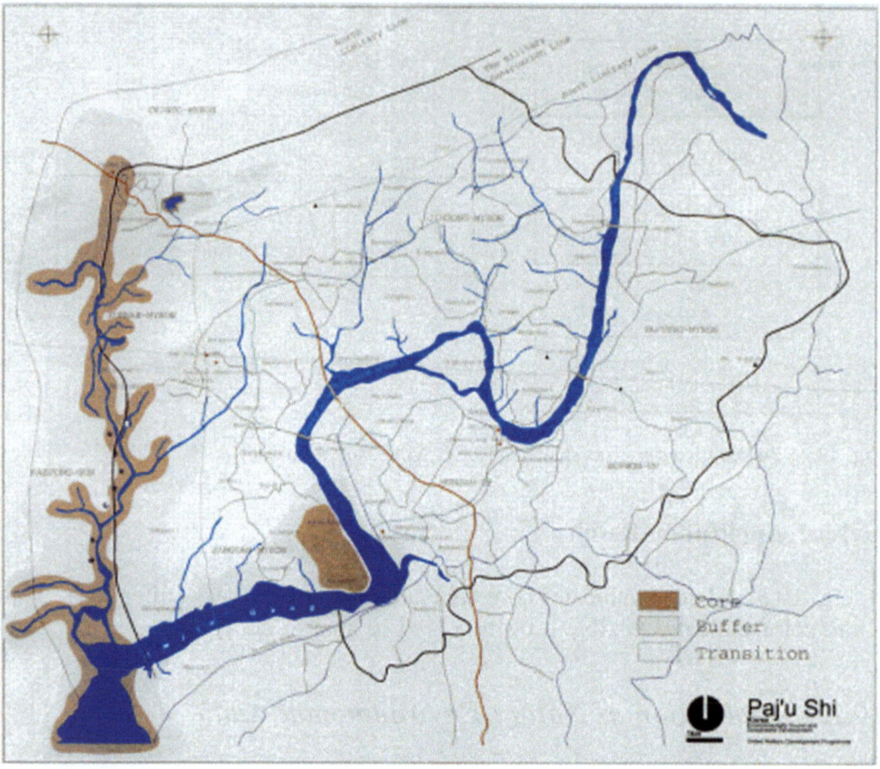

Fig. 10.3 The application result

Decide on sites within a target area that meet various reserve criteria
Compile and evaluate resources.

10.1.4.2 Application Results

In the target area, natural resources-based conservation value has not been segmented earlier. Before performing nature environmental audit, survey, analysis, and evaluation of important species and habitats found extensively in the target area need to be conducted first.

Therefore, natural areas were categorized first. Although natural area zoning is not a process that must be done in nature environmental audit, it is a unit to be surveyed in audit.

In Korea, there has been no nationwide segmentation by natural features such as geology, terrain, a mountain range, and species. In order to apply the model in Paju CCZ and DMZ, a unit including natural and cultural features has been selected as natural area.

Based on such features, it has been divided into three natural areas—the Imjin river basin, the Sachon basin, and hills and fields—that have similar natural features. The segmentation has an advantage that this may be an integrated unit in developing management strategy also.

The types of major habitats may be divided into forest, pond and reservoirs, wetland, grassland, river and streams, reed fen, willow community, shrubs, and rice paddies close to water system (Fig. 10.4). These may be found in a certain natural area as a distinguishing feature. However, similar habitats may be distributed throughout many natural areas.

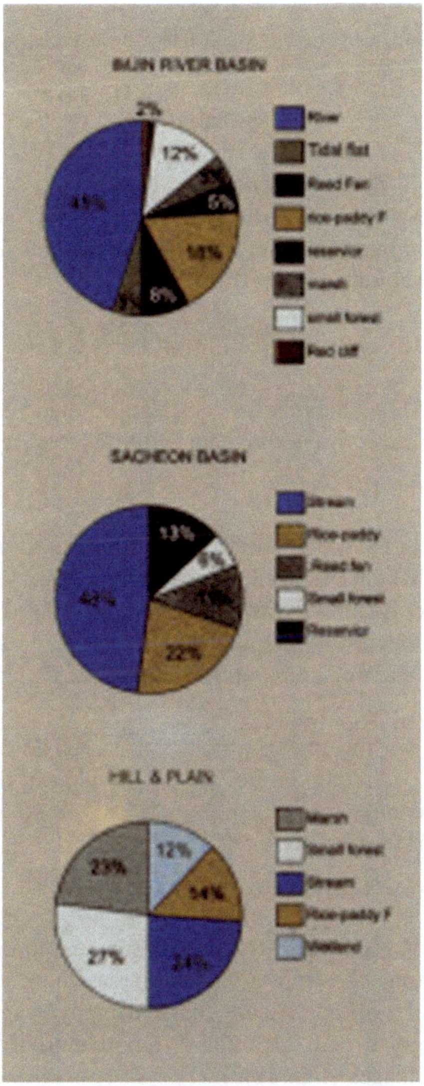

Fig. 10.4 Land us status of the divided study sites

Table 10.1 Regional features and key habitats for the divided study sites

	Regional feature	Key habitats
Imjin river basin	A basin ecosystem centered around the Imjin river; good living environment for wild life including tidal flat and reed community, extensive wetland, reed fen, and riverside corridor; include good wetland habitats where 37 migratory bird species come each year	Imjin river, tidal flat, reed community, arable land around wetland, scrub, and rock
Sacheon basin	MDL goes through the middle of Sacheon that branches off from the Imjin river; various hydrophytes are found along each depth	Streams, arable land around wetland, reed fen, scrub, withered forest
Hills and plains	Low-hill and relatively flat area to the north of Imjin river; most land is used for cultivation; valuable grassland and scrub are included	Grassland, scrub, riverside, arable land around wetland

When areas of major habitats in each natural area are compared, although there are slight differences, rivers and wetland/grasslands are identified as major habitats in overall. Reservoir and rice paddies near wetland are known to be representative habitats (Table 10.1).

10.1.4.3 Assessment of Target Area

Major habitats in each natural area were assessed for their diversity, rarity, naturalness, potential for wildlife, fragility, and intrinsic appeal.

(a) **Diversity**

High-diversity area usually refers to an area where conditions and types of habitats as well as species are diverse. In here, only the diversity of species or community was considered. In general, this may be indexed based on number of species and population using Shanon's index. This study has based on the population of amphibians, reptiles, birds, and mammals.

(b) **Rarity**

Among entire species found in the study area, species that are rare and endangered globally and locally were graded. Grading is based on the number of rare species.

(c) **Naturalness**

As a base for determining naturalness of a habitat, a high-naturalness area translates into a valuable area where diverse animals and plants inhabit.

Although there may be many ways to evaluate naturalness, this study used green area naturalness reported in Korea and analysis result of NDVI satellite images (Fig. 10.5).

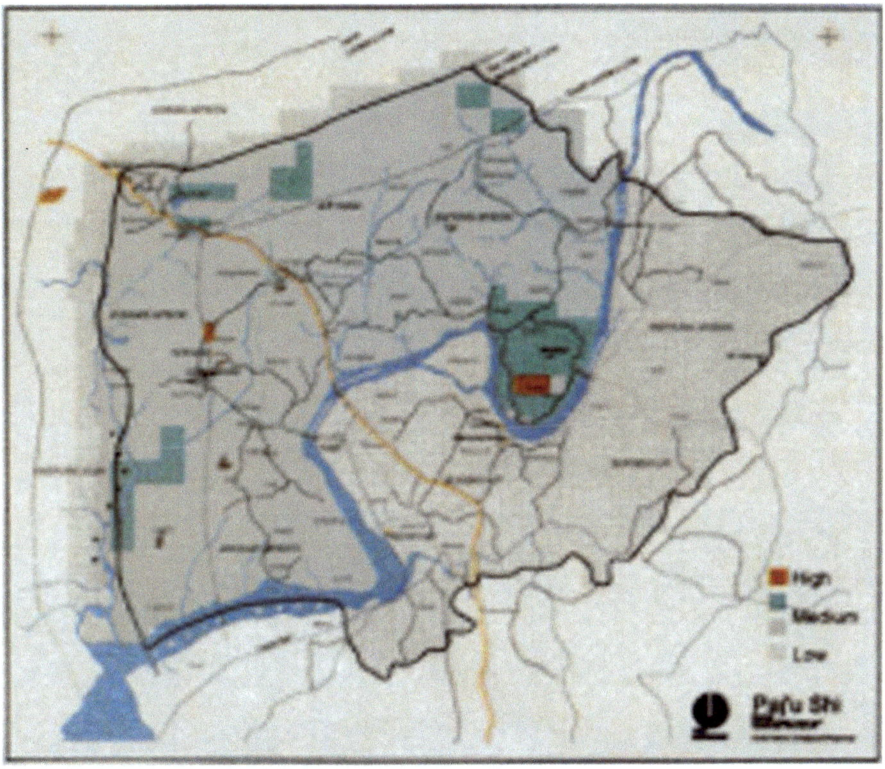

Fig. 10.5 The map of assessment for 'naturalness'

(d) **Potential for Wildlife**

It represents a capacity of a certain habitat to breed a certain species. In this study, spawning grounds of amphibians and reptiles were used.

Usually, radius of habitats of amphibians and reptiles is 0.5–1 km from a spawning ground has potential to breed a certain species. In particular, expected potential stream was calculated based on DEM figure of target area (Fig. 10.6).

(e) **Fragility**

This describes an area where healthiness is declining or already weak. In here, it refers to an area that is or likely to be damaged easily by an external impact.

Assessment was based on the distance to a road in/near the target area or to an area where ecologically negative development is underway, its area, and its density.

(f) **Intrinsic Appeal**

As a fundamental value inherent in the landscape of natural area, this assesses how similar it is to an ideal nature. As the assessment factors are qualitative, this

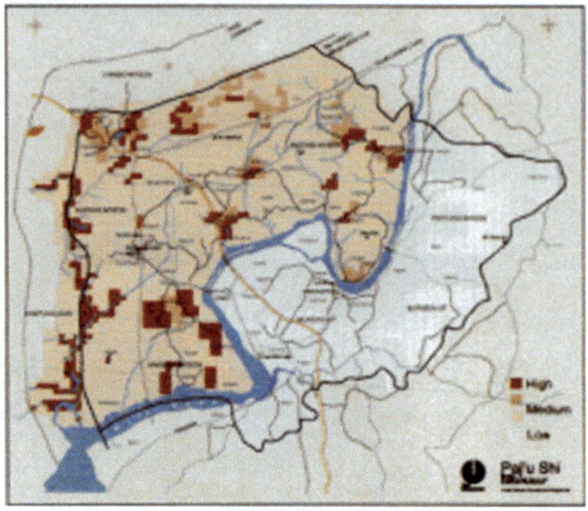

Fig. 10.6 The map of assessment for 'potential for wildlife'

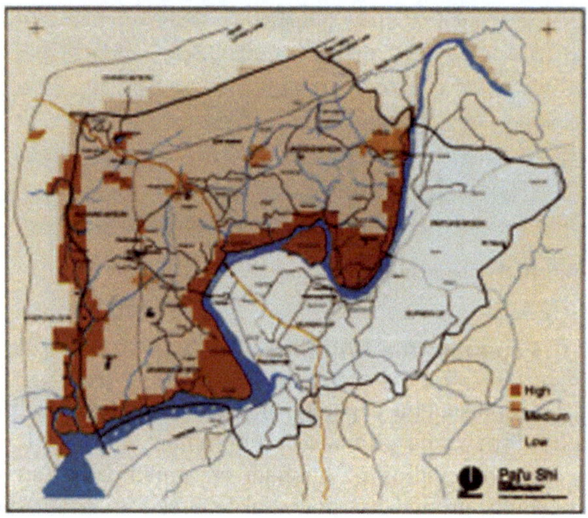

Fig. 10.7 The map of assessment for 'intrinsic appeal'

study used qualitative assessment method by compiling opinions of experts and advisers participating in the study (Fig. 10.7).

10.1.4.4 Synthesis for Conservation Value

Through an algorithm of drawings that were assessed by each assessment factor, conservation value of the study area has been divided into CNC, SC, CNA, and PEZ (Fig. 10.8).

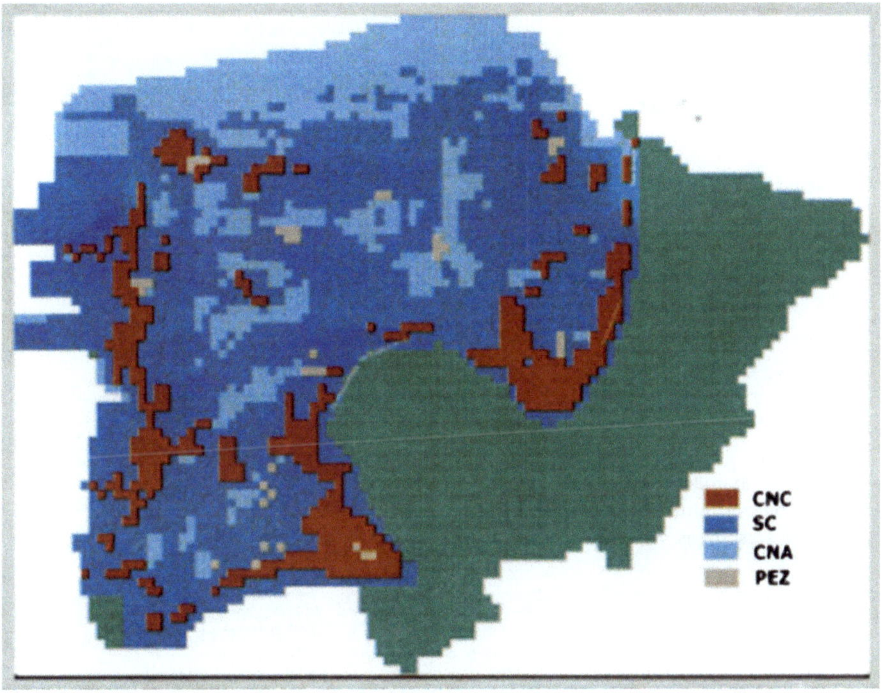

Fig. 10.8 The overlay map of each assessment factor

10.1.5 Application of Bioregional Approach Model

10.1.5.1 Process

Figure 10.9 shows the bioregional approach process used in this particular study.

The features of the above-mentioned process would be steps 2 and 3. In the process, step 2, understanding of the area, segments regional places by physical elements such as waters and terrain based on bioregional features and defines regional and intrinsic resources forming the features of such region. In step 3, drawing areas segmented from bioregion to place unit, photos of each district, and testimonies by local residents define characteristics and features of each region. The relationships between such resources and each region from the landscape unit of each place unit represent natural and cultural essence and show a process in a larger bioregional location.

What is important is to understand the value and attitude (lifestyle) of people living in the area as well as to have thorough visits. In other words, recognition of local residents actually living there and consideration of segmentation of established space as well as segmentation of physical aspects need to be performed simultaneously (Table 10.2).

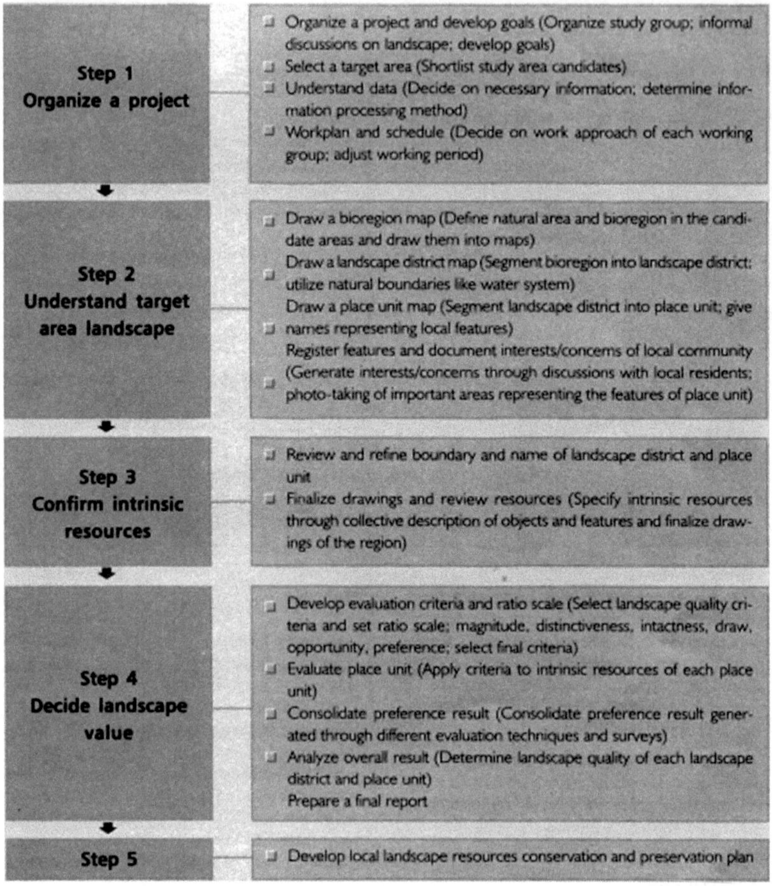

Fig. 10.9 The bioregional approach process

Since such segmentation is performed while considering the demographic and social aspects as well as physical features of the area in overall, in order to ensure the accuracy of segmentation, it should be conducted based on the review of extensive basic knowledge and data on a target area and the field experience.

Therefore, based on such concepts and procedures, this study performed a further-developed approach by fully considering elements often ignored in a traditional planning process.

10.1.5.2 Application Results

Using the bioregional approach to achieve the objectives of this project within the theoretical framework required to do a site analysis based on the intrinsic resources of the bioregion. Intrinsic resources are the economic, environmental, and

Table 10.2 The concept and characteristics of bioregion, landscape district, and place unit

Bioregion	- The highest unit in bioregional approach units - Segmented by unique climate, terrain, vegetation, river, and land types in a region - Segmentation should undergo the understanding of features and boundary recognized by local residents through contacts with and participation by local residents - In landscape of a target area, provide a conceptual framework of conservation value evaluation and plan and management
Landscape district	- Sub-unit of bioregion - Segmented by waters and mountain; scope recognized by an observer - Local residents often have a familiar name for the region; it is sometimes perceived as a certain region Preliminary segmentation based on terrain, animal and plant inhabitation status, waters, and land use pattern; culture and lifestyle are considered additionally through field visits
Place unit	- Sub-unit of landscape district; the lowest unit in bioregional segmentation - A space surrounded by areas with unique visual features - Visually separable at once - Boundary is determined by vegetation, landform, and ridge

educational treasures of the land which make places distinct and experiences memorable-scenic, natural, historic, archeological, cultural, and recreational resources. The land is made up of an uninterrupted tapestry of small places (of place units) which are the homes of these resources. Each place unit can be inventoried for the six resources shown in Fig. 10.11 and be incorporated into management strategies for growth and change.

It is important to note that interrelationships often exist between one or more intrinsic resources. For instance, typical outdoor recreation may occur in areas with both high natural and scenic resources. The benefit of this model of inventory and analysis is that it allows complex cataloging of place units which correspond to a distinctive of specific combination of experiences for the visitor.

Scenic Resource contains views of natural or manmade features that are memorable or unique and may include panoramic seascape, pastoral rural settings, picturesque structures, coastlines, rain forests, rivers, and bays. These vital views and landscapes best represent the intrinsic character of a place or community. Often this is more subjective than the other qualities. Scenic resources often heightened visual experiences because of their strikingly distinct and pleasing elements which include landform, water, vegetation, as well as manmade development.

Natural Resource focuses on features of the visual environment that are in a relatively undisturbed state. These pristine physical resources, such as forests, geological formations, wetlands, nesting areas, river shoreline, creeks, and streams, represent the greatest natural beauty of a place. These areas should not be disturbed by humans.

Historic Resources record and create an appreciation of past human activities. These structures often tell stories of a place as well as of the inhabitants of a landscape. In that manner, cemeteries, battlegrounds, churches, and town structures reflect our social interaction with landscape and the course of historical development. Traditional architecture and settlement patterns educate and reveal the character of our landscape and the intention behind our actions.

Archeological Resources contain physical evidence of historic or prehistoric human life or activity that is capable of being invented or interpreted. Identified through ruins, artifacts, structural remains, and other physical evidence, they have scientific significance that educate the viewer and tie our present day activities with the past. Native American, colonial settlement, industrial revolution, early tourism, and agrarian themes are common examples.

Cultural Resources are the living expressions or the traditional lifestyle and customs of a distinct group of people. They are celebrated by both lavish ceremonies and everyday rituals such as traditions, unique indigenous industries, food, music, dance, rituals, speech, and markets. These activities provide insight into how the traditional culture of an area may be further integrated in the future. The working landscape is of special importance, ensuring the continued existences of agrarian- or rural-based traditions and allowing for the culture's continued survival.

Recreational Resources are those aspects of the landscape that support outdoor activities and include hiking, rafting, camping, kite flying, bird watching, photography, and pleasure driving. Both passive and active recreational opportunities can be accommodated. These recreational qualities do not propose intensive development for private recreation, but offer distinct and exceptional travel experiences for the general visitor.

Overall land use plan and each resource assessment are presented Figs. 10.10 and 10.11, respectively.

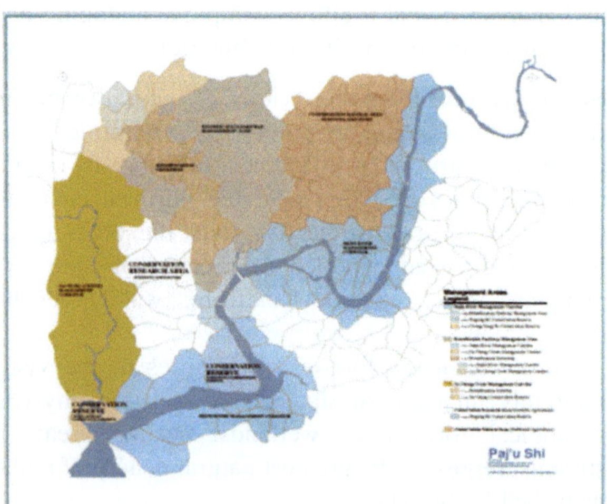

Fig. 10.10 Overall land use management plans

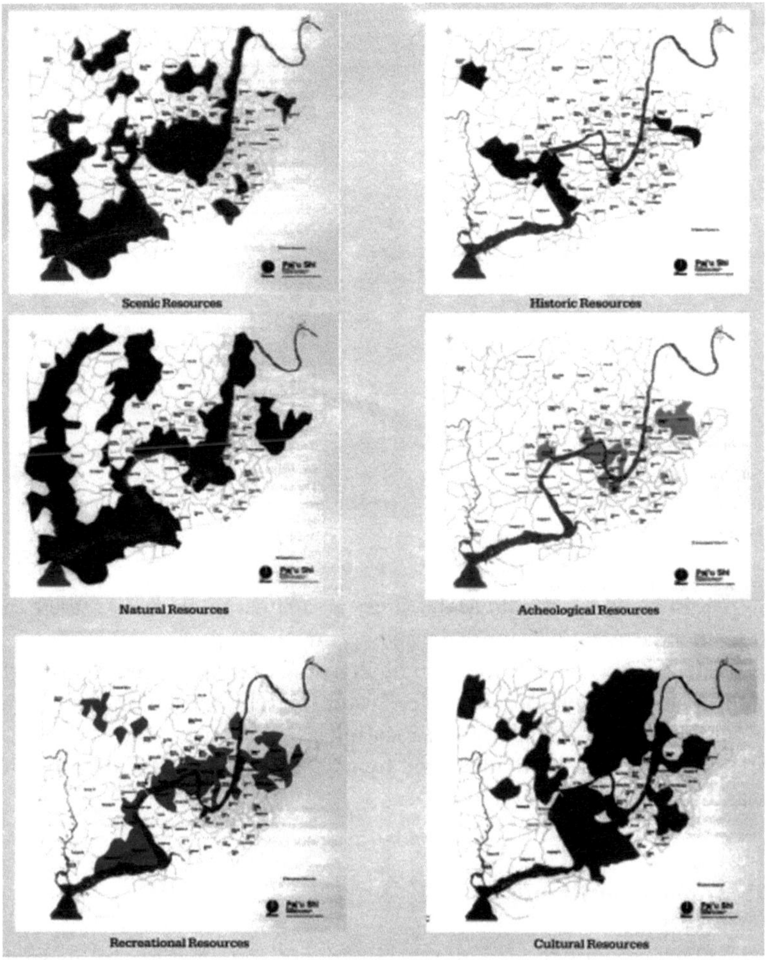

Fig. 10.11 The maps of assessment for each resources

10.1.6 Application of Land Suitability and Carrying Capacity

10.1.6.1 Process

Application process of suitability analysis model was defined in three steps as shown below.

STEP 1
Setting Objectives
STEP 2 ↓
Setting Criteria
STEP 3 ↓
Assessment

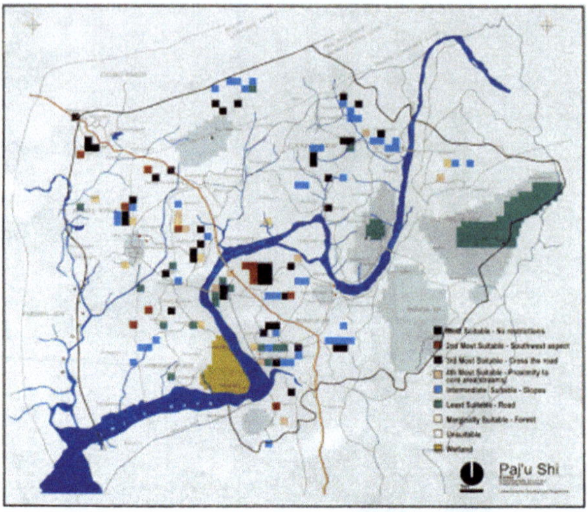

Fig. 10.12 Result of land suitability and carrying capacity analysis

In Step 1, objectives suitable to the study area situation are set. Some examples are eco-residential area, eco-tourism, and eco-research park.

Step 2 sets criteria suitable for each objective. Criteria have been set for factor: topography, slope, direction, altitude, water system, vegetation, wild animal habitats, drainage, and transportation location.

In Step 3, a suitable site is identified by assessing each with earlier established criteria and overlapping drawings.

10.1.6.2 Application Results

Established factors were divided into details according to features of the study area and weights were given. In case of water system in particular, it was set with a boundary of 500–1,000 m in water supply aspect and preservation of habitats surrounding the water system aspect (Table 10.3). The result of land suitability and carrying capacity analysis is shown in Fig. 10.12.

Transportation

Table 10.3 Criteria suitable for selected factors

Item		Criteria	
Distance between producer and consumer, high economics, easy to commute	0–500 m	4	Convenient
	500–1,000 m	3	↕
	1,000–1,500 m	2	
	1,500 m	1	Inconvenient

Topography

Item		Criteria	
Suitable land with firm ground and plain, lowland plain, hill, mountain	Lowland plain	3	Flat land
	Hill	2	↕
	Mountain	1	Mountain area

Slope

Item		Criteria	
Gentle slope area	0–8 %	4	Gentle
	8–15 %	3	↕
	15–25 %	2	
	25 %	1	Steep

Water system

Item		Criteria	
Easy availability of water, distancet from biotope	0–500 m	4	
	500–1,000 m	3	Degree of water
	1,000–1,500 m	2	Opportunity
	1,500 m –	1	

Drainage

Item		Criteria	
Good drainage	Good	3	Good drainage
	Slightly good	2	↕
	Poor	1	Poor drainage

Vegetation

Item		Criteria	
Categorization by state of vegetation	Agricultural land	4	
	Secondary grassland	3	Poor forest
	Afforest area	2	↕
	Secondary forest	1	Good forest quality

Aspect

Item		Criteria	
Traditional location condition: south	Southeast	4	
	Southwest	3	Good energy efficiency
	East	2	↕
	West and north	1	Poor energy efficiency

Altitude

Item		Criteria	
Set the lowest place in the study area as 0	0–100 m	4	Advantageous for residential
	100–200 m	3	And facility location
	200–300 m	2	↕
	300–400 m	1	Disadvantageous for residential and facility location

10.1.7 Total Assessment and Zonation

In developing land use strategy, the following zonation process needs to be performed as a preceding process. Its process is as follows:

Analysis results of each applied model are built into a database. Factors interact with others and generate a new result. Then, the result becomes an input to another model (Fig. 10.14).

The review of zonation developed at home and abroad shows that although terms may be different; it is usually segmented depending on the level of conservation value.

When it is reviewed by zoning entity and governing laws, zones are largely divided into an area that has valuable natural environment and need to be conserved, an area that may be developed because natural environment is relatively less valuable but has an area that is environmentally very important but was destroyed somehow and needs to be restored.

In border area zonation by the Globalization Initiative Committee, entire DMZ is set as reserve area. CCZ has been divided into north CCZ and south CCZ and designated as conservation area and semi-conservation area, respectively. Area south of CCZ is set as reorganizing area. However, the result of field survey on DMZ by the team shows that such zonation is based on a precondition that

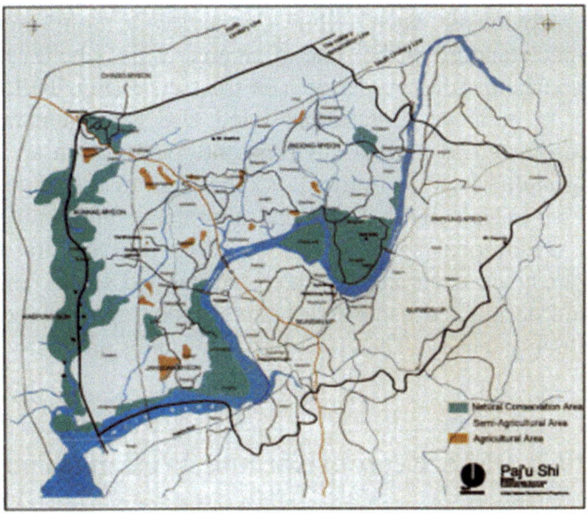

Fig. 10.13 The zonation map considering land use management law

conservation value will decline from DMZ to the south of CCZ as the area become distant.

STEP 1
Synthezing models
STEP 2 ↓
Setting Criteria of land use
STEP 3 ↓
Zonation

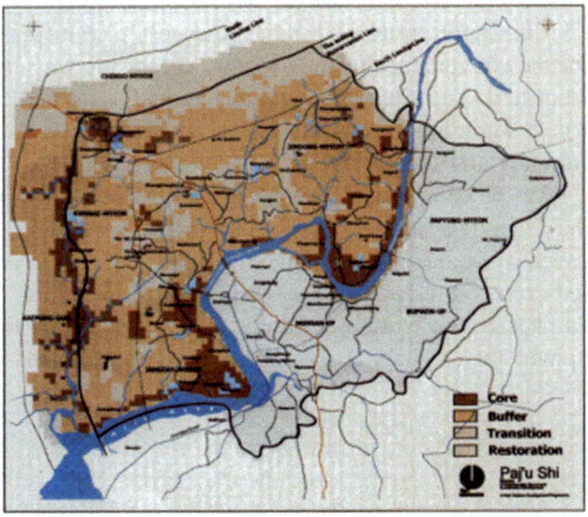

Fig. 10.14 The zonation map by overlaying of each model

Obviously, such a linear zonation is convenience-driven without accurate understanding of the state of an area. This is believed to be far from conservation management of border area in true sense (Fig. 10.13).

In pursuant to National Land Use Management Law in Korea, the area is designated as natural environment conservation area, semi-agricultural area, and agricultural area. Entire DMZ belongs to the natural environment conservation area. CCZ is divided into agricultural area and semi-agricultural area (Fig. 10.14).

Traditional zonation was based on the "Natural Land Classification Survey," which surveyed and classified natural and social conditions. An area was divided and designated using five factors: topography, slope, soil, traffic index, and population density. The zonation was usually productivity driven.

10.2 DMZ Ecotourism and Wetland Tour

10.2.1 Site Selection

10.2.1.1 Zoning of DMZ and the Boundaries for Ecotourism

In selecting the sites for ecotourism, the bioregional approach was adopted. This approach was utilized because the DMZ has distinct characteristics and because of the restrictions the ecological study had to abide with. The zone was divided into three parts: the western ecotourism circle, the central ecotourism circle, and the eastern ecotourism circle (Fig. 10.15). The sites were specifically selected in such a way that the educational component of the program is maximized and at the same time convenient to transportation and other amenities. Besides being a natural wonder, the educational criteria in the identification of the tourist spots should include the historical significance and cultural value of the sites.

It is important to note that the access to transportation is the first requirement for the convenience of the tourists. So, for the western ecotourism circle, Paju was identified. The site will highlight the historical and cultural value of Paju for it is where Panmunjom is situated. In addition, Paju also sustains habitats of natural wonders and creatures.

For the central ecotourism circle, Cheorwon was identified. This is a site where the Gyeongwon line passes. The primary consideration is the historical significance of the site. It was once the capital of Goguryeo. The main attraction is the red-crowned crane, one of the natural wonders of the DMZ. The tour can emphasize the features of the area in terms of the factors attributed to the development of the habitat that is favorable for the red-crowned crane, among other

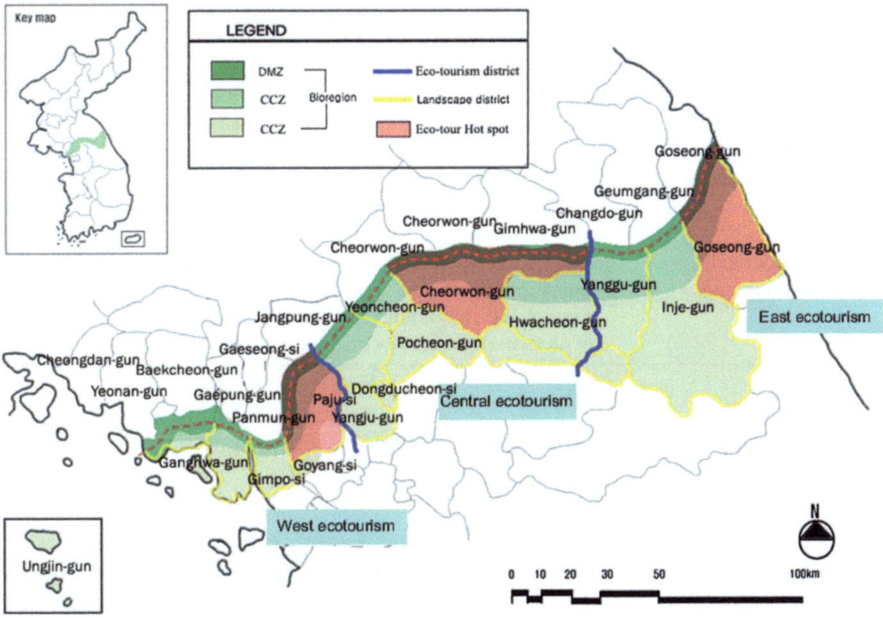

Fig. 10.15 Zonation for ecotourism

creatures that can be observed. The educational component may include the Bird Feeding Program of the government and the active participation of the local farmers.

For the eastern ecotourism circle, the Goseong was identified. The transportation in this site is the Donghae line. This tour will highlight Mt. Geumgang and the connectivity of the forest, wetland, grassland, ponds, and reservoir ecosystems including wildlife and the attributes in the development of the habitats. Programs on restoration and conservation of the habitats are highly educational to tourists, especially those that can be duplicated in their own countries.

Above all, the ultimate learning of the tour should be on the effective conservation and restoration management the Korean experience could share to the world.

10.2.2 Plan for Ecotourism Route

The routes and methods for the tour were designed in accordance to the significant characteristics of the ecotourism circles selected. For details, please refer to Table 10.4.

Table 10.4 Planning method of ecotourism route per foothold area

Classification of foothold area	Planning method
Western ecotourism circle—Paju	Select the sites with various species based upon the PBAs method and make the site as the tourism route
Central ecotourism circle—Cheorwon	Categorize resources that can be classified as the ecotourism resources and develop programs using those resources. And based upon those programs, plan for route by subject, season, and lead time
Eastern ecotourism circle—Goseung	Classify the ecological resources of Goseong into core, buffer, and transition area based upon the UNESCO MAB and plan for the route using storytelling scenario techniques focusing on evolutionary process of species and restoration course

10.2.3 Ecotourism Resources of DMZ

Paju

Compared to other regions, Paju is most abundant with historical and cultural resources potential for tourist attraction. The tunnel along the Gyeonui Line is a historical memory of Korean division. The remains of Panmunjom are still visible, and till now remain as a painful reminder of the effects of war. It even appeared on several movies. Another attraction is the Tongilchon within CCZ whose local economy was revived by a humble bean (Figs. 10.16, 10.17, 10.18, 10.19, 10.20, 10.21, 10.22, 10.23, 10.24, 10.25, 10.26, and 10.27).

Cheorwon

Cheorwon in Gangwon province boasts on its plains such as the Yeokgok, Namdaecheon, and Cheorwon plain that are highly abundant with ecological resources. Various kinds of species and specialized habitats for *Grus japonensis* and *white-naped crane* can be viewed in the plains. Like Paju, Cheorwon DMZ is

Fig. 10.16 The forest vegetation near the Panmunjom within DMZ

Fig. 10.17 The forest vegetation near the roadbed of existing Gyeonggi line

Fig. 10.18 Swan Goose

Fig. 10.19 The view of wetland in the upper stream of Sacheon river

Fig. 10.20 The view of ecosystem for the wetland of Eoryong reservoir

Fig. 10.21 White-naped crane

Fig. 10.22 Narrow-mouth frog

Fig. 10.23 Rat snake

Fig. 10.24 *Acorus calamus var. angustatus*

Fig. 10.25 Penthorum chinense PURSH

Fig. 10.26 *Typha angustata* and *P. australis* community

Fig. 10.27 Tributary wetland of Sacheon stream

also a habitat for many kinds of migratory birds. There are also wetlands in Cheorwon formed around Tokyo and Gangsan reservoirs that harbor diverse species.

Majority of historical and cultural ecotourism resources of Cheorwon districts are associated to the tragedy of Korean division. The Russian—style headquarter of the North Korean Labour Party is located within the CCZ which was built in 1946 when Cheorwon was still occupied by North Korea. It is now used for cultural events such as concerts and musical presentations. Since 2001, it has been protected by the government and is designated as a cultural heritage. Other significant cultural heritage in Cheoreon includes the Woljungri station, the Dopian Buddhist temple—a three-story stone pagoda, and the war monument at Baekma highland—a remembrance of the Korean victory after 24 bloody battles (Figs. 10.28, 10.29, 10.30, 10.31, 10.32, and 10.33).

Fig. 10.28 *Grus japonensis* and *Grus vipio* observed in Cheorwon plain (CCZ)

Fig. 10.29 *Chelidonium majus var. asiaticum* (Hara) Ohwi and *Spiraea salicifolia L.* community in Cheorwon wetland

Fig. 10.30 Forest within DMZ

Fig. 10.31 Goseokjung pavillion area

Fig. 10.32 Headquarter of North Korean Labour Party

Fig. 10.33 Sundam valley

Goseong

The Rare lagoons and coastal wetlands are in Goseong. Its sandy beaches are added attraction to the tourists. Goseong is regarded as the best place for education, cultural, and sentimental experiences. The site is very distinguishable from other tourist locations because of the features of its inland ecosystem, brackish water

Fig. 10.34 Gamho and DMZ observed at Geumgangsan observatory

Fig. 10.35 Marsh wetland in front of unification observatory

zones, seashore wetlands, and sandy beaches. With the Taebaek mountains range at the western background and with the vast sea in the east, Goseong is a place every tourist would like to be. The natural ecosystem in the area provides knowledge on several fish and aquatic species that thrive in a brackish water habitat, while communities of *Rosa rugosa var*. Rugosa give color to the eerie white beaches.

Moreover, it is the best place for observing the estuarine wetland ecosystem and its various kinds of fish species. In upstream, the tourists can observe freshwater fish while in downstream, they can observe brackish water fish.

The research group has been monitoring the habitat and species within the area since 2009. They found out that wetlands are reestablished and the ecological corridor is formed around Donghae line. Nature has healed itself and is performing well as habitat for the species normally inhabiting the region.

Fig. 10.36 Ecological bridge (eco-bridge) built for the preservation of wetland located in front of unification observatory

The Unification observatory with an educational exhibition hall is another historical and cultural resource of Goseong. In this observatory, tourists are able to enjoy a wider view of Mt. Geumgang and Haegeum river and can fully appreciate the natural landscape of the entire South Korean part of DMZ. Moreover, they can also see the truce line installed along the coastal line, a very painful indication of a divided country. The view may be regarded as a historical and cultural asset for Goseong, but the experience would be very sentimental as well (Figs. 10.34, 10.35, and 10.36).

10.2.4 Development of DMZ Ecotourism Program

Based on the ecotourism resources suggested above, a plan is introduced in this section specifying the routes, boundaries, or border land and proposed specific activities (Table 10.5). The tourism zone is classified into six categories and they include the Peaceful Unification zone, Natural ecosystem zone, Mountain climbing zone, Marine cruise zone, Water's edge recuperation zone, and historical and cultural zone. The classification is done for the purpose of ensuring the educational outcomes expected to be achieved by tourists in every site they would visit.

10.2.5 Planning for DMZ Ecotourism

Western ecotourism circle: Paju.

Ecotourism concept for Paju

The specific target sites in Paju shall be highlighted as a peaceful unification zone introducing resources such as the battlefield, Panmunjom, and the Unification observatory. For the natural ecosystem component of the tour, major ecological inhabitants including habitat for migratory birds and the estuarine wetland shall be presented giving emphasis on recuperation tourism using other resources such as the water edges of the estuary of Han and Imjin river in connecting the historical and cultural values of the DMZ. Figure 10.37 illustrates the plan for ecotourism designed for Paju area.

The Prime Biodiversity Area (PBA) method was applied in evaluating and identifying the specific habitats and species to be included in the ecological tour in Paju area. The steps used for developing the ecological route are as follows:

Step 1: *Organization and analysis of the species in Paju*

In order to plan for the ecological route in Paju area, the habitats present in the area are organized and the details of the respective species are used as database for the PBA evaluation. Evaluating the present status of the ecotourism resources is an

Table 10.5 Ecotourism program and route

Classification	Activity program	Route
Peaceful unification zone	Experience activity on DMZ truce line Experience activity on North Korean culture Educational activity on nostalgia and home town experience activity Experience activity on war	DMZ truce line Course Course for experiencing North Korean culture Education Course on unification culture Course for nostalgia and home town experience Course for war experience
Natural ecosystem zone	Wild animals observation and educational activity Natural in habitat observation and educational activity Experience activity on wildness Biological evolution and restoration experience activity Observation activity on natural landscape Exploration activity on geographical features	Course for wild animal exploration Course for observation of natural habitats Wildness experience course Course for inhalation of fresh air Course on biological evolution and restoration Observation course on natural landscape Course on geographical features
Mountain climbing zone	Experience activity on mountain climbing Experience activity on observation activity on natural landscape Wild animals observation and educational activity on Astronomy observation and educational activity	Course for mountain climbing Wildness experiencing course Observation course on natural landscape Wildlife animals exploration course Course for galaxy and star observation
Marine cruise zone	Marine cruise activity Observation activity on coastal landscape Coastal recuperation activity Activity for physical therapy	Marine cruise course Course for coastal landscape observation Course for recuperation in seaside Course for leisure activity in seaside Course for physical therapy
Water's edge recuperation zone	Observation activity on natural landscape Recuperation and experience activity at the edge of water Activity for physical therapy	Course for natural landscape observation Course for recuperation at the edge of water Course for leisure activity at the edge of water Course for physical therapy
Historical and cultural zone	Observation and educational activity on history and culture Observation and educational activity on culture of Goryeo dynasty Observation and educational activity on DMZ unification culture	Observation course on history and culture Observation and educational activity Course on Goryeo dynasty Culture observation course on DMZ unification

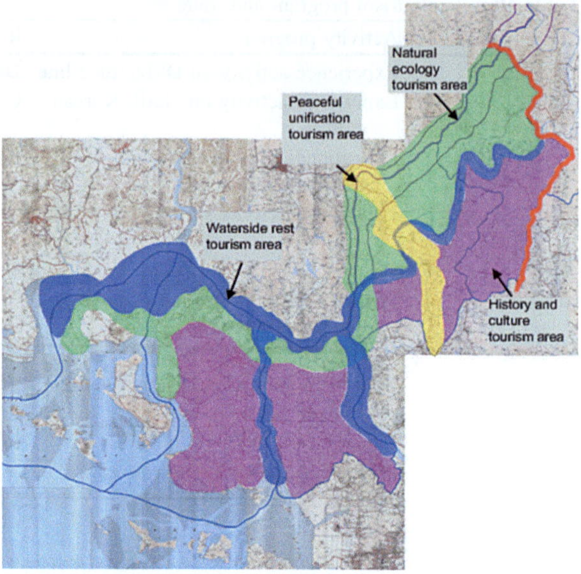

Fig. 10.37 Step 1: plan for ecotourism in the western part of DMZ and its borderland

Table 10.6 The list of species observed in DMZ as of 1 June, 2006

	Outlet water way of Daesung reservoir	Daesung reservoir wetland
Habitat	It's an outlet waterway of Daeseong reservoir observed with wetland vegetations including *Phragmites japonica, Salix purpurea var. japonica, Salix glandulosa* SEEM	It's a wetland area where *P. australis*, dominating the area, is going under the transformation into the community of willow. It is worth for further investigation
Vegetation	*Phragmites japonica, Salix purpurea var. japonica, Salix glandulosa* SEEM	*P. australis*, willow
Mammals	*H. inermis*, Leopard	*P. bengalensis*
Reptiles and amphibians	*G. rugosa*	Red-sided water snake
Birds	Little egret, large egret, white-winged black tern, little grebe, *Aix galericulata*	Black-capped kingfisher habitat
Insects		
Crustacean		River snail
Fish	Korean rose bitterling, Korean spotted sleeper, Japanese eel, *Cobitis lutheri*	Korean spotted sleeper, Korean bullhead, *Carassius auratus, Rhinogobius giurinus, Pseudoperilampus notatus*

essential step in planning the course of wildlife exploration. Table 10.6 shows the list of habitats and species in Paju Daeseong reservoir based on the series of investigations conducted in DMZ/CCZ.

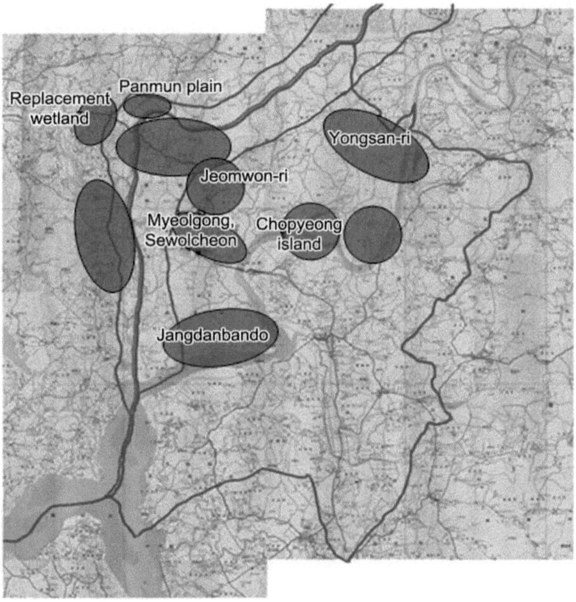

Fig. 10.38 Step 2: distribution of species in Paju region

Step 2: *Evaluation on the distribution of species*

Based on survey of species, the distribution of species is shown in Fig. 10.38.

Step 3: Analysis on habitat patch network

The result of the analysis on habitat patch network is illustrated on Fig. 10.39. Main water system appears around the Imjin river, while subwater system observed around Sacheoncheon and its tributary. The green belt is in north to south direction.

Step 4: Evaluation on potential PBAs

Eight regions including Jangdan peninsula, Yongsan-ri, Gyeonggi Line area, Jeomwon-ri, replaced wetland, Chopyeong island, Myeolgoingcheon, Sewolcheon, and Panmunbeol are selected as the target area for PBAs evaluation and the result shows that Chopyeong island and Panmunbeol have the most diversity in species, therefore with highest potential for ecotourism and preservation value. The result of this evaluation can also be seen in Table 10.7.

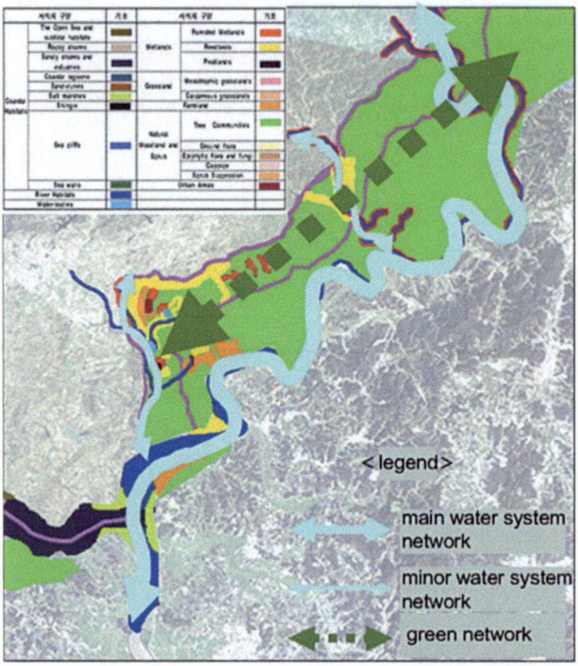

Fig. 10.39 Step 3: the result of analysis on habitat patch network

10.2.5.1 Ecotourism Resources by Major PBAs Region

The present status of the eight potential areas enumerated in the preceding section is described in Table 10.6.

10.2.5.2 Designing Wild Animal Exploration

Based on the PBAs, the status of the ecosystem of Paju is good. Thus, the wild animal exploration course was designed. The route targeted the DMZ and CCZ including the borders of Paju to provide an opportunity for tourists to observe the habitats in the wild and trace the imprints of wild animals. The interest of the tourists shall be drawn toward appreciating the wilderness of the wetlands and the forests. The planned wild animal exploration route targeting the overall Paju area is shown in Fig. 10.40.

Central ecotourism zone: Cheorwon

Ecotourism concept for Cheorwon

The first step initiated by the research team was the classification of the ecotourism resources found in Cheorwon. The central tourism circle can be classified into the following parts:

Table 10.7 The result of PBAs evaluation on Paju

Classification	PBA 1 Jangdan peninsula	PBA 2 Yongsan-ri	PBA 3 Gyeonggi line area	PBA 4 Jeomwon-ri	PBA 5 Replaced wetland	PBA 6 Chopyeong island	PBA 7 Myeolgongcheon/ Sewolcheon	PBA 8 Panmunbeol
Size of habitat	**	**	*	**	***	***	***	***
Ecological unit	***	***	**	*	***	***	***	***
Ecological structure	**	X	**	*	***	***	***	***
Size of target species	***	**	**	*	**	***	**	***
Abundance of preferred species	***	**	**	X	**	***	**	***
Diversity	***	***	**	*	***	***	***	***
Type of habitats	*	***	**	*	**	***	**	***
Potential	***	**	**	**	**	***	**	***
Mosaic/Eco tone	**	**	*	X	**	***	**	***
Adaptability to habitats	**	***	**	X	**	***	**	***
Scale for evaluation time	**	*	*	**	***	***	***	***
Existence of target species	*	*	X	X	***	***	***	***
Conservation priority habitat	***	**	X	*	***	***	***	***
Threatening factors	*	X	*	X	**	**	**	**

*** Very satisfied, ** Satisfied, * Moderate, ^{X} Unsatisfied

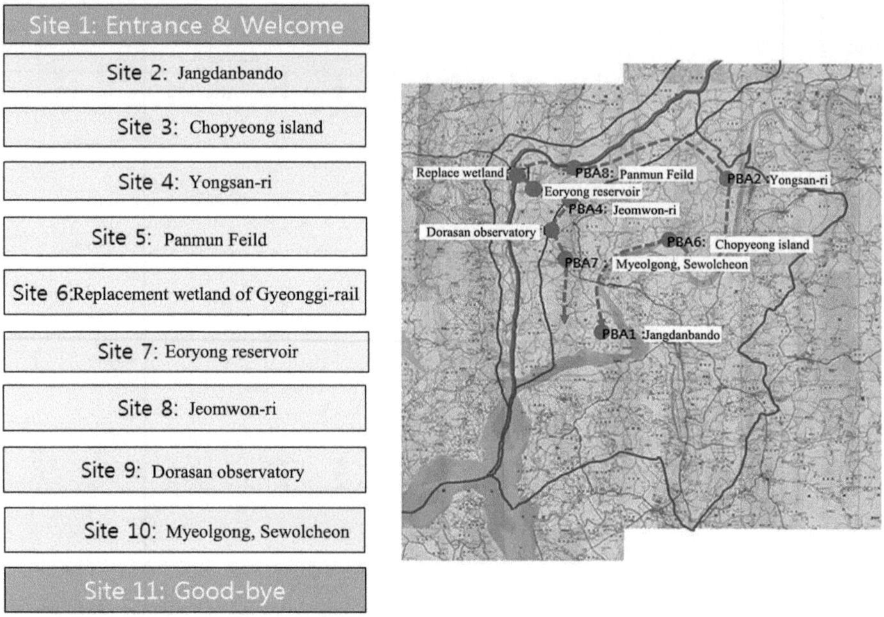

Fig. 10.40 Ecotourism route targeting the whole area of Paju

(a) Natural ecotourism zone: DMZ and major ecological habitats are connected.
(b) Peaceful unification tourism zone: Restoration and conservation areas along the Gyeongwon line are included.
(c) Recuperation zone at the edges of water: Hantan river, Bukhan river, and Paro lake are connected.
(d) Historical and cultural tourism zone: cultural resources of Goryeo dynasty is connected with that of unification.
(e) Mountain climbing tourism zone: Mt. Daesung and Baekam are connected.

The planned design for the ecotourism at Cheorwon proposed by the research team is illustrated in Fig. 10.41. The design is basically based on existing ecotourism resources reflecting their respective geographical features.

10.2.5.3 Ecotourism Plan for Cheorwon

After classifying the ecotourism resources at Cheorwon, a program was developed based on those resources. The route, on the other hand, was designed according to the route by subject, route by season, and route by leading time as follows:

(a) **DMZ Ecotourism resources in Cheorwon**

The resources are then reclassified as ecological, natural, historical, and cultural in order to make the ecological experience educational and meaningful for tourists.

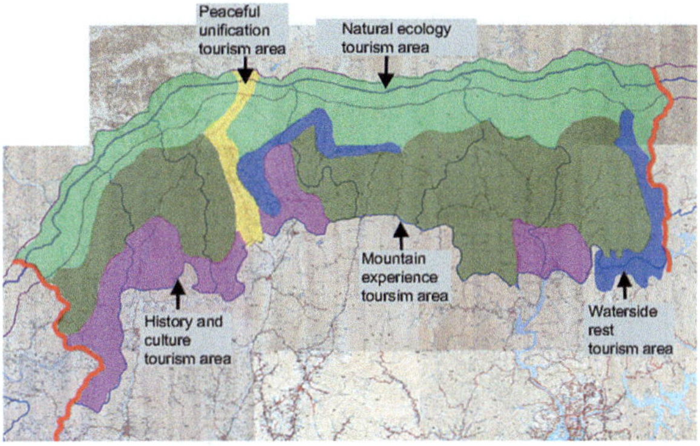

Fig. 10.41 Central part of DMZ and plan for ecotourism in borderland

(b) **Overall route**

The overall route planned for the ecological experience is presented in Table 10.8 accompanied with Tables 10.9, 10.10, 10.11 & 10.12 which contain the route plans for ecotourism in Cheorwon DMZ, by subject, leading time and season.

(c) **Route by subject**

The route per subject can be classified into four categories based on the existing ecotourism resources namely: (1) war experiencing course, (2) natural ecosystem experiencing course, (3) sightseeing course, and (4) historical and cultural experience course. This is shown in Table 10.9 where each of these subjects is specifically described in details. Ecotourism experience is an essential index for

Table 10.8 Overall route plan for ecotourism in Cheorwon DMZ

Overall route	Leading time	Course
DMZ/CCZ zone	Overall Course in DMZ/CCZ	Goseokjung → Jiktang waterfall → Jusangjeolri → Hantan river → Yangjiri → Tokyo reservoir → Dongsong reservoir → Woljungri station → Holy land of Taebong → Cheontong → Ice cream uplands → Baekma highland → Headquarter of Labour Party → Dopian Buddhist temple → Hak reservoir
DMZ/CCZ zone and other region in Cheorwon	Comprehensive course	Goseokjung → Sundam → Sungil bridge → Jiktang waterfall → Jusangjeolri → Hantan River → Dolmen → Earthen wall → Yangjiri → Tokyo reservoir → Dongsong reservoir → Woljungri station/ Cheolui triangular observatory → Holy land of Taebong → Cheontong → Ice cream uplands → Baekma highland → Headquarter of Labour Party → Dopian Buddhist temple → Hak reservoir → Grave stone for general Kim Eung-ha → Sangmeori limestone cave

Table 10.9 Route plan for ecotourism in Cheorwon DMZ by Subject

Subject	Leading time	Course	Program
War experience course	War experience course	Goseokjung → Seungil bridge → Battle monument of sniper ridge fight → Seungri observatory → Underground tunnel no 2 → Cheolui triangular observatory → Ice cream uplands → Baekma highland	Eduction activity on unification
	DMZ truce line experience course	Goseokjung → Baekma highland → Woljungri and Chului triangular observatory → Underground tunnel no 2 → Seungri observatory	Eduction activity on unification
	DMZ displaced person experience course	Goseokjung → Tokyo reservoir → Woljungri station/Cheolui triangular observatory → Ice cream uplands → Baekma highland → Headquater of Labour Party → Hak reservoir	Nostalgia activity
Nature and ecology experience course	Bird experience course	Goseokjung → Yangjiri → Tokyo reservoir → Dongsong reservoir → Sanmyeong lake → Cheontong → Ice cream uplands → Daemari → Headquarter of Labour Party	Observation activity on ecology
	Spontaneous healing course	Goseokjung → Sambuyeon waterfall → Maewol tower → Jusangjeolri → Hantan river → Cheontong → Daemari '→ Hak reservoir → Sangmeori limestone cave	Psychotherapy activity
	DMZ inhabitat experiencing course	Goseokjung → Hantan river → Yangjiri → Tokyo reservoir → Dongsong reservoir → Woljungri station → Cheontong → Daemari → Headquarter of Labour Party → Hak reservoir	Educational activity on ecology
	DMZ wetland experience course	Goseokjung → Yangjiri → Tokyo reservoir → Dongsong reservoir → Woljungri station → Sanmyeong lake → Cheontong → Daemari artificial rice paddy wetland) → Headquarter of Labour Party(forested wetland)	Observation activity on ecology
	DMZ wild animals experiencing course	Goseokjung → Hantan river → Yangjiri → Tokyo reservoir → Dongsong reservoir → Cheontong → Ice cream uplands → Daemari → Headquarter of Labour Party → Hak reservoir	Observation activity on ecology
	DMZ sightseeing course	Goseokjung → Jiktang waterfall → Jusangjeolri → Woljungri station → Ice cream uplands → Baekma highland → Hak reservoir → Sangmeori limestone cave	Psychotherapy activity
Sightseeing course	Historical and cultural course of DMZ	Goseokjung → Seungil bridge → Dolmen → Earthen fortress wall → Woljungri station/Cheolui triangular observatory → Holyland in Taebong → Headquarter of Labour Party → Dopian Buddhist temple → Gravetomb of general Kim Eung-Ha	Educational activity on history
Historical and cultural experience course	Unification and cultural course of DMZ	Goseokjung → Seungil bridge → Battle monument of sniper ridge fight → Seungri observatory → Underground tunnel no.2 → Woljungri station/Cheolui triangular observatory → Baekma highland → Headquarter of Labour Party	Educational activity on history

Table 10.10 Route plan for ecotourism in Cheorwon DMZ per leading time

Leading time (h)	Course	Program
1 day 3 h	Goseokjung → Jiktang waterfall → Columnar joint → Hantan river → Yangjiri → Tokyo reservoir	Nostalgia activity
	Goseokjung → Yangjiri → Tokyo reservoir → Cheontong → Headquarter of Labour Party	Observation activity on ecology
	Goseokjung → Columnar joint → Hantan river → Tokyo reservoir → Dongsong reservoir	Psychotherapy activity
6 h	Goseokjung → Jiktang waterfall → columnar joint → Hantan river → Tokyo reservoir	Educational activity on ecology
	Goseokjung → Headquater of Labour Party → Ice cream uplands → Woljungri station/Cheolui triangular observatory → Baekma highland	Psychotherapy activity eduction activity on unification
	Goseokjung → Jiktang waterfall → Cheontong → Sanmyeong lake → Dongsong reservoir → Tokyo reservoir → Hantan river	Psychotherapy activity
	Goseokjung → Hantan river → Yangjiri → Tokyo reservoir → Dongsong reservoir → Cheontong → Ice cream uplands → Daemari → Headquarter of Labour Party	Observation activity on ecology
9 h	Goseokjung → Hantan river → Yangjiri → Tokyo reservoir → Dongsong reservoir → Woljungri station → Cheontong → Ice cream uplands → Baekma highland → Headquarter of Labour Party → Sangmeori limestone cave	Educational activity on ecology

quality and effective tourism. The routes designed by subject are illustrated in Fig. 10.41.

(d) **Route by leading time**

Cheorwon is the region that can be explored in a day's trip. Time bound routes are then proposed; 3 h route plan, 6 h route plan, nine h route plan, one night route plan and 2 day route program. Route designs were based on the ecological features.

(e) **Route by season**

Cheorwon is a natural beauty and maintains remarkable landscapes with streams, forests, and plains, all in one region. Distinct changes are observed in Cheorwon in different seasons. Therefore, another set of routes were designed per season (Tables 10.10, 10.11, 10.12).

Table 10.11 Route plan for ecotourism in Cheorwon DMZ per leading time

Leading time	Course	Program
1 day 9 h	Goseokjung → Dolmen → Battle monument of sniper ridge fight → Earthen wall → Underground tunnel no. 2 → Woljungri Station/Cheolui triangular observatory → Holy land of Taebong → Baekma highland → Headquarter of Labour Party → Dopian Buddhist temple	Educational activity on history
	Goseokjung → Columnar joint → Hantan river → Tokyo reservoir → Woljungri station → Sanmyeong lake → Cheontong → Ice cream uplands → Daemari → Headquarter of Labour Party → Sangmeori limestone cave	Eucational activity on ecology
1 day and 2 nights	Goseokjung → Seungil bridge → Dolmen → Earthen wall → Battle monument of sniper ridge fight → Seungri observatory → Underground tunnel no. 2 → Woljungri station/Cheolui triangular observatory → Holy land of Taebong → Baekma highland → Headquarter of Labour Party → Dopian Buddhist temple → Tombstone of general Kim Eung-Ha	Educational activity on history and unification
	Goseokjung → Seungil bridge → Columnar joint → Hantan river → Yangjiri- > Tokyo reservoir- > Undergrand tunnel no. 2 → Dongsong reservoir → Woljungri station/ Cheolui triangular observatory → Sanmyeong lake → Cheontong → Baekma highland → Ice cream uplands → Headquarter of Labour Party → Hak reservoir → Sangmeori limestone cave	Educational activity on ecology and unification

10.2.5.4 Eastern Ecotourism Circle: Goseong

Ecotourism concept for Goseong

Considering its geographical location of the ecotourism resources, the location is classified into four

(a) Peaceful unification zone: resources along the Donghae Line are preserved and restored.
(b) Natural ecotourism zone: DMZ is connected with the major ecological habitats including coastal wetland and estuarine.
(c) Marine cruise zone: mainly in the eastern coast.
(d) Mountain climbing zone: Mt. Geumgang, Hyangro-bong, and Mt. Seolak are connected.
(e) Recuperation zone in the edge of water: mainly around the Naerin stream.
(f) Historical and cultural zone: historic and cultural resources of Goryeo dynasty and prehistoric sites are connected.

Table 10.12 Route plan for ecotourism in Cheorwon DMZ by season

Subject	Leading time	Course	Program
Experience on nature and ecology	Spring course	Goseokjung → Tokyo reservoir → Dongsong reservoir → Woljungri station → Sanmyeong lake → Cheontong → Ice cream uplands → Daemari	Educational activity on ecology
Experience on nature and ecology	Summer course	Goseokjung → Jiktang waterfall → Columnar joint → Hantan river → Headquarter of Labour Party	Educational activity on ecology
Experience on history and culture	Autumn course	Goseokjung → Dolmen → Battle monument of sniper ridge fight → Underground tunnel no.2 → Woljungri station/Cheolui triangular observatory → Holy land of Taebong → Baekma highland → Headquarter of Labour Party → Dopian Buddhist temple	Educational activity on history
Experience on nature and ecology	Winter course	Goseokjung → Hantan river → Yangjiri → Tokyo reservoir → Dongsong reservoir → Cheontong → Ice cream uplands → Daemari → Headquater of Labour Party	Educational activity on ecology

10.2.5.5 Ecotourism Plan for Goseong

The ecotourism resources of Goseong are classified into: core, buffer, and transition zone, based on UNESCO/MAB. The sites are specifically selected in such a way that observation of biological evolutionary process and restoration can be observed. The Information, Education, and Communication (IEC) approach proposed is "storytelling."

a. **Ecotourism resources of Goseong, DMZ**

The ecotourism resources of Goseong can be classified into five categories: (1) Unification education activity, (2) Observation and education activity on ecological wetland, (3) Historical and cultural activity (4) Unification and cultural activity, and (5) Observation and education activity on biological evolution process.

Moreover, the ecotourism resources of Goseong can be classified as Core, Buffer, and Transition zone according to UNESCO/MAB as shown in Table 10.14.

b. **Ecotourism scenario**

Based on the ecotourism resources classified on Table 10.13, an ecotourism scenario is designed and shown in Table 10.14. According to the planned course, the tourists will be given a chance to observe the replacement wetland habitats,

Table 10.13 DMZ Ecotourism resources of Goseong classified by the UNESCO/MAB

Area for ecotourism	Ecotourism program	Core zone	Buffer zone	Transition zone
Unification observatory	Education activity on unification			•
Coastal wetland in the eastern sea	Observation and education on ecological wetland	•	•	
Calystegia soldanella and Lathyrus japonicus community in the Donghae railroad	Observation and education on ecological wetland	•	•	
Sacheon region	Observation and educational activity on wetland	•	•	
Anho reservoir	Observation and appreciating activity of wetland	•	•	
Coastal sanddune wetland	Observation and educational activity on coastal wetland	•	•	
Community of sweetbrier	Observation and educational activity on species	•	•	
Gyeongwon line zone	Historical and cultural activity		•	•
Community of forest wetland	Observation and appreciating activity of forested wetland	•	•	
Myeonghori area	Observation and appreciating activity of wetland		•	•
Myeongpari area	Observation and appreciating activity of wetland		•	•
Truce line zone	Unification and cultural activity		•	•
Railroad bridge of Mt. Geumgang	Historical and cultural activity	•	•	
Forest near stream	Observation and education on ecological wetland	•	•	
Gamho region	Observation and education on ecological wetland	•	•	
Mt. Geonbong region	Forest experiencing activity	•	•	
Replacement wetland	Observation and education activity on biological evolutionary process		•	•
Ecological corridor for living creatures	Observation and education activity on biological evolutionary process		•	•
Dolmen	Historical and cultural activity		•	•
Hwajinpo	Historical and cultural activity		•	•
President Lee Seung Man village	Historical and cultural activity			•
Lee Ki Bong Village	Historical and cultural activity			•
Kim Il Sung Village	Historical and cultural activity			•
Exhibition hall for historic security	Unification and cultural activity			•
Dolmen historic site	Historical and cultural activity			•

Table 10.14 Ecotourism scenario in Goseong

Site 1 : Entrance & Welcome		
	By receiving the instructions and matters to be noticed prior to the visit, we can avoid any damages to its environment and promote its sustainable use. Also we can record the species observed in each habitat on the observation sheet received. Receiving the observation note on species detected per inhabitat ▸ ◂ Receiving instructions and matters to be atteneded before making the trip	
	Daegang-ri/Myungho-ri, Hyunnae-myeon, Goseong county Exhibition hall for inforamation ▸ ◂ Explanation board on species	

▾

Site 2 : Songjiho			
		We can do marine activity in this area and also receive explanation on ecological succession so we can observe species. Receive explanation on species ▸	
	Explanation board ▸ ◂ Observing species		

▾

Site 3 : Hwajinpo			
		We can receive the education on marine species and participate in the marine activity. Also we can take some rest by taking a walk Trail ▸	
	Marine activity Pergola in nature ◂ Environmental education in marine ecosystem exhibition room		

▾

Site 4 : Estuarine wetland of Donghae Line			
		By observing and doing marine activity in estuary wetland, we can record the species observed in the area and appreciate the marine environment Seashore trail ▸	
	◂ Marine activity ◂ Record species observed	Environmental education & exhibition room ▸	

sweetbrier restored site, coastal sanddune wetland and coastal wetland, and also be able to appreciate the process of restoration and evolution that transpired within the area. Then they are allowed to observe wildlife, while the nature and characteristics of the creatures being observed are explained by the ecoguide. The beautiful landscape of the area alone gives comfort and rest for visitors.

10.3 Ecological Restoration

One long standing approach in reducing the negative impact of fragmentation and isolation of populations is to connect habitat fragments with corridors of similar habitat. Despite the logic of the approach, evidence of the effectiveness of habitat corridors to promote movement of animals between habitat patches has been scant (Molles Jr 2010). Experimental evidence in particular, has been lacking. Recently, however, a number of experimental studies have filled many of the gaps in understanding the effects of habitat corridors on the movement of animals.

Kim (2004) and his colleagues studied the influence of ecobridges on the movements of animals associated with early successional habitats. Their study sites were "patches of open habita" created right over the railroad of Donghae line of Goseong county, Gangwon province. Figs. 10.42 and 10.43 illustrates the map

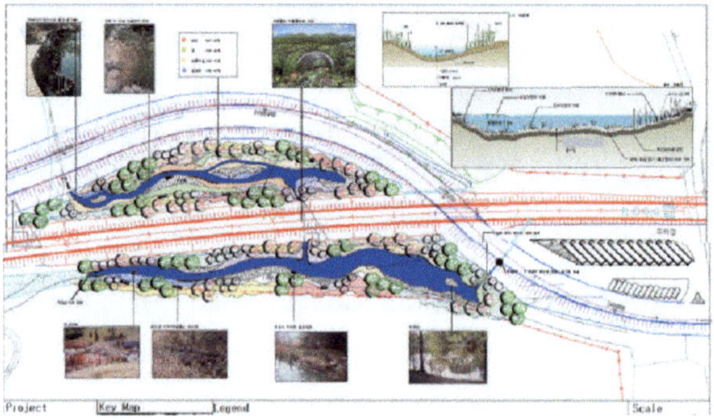

Fig. 10.42 Plan of replacement wetland

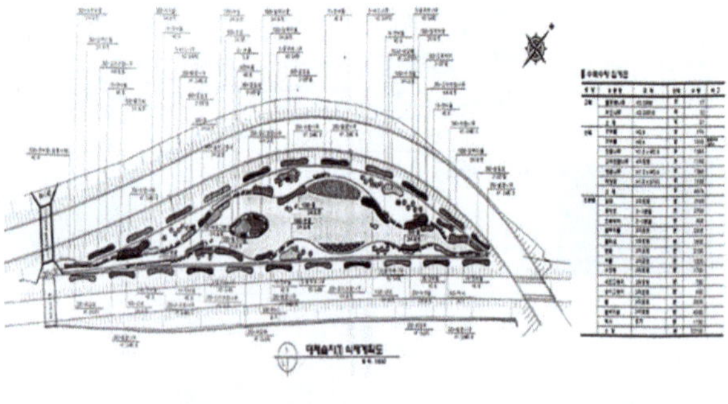

Fig. 10.43 Planting plan of replacement wetland

of the general plan and plating plan of replacement wetlands along the Donghae line.

10.3.1 Construction of Replacement Wetlands

10.3.1.1 Site Selection

The selected sites expand from Gyeonggi Line to the project area of North–South link road and from Donghae Line to the project area of North–South link road.

10.3.1.2 Process

The construction process of replacement wetlands is very similar to that of the general ones. However, in order to judge the success of the replacement wetlands, it is essential to establish the strategies, the objectives, and the standard of practices before the actual construction begins. Moreover, the conformity assessment of the habitat must play a central role when site selection process is preceded. The process of developing construction techniques for replacement wetlands is illustrated in Fig. 10.44.

10.3.2 The Replacement Wetland Created in DMZ and CCZ

10.3.2.1 The Replacement Wetland in Gyongui Line and the Projet Area of North–South Link Road

This area showed high conservation value when evaluated and deserved to be preserved in its natural state. Meanwhile, due to the development project of the inter-Korean exchange and cooperation program, the alternative management action is the creation of wetlands.

First, the objective and standards of practices were established regarding the wetland that would be created. The objective of the creation of wetlands along the Gyeonggi Line was to restore the wetlands affected by the developments to provide a suitable habitat for species that were lost and displaced especially the *Nannophya pygmaea* Ramder as the primary targeted species, as well as the *Kaloula borealis* and *Elaphe schrenckii*. The standard practice was the provision of a spawning pond with open water surface and aquatic plants.

The next step was the selection of the site for the creation of wetland. The selection of probable sites undergone two processes; first, the evaluation was on the indoor environment and second, the evaluation was focused more on site survey and analysis. A long list of proposed sites were evaluated which was

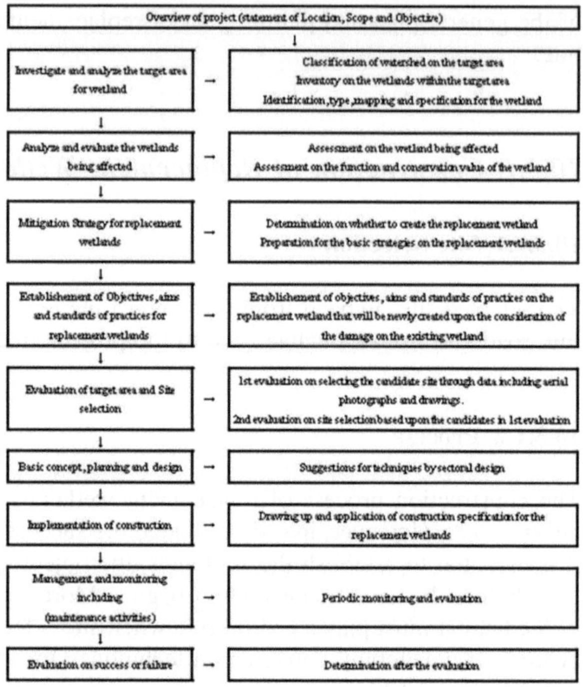

Fig. 10.44 The process of construction for replacement wetlands

eventually cut down into two final sites to be decided upon and finally a site was identified. After the selection, it went through the process of basic concept, planning, and design based on the following two objectives: (1) Creation of wetland for water purification, and (2) Creation of wetland using target species. The created wetland is referred to as "replacement wetland" in this chapter.

A "prototype wetland" was identified in the same area to serve as a reference site and which shall be duplicated in the "replacement wetland" to be created. Naturalization is applied in some areas to induce natural competition and ecological succession.

Results

Completed Features of the Replacement Wetlands

The creation of the wetlands in Paju area began based on the plan completed in October 2003. The replacement wetland created in Paju is shown in Fig. 10.45. The replacement wetland was created with porous materials, wooden bridges, and piled stones at the water's edges to serve as bank barriers and at the same time serve as spawning and playing areas for creatures. It was made sure that the replacement wetlands are connected to the natural wetlands in the area.

Fig. 10.45 Detailed features of the replacement wetlands

The result of the monitoring on habitat

The monitoring on the replacement wetlands of Gyeonggi Line was conducted in the year 2003–2006. There was no vegetation right after its creation but as time passed, herbs started to appear and by year 2006, marshy vegetation were already observed around the replacement wetlands. As the vegetation stabilized the habitat, the number of species in the area gradually increased.

The result of the monitoring on number of species

As expected, no species were seen 2 months after the creation of the replacement wetland until winter of 2003.

But after 9 months, species such as *Colias erate*, carpenter bee, Copper Butterfly, *Neocaridina denticulata*, *Abbottina rivularis*, *Lefua costata*, *Pseudorasbora parva*, *Aphyocypris chinensis,* and *Glandirana rugosa* were found.

Table 10.15 The classification of number of species found by date

Classification	Vegetation	Mammal	Bird	Insect and crustaceans	Fish	Reptiles and amphibians
July 2004	10	5	36	12	10	5
September 2004	10	5	37	12	9	5
January 2005	–	7	32	–	–	–
March 2005	–	6	32	5	11	3
July 2005	18	9	34	14	10	3
September 2005	18	5	27	14	10	3
November 2005	18	7	31		10	–

After 11 months, *Ophraella communa LeSage*, *latiabdominis Miyamoto*, *A. rivularis*, *L. costata*, *P. parva*, *A. chinensis*, *G. rugosa*, *Salmosa*, *Phasianus colchicus,* and *Hydropotes inermis* were observed.

After 19 months, 1 Rana rugosa and Agkistrodon brevicaudus were observed in addition to the species found in month 11.

After 21 months, *L. Miyamoto*, *C. erate*, carpenter bee, Copper Butterfly, *N. denticulata*, *A. rivularis*, *L. costata*, *P. parva,* and *A. chinensis* were found.

On the 23rd month, *atiabdominis Miyamoto*, *N. denticulata*, *A. rivularis*, *L. costata*, *P. parva*, *A. chinensis*, *P. colchicus,* and *H. inermis* were found.

On the 25th month (Nov 30th, 2005), *atiabdominis Miyamoto*, *A. rivularis*, *L. costata*, *P. parva*, *A. chinensis*, *P. colchicus,* and *H. inermis* were found

And finally after 37 months from its creation, various kinds of vegetation including marshy vegetation covered up most of the wetland where the open surface could hardly be found.

Moreover, the replacement wetlands were connected to the tributary of Noth Korea's Sacheon river. The species living in the mouth of Imjin river came all the way down to the replacement wetlands brought by current. As a result, *Neocardina denticulata* and *Eriocheir sinensis* were already found inhabiting in the wetlands including; *Zacco platypus*, *Rhynchocypris oxycephalus*, *sinensis* and *Odontobutis intereupta*. In the surrounding vegetation, *Tenodera angustipennis*, *Accipiter gentilis* and *Ophraella communa LeSage* were found. Various kinds of birds and mammals like the *H. inermis* and *Prionailurus bengalensis* gathered in this area in search of water and prey. Fish is abundant and increasing in number such as; the *Rhodeus uyekii*, *Abbottina springeri Banarescu et Nalbant*, *Squalidus gracilis majimae* and *Odontobutis interrupta* which are regarded as endemic species. The abundance of *Gammarus sobaegensis Ueno* and *planktons* found in the wetlands are proofs that the wetlands are free from pollutants (Table 10.15).

The types of species found on the replacement wetlands of Gyeonggi Line are shown in Fig. 10.46.

Fig. 10.46 Species observed in the replacement wetland of Gyeonggi line

10.3.2.2 The Replacement Wetland in Donghae Line and the Projet Area of North–South Link Road

Completed features of the replaced wetlands

The replacement wetlands in Paju area were completed in November 2004 and the process of the construction is shown on the following pictures below.

Fig. 10.47 The process of the construction of the replacement wetland in paju area

The result of the monitoring on habitat

With the noticeable increase of the marshy vegetation, the wetlands are naturally developing suitable habitats for larger wildlife like the mammals, birds and reptiles. In order to verify if the replacement wetlands are already stable habitats for these species, an ecological study was performed by experts of every field, four times a year. During the monitoring, various kinds of wildlife were observed including; *Nyctereutes procyonoides*, *H. inermis*, Korean fire-bellied toad, *Orthetrum albistylum*, *Ardeidae*, and *Anas poecilorhyncha*. Moreover, the discovery of a protected species—*Lethocerus deyrollei* in marshy vegetation just proves that the replacement wetland is functioning at its best and as a perfect habitat for larger wildlife.

Expectedly, almost no vegetation were found right after its creation in November 2004 but after 3 months (March 2005), some of *N. procyonoides* and *H. inermis* were found at the water's edge. In June 2005, 6 months after its creation, an increased number of vegetation was noted and even expanded to the near water way forming like stone piles and islands in the wetlands.

In August 2005, 10 months after its creation, not only herbaceous plants were observed but also marshy vegetation such as *Zizania latifolia*, *Phragmites australis,* and *Typha orientalis* C. Presl. Moreover, the presence of reptiles and amphibians including Korean fire-bellied toad were confirmed by the researchers.

In October 2005, a year after its creation, the wetlands had lush vegetation and small mammals including *N. procyonoides H. inermis* and insects like *O. albistylum* were found as well.

Table 10.16 The status of monitoring on replacement wetlands in different times

March 30. 2005	
August 31, 2005	
October 24, 2005	
December 27, 2005	
November 23, 2007	
July 3, 2008	
October 7, 2008	

In December 2005, the marshy vegetation seemed stabilized although a difference in the quantity of water transported into the replacement wetland was noticed. It was already carrying out successfully its function as habitat for wildlife, both as vegetation and stabilizer of the soil (Table 10.16).

10.3.2.3 The Result of the Monitoring on Number of Species

Since the construction of the replacement wetlands in Donghae Line, monitoring continued and found many species inhabiting there. Reptiles and amphibians like Korean fire-bellied toad, Dybowski's Brown Frog, *Hyla japonica* were found in front of the sluice way. In light of the footprints of *H. inermis* and *Sus scrofa* and droppings of *N. procyonoides* discovered in the area, it is a significant proof that the replacement wetland is functioning as water source for mammals. Moreover, it is also presumed to be a resting place for birds since *A. poecilorhyncha* is already a common sight in the spot.

In the replacement wetlands created via waterway, *Juncus effusus* var. decipiens, *Scirpus triangulatus*, *P. australis*, *T. orientalis* C. Presl, and *Scirpus maritimus* L. are found, proving that it is already in a stable state.

Droppings of *H. inermis* were discovered and *S. scrofa* were seen bathing in the surface waters of the wetlands. These sights validate that the wetland supplies water for mammals.

G. rugosa as the representative of reptiles and amphibians and, Lesser Purple Emperor, yellow swallowtail butterfly, *Ranatra chinensis Mayr* as that of insects were found as well. In the wetlands formed by the construction of the waterway, various kinds of fish were observed including endangered species such as *Lethenteron camtschaticum* and *Pungitius sinensis*. Even the *Cobitis pacifica* and *Moroco lagowskii* which are often found in cold water were also found. The floating goby, a fish normally found in brackish water was also found in the wetland. The species discovered in the wetlands are all illustrated in Fig. 10.48.

10.3.3 Eco-Bridge

The Ecobridge plays a pivotal role in providing the nature with the ecological network.

10.3.3.1 Site Selection

In order to restore the ecosystem damaged by the construction of the railroad and the road, two regions were selected as target areas: (1) the Gyeonggi line and project area for North–South connecting road, and the (2) Donghae line and

Fig. 10.48 Species observed in the replace wetland of Donghae line

project area for North–South connecting road. There is a total of five ecobridges within the Paju DMZ four in DMZ and one in CCZ region.

In the Donghae line, five sites were selected including three areas where the forest and the coastal ecosystems are disconnected due to the construction of the railroad within the CCZ and two areas where the forest and wetland ecosystems are disconnected due to the road construction.

10.3.3.2 Design Process for the Eco-Bridge

To create an ideal eco-bridge, complete understanding on the corridor of animals and on the impact of development (construction of roads, etc.) are necessary for designers. In designing the construction, the location, selection of types, and determination on dimensions and methods for anti-collision are seriously considered. The work does not end there, after the completion of the construction, focus is diverted on finding effective ways to maintain, manage, and monitor the site.

However, no preliminary predictive assessment could be adopted for both Gyeonggi Line and Donghae Line in this case. Therefore as a supplemental measure, experts in every field ran a survey on the overall ecology including species composition, distribution, seasonal changes, spawning ground, habitat, activities of daily living, vegetation, and macrowater environment of wildlife in the target area. This is for the purpose of predicting and minimizing the damage on ecosystem, while constructing the road and railroads. Figure 10.49 illustrates the design and construction process of the eco-bridge.

10.3.3.3 Eco-Bridges in Gyeonggi Line

Completed features of the eco-bridges

The eco-bridge was created along the Gyeonggi line within CCZ in order to reconnect the forest habitat disconnected by the existing railroad. Most of the surrounding vegetation were damaged during the construction of the road but were restored after the completion of the eco-bridge. Right after the construction, development of vegetation was already detected. As time passed, the ecosystem was able to recover and balanced by the increasing vegetation. However, the pine trees were not adapting well as observed during continued monitoring. Two CCTVs were installed on the eco-bridge in order to monitor the use and behavior of wild animals.

The result of the monitoring on habitat

Due to the scanty vegetation in the initial stage, bare grounds were seen. But after a year, in the summer of 2004, growth of herbaceous plants was detected on the eco-bridge. In autumn of 2004, both herbaceous plants and bushes were already seen and, in winter of the same year, communities of Saw tooth oak were observed.

By May of 2005, almost all kinds of vegetations including communities of *Quercus acutissima Carruth.*, *Lespedeza bicolor,* and Korean azalea were found.

The result of the monitoring on the number of species:

The entire area covered by the Paju Gyeonggi Line is located at the bottom of CCZ and the eco-bridge created within the ecological network connecting the forest habitat, agricultural land and palustrine wetlands well developed.

Moreover, the eco-bridge was frequently used by mammals and birds such as; *H. inermis*, *N. procyonoides,* and *S. scrofa*. Their footprints and droppings on the eco-bridge were seen during the monitoring activities. The movement of small mammals including *H. inermis*, *S. scrofa,* and cat are validated through infrared cameras. The species observed on the eco-bridge of Gyeonui line are illustrated on Table 10.17.

The vegetation on the eco-bridge include: *Taxus cuspidata* (planted), *Q. acutissima Carruth.* (planted), and pine trees. Species of *Q. acutissima Carruth* are the

Table 10.17 The species detected on the eco-bridge of Gyeonggi line

November 8 2004	 Footprint of Hydropotes inermis	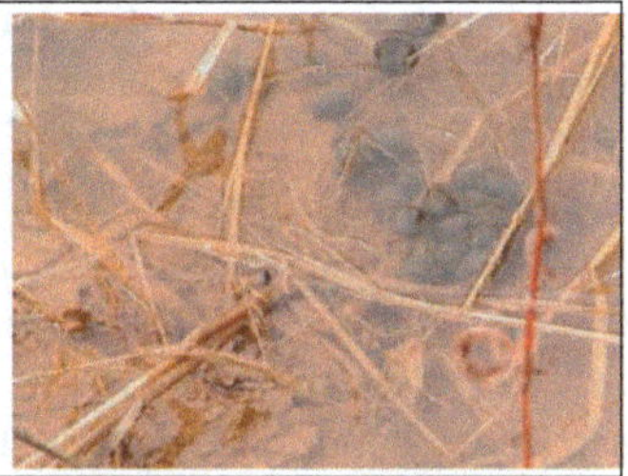 Droppings of Hydropotes inermis		
January 17 2005	 The bird nest of pale thrush	 Droppings of Hydropotes inermis	 Droppings of bush dog	 Tunnel of Mogera wogura
March 21 2005	 Footprint of Hydropotes inermis detected on the eco-bridge	 Footprint of Prionailurus bengalensis detected on the eco-bridge		
May 16 2005	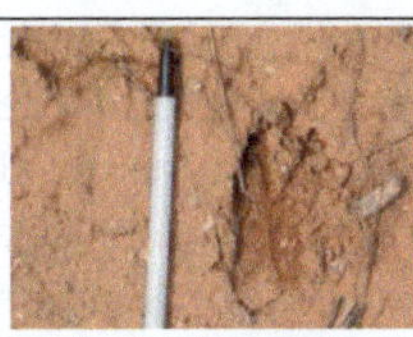 Footprint of Sus scrofa	 Droppings of Nyctereutes procyonoides.	 Silver-studded blue	

most dominant in the area. *Quercus mongolica Fisch.* ex Ledeb., *Acer tataricum subsp. ginnala* (Maxim.) *Wesm.*, *Quercus aliena Blume*, *Ailanthus altissima*, *Meliosma myriantha Siebold,* and *Zucc.* were found as forest vegetation. Footprints of *H. inermis*, cat, and *N. procyonoides* confirmed the existence of these three mammals in the eco-bridge.

In January 17 2005, *T. cuspidata* (planted), *Q. acutissima Carruth*, and *Q. acutissima Carruth* were noted. *Q. mongolica Fisch.* ex Ledeb., *A. tataricum subsp. ginnala* (Maxim) *Wesm., Q. aliena Blume*, *A. altissima,* and *M. myriantha Siebold* and Zucc. were the forest vegetation observed in the eco-bridge. Moreover, for mammals, footprints of *H. inermis*, bush dog, and *Mogera wogura* were detected and, as for birds, pale thrush was observed.

In March and May of 2005, *T. cuspidata* and *Q. acutissima Carruth* were the observed vegetation while *Q. mongolica Fisch.* ex Ledeb., *A. tataricum subsp. ginnala* (Maxim.) *Wesm.*, *Q. aliena Blume*, *A. altissima,* and *M. myriantha Siebold* and *Zucc* were the forest trees noted. For mammals, footprints of *H. inermis*, *P. bengalensis,* and cat were found.

10.3.3.4 Eco-Bridges in Donghae line

Completed features of the eco-bridge

The eco-bridge was created along the Donghae line within CCZ of Goseong county where the forest and coastal ecosystems were disconnected due to the construction of the railroad. This eco-bridge plays a pivotal role as the ecological corridor of mammals, reptiles, and amphibians. The ecological pond on the eco-bridge was primarily developed as wetland and with the formation of the marshy vegetation, and it has balanced well with the surroundings. A CCTV was installed on the bridge to monitor the movement of inhabiting wildlife. The detailed stages of the construction of the eco-bridge are illustrated in Table 10.18.

The result of the monitoring on habitat

In the winter of 2006, the ecological pond was installed but has not yet stabilized a habitat as not enough vegetation were detected. But in April 2007, some marshy vegetation started to emerge such as the *T. orientalis* C. Presl, and insects were also discovered near the ecological ponds. In June 2007, it was found to be stable as a wetland. Some mammals, reptiles, and amphibians were found. In August 2007, the growth of vegetation including *T. orientalis* C. Presl was found to be generally in good condition but other species were rarely detected during rainy season because of the overflowing water. The creatures need not go to the eco-bridge though, to secure vegetation during rainy seasons because it is available anywhere. In November 2007, vegetation found to be stably rooted on the bridge had developed into a marshy wetland. Increased number of water beetles was detected.

The result of the monitoring on the number of species

For the vegetation, *T. orientalis* C. Presl, Quercus variabilis Blume, broad-leaved tree, tomato seedlings, and wild rosebush were detected.

Table 10.18 The change of eco-bridge in Donghae line

During the groundwork (June 28, 2005)		December 26, 2006	
During construction (October 24 2005)		June 19,2007	
Right after the completion (June 19, 2006)		November 23, 2007	
March 28, 2008		July 3, 2008	
October 7, 2008		April 23, 2009	
August 27, 2009			

Table 10.19 The species observed on the eco-bridge of Donghae line

October 24 2005				
	Solenoglyph	Footprint of Sus scrofa		
December 26 2006				
	Droppings of Hydropotes inermis	Droppings of Sus scrofa		
April 9 2007	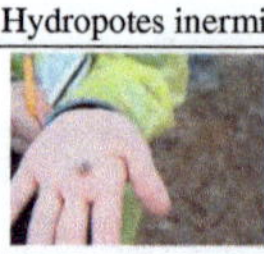			
	Notonecta triguttata Motschulsky	Ranatra chinensis Mayr	Droppings of Nyctereutes procyonoides	Droppings of Hydropotes inermis
June 19 2007				
	Typha orientalis C.Presl	Wild rose bush	Quercus variabilis Blume	

	Droppings of Hydropotes inermis	Rana nigromaculata	Viper Snake	
August 7, 2007				
	Tomato seedlings	Footprint of Hydropotes inermis	Droppings of Nyctereutes procyonoides	Mogera wogura tunnel
November 23, 2007				
	Typha orientalis C.Presl	Droppings of Hydropotes inermis		

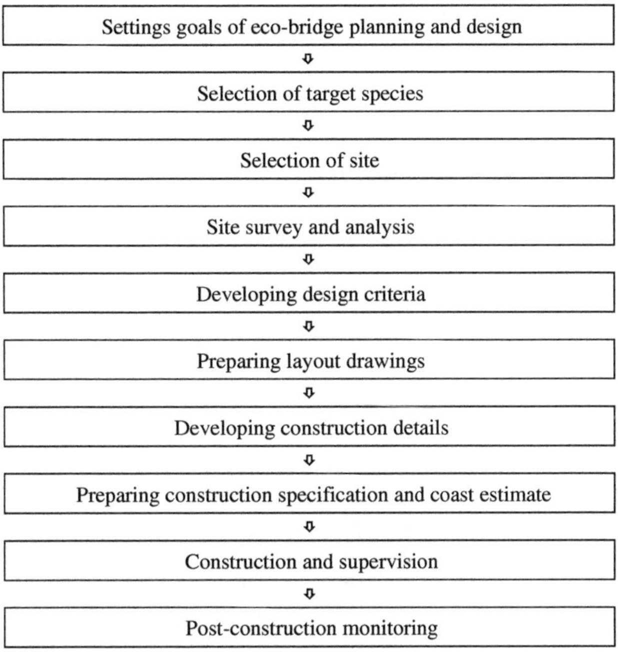

Fig. 10.49 Design and construction process of the eco-bridge

The following species were observed: (1) Mammals; footprints of *H. inermis*, *S. scrofa*, *N. procyonoides*, Mustela sibirica, *M. wogura,* and cat. (2) Reptiles and amphibians; *Rana nigromaculata*, Korean fire-bellied toad, Dybowski's Brown Frog, *Rhabdophis tigrinus*, *Salmosa,* and Water snake. (3) Insects in the ecological pond; *Notonecta triguttata Motschulsky*, *R. chinensis Mayr*, Korean Water Scorpion, *Aquarius paludum* (Fabricius), and *Eretes sticticus*. The details of species detected on the eco-bridge are shown in Table 10.19).

10.4 Trans-boundary Cooperation

10.4.1 One Ecological Unit: Biodiversity Management and Restoration

The DMZ is a politically drawn boundary demarcating the boundaries of South and North Korea. But the ecological boundary crosses the dimensions of both. The DMZ is one interconnected ecological unit formed at both sides of the MDL (Military Demarcation Line) with a width of 2 km in the south and 2 km on to the north. Therefore, the DMZ offers intriguing possibilities for promoting

After the groundwork (before the groundbreaking of eco-bridge construction in August 2001)

During the construction (November 23, 2001)

After the completion (May 16, 2005)

Fig. 10.50 The process of the construction of the eco-bridge in Gyeonggi line. **a** After the groundwork (before the groundbreaking of eco-bridge construction in August 2001), **b** During the construction (November 23, 2001), **c** After the completion (May 16, 2005)

conservation for the transboundary ecosystems, as well as for the transfrontier cooperation and peace of North and South Korea. To protect extensively the largely unspoiled biotopes, a border-spanning composite biotope system should be set up cooperatively by both countries. Details of the ecological unit created around the MDL are shown in Figs. 10.51, 10.52 and 10.53 including images of wetlands and important areas found in the DMZ.

The Korea Institute of Geoscience and Mineral Resources developed a 3-D image of a potential course to be restored as South-North railroad. The image was developed by combining photos from satellite Arirang No.1 and GOS of France and was released to the public on 3 August 2000. Yellow line is 12 km long, which runs from Munsan to the southern boundary of DMZ, and red line is 8 km long, which is the northern boundary of DMZ to Bongdong (Chosun Ilbo Photo DB, 2000).

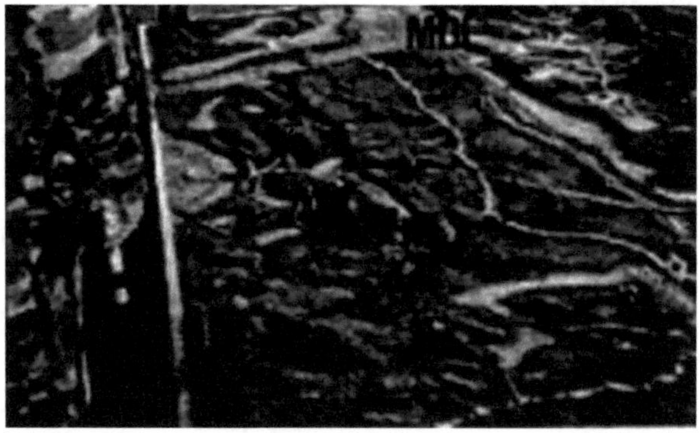

Fig. 10.51 Western DMZ and MDL

Fig. 10.52 Eastern DMZ and MDL

Fig. 10.53 Landsat image of western DMZ topography and MDL

10.4.1.1 Migratory Bird's Flyway

The DMZ is a regular flyway of migratory birds in Northeast Asia including the red-crowned crane, white-naped crane, and the hooded crane. Therefore, there is a need to manage the area through joint efforts by the DPR Korea and the ROK to promote ecological sustainability and contribute in the conservation of global

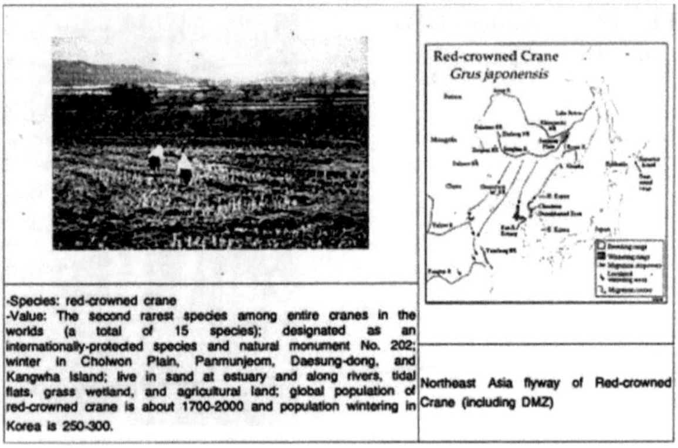

Fig. 10.54 Northeast Asia Flyway of Red-crowned Crane and its photograph and Value

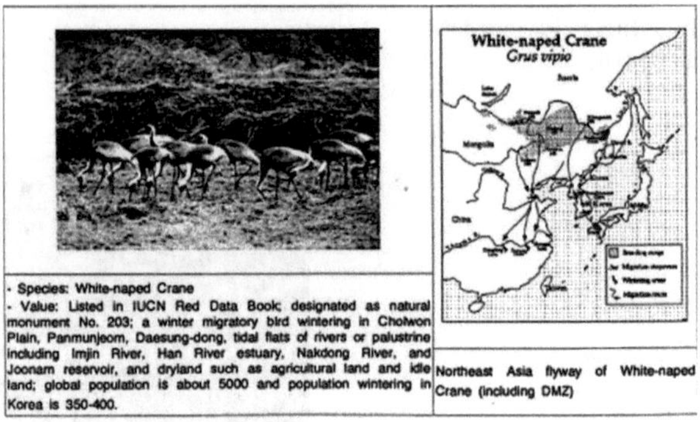

Fig. 10.55 Northeast Asia Flyway of White-naped Crane and its photograph and Value

Fig. 10.56 Northeast Asia flyway of hooded crane and its photograph and value

biodiversity. The final destination of the cranes in the DMZ is the Cholwon and Panmunjom areas. But the red-crown cranes are also found settled at an estuary where the Imjin river and the Han river join the West Sea, whereas the white-napped cranes and hooded cranes are also found. Figures 10.54, 10.55, and 10.56 show the Northeast Asia flyways of the cranes.

10.4.1.2 Biodiversity and Economic Cooperation

The DMZ is a place where the two Koreas can share biodiversity-based economic benefits.
Eco-Tourism

(a) Potentials of Nature and Landscape. As described in the previous chapter, wetlands are well developed around the five major rivers running through south and north of the DMZ. Habitats of rare animals, plants, and migratory birds increase the opportunities for ecotourism. Therefore, if the DMZ of both sides is combined and developed as an integrated eco-tourism system, the status of the DMZ is expected to be improved much greater.
(b) Potential of Cultural Resources. The DMZ has high-quality cultural tour products offering good opportunities for cultural tourism. However, the DMZ in both south and north need to be integrated in order to be registered at the Trans-boundary UNESCO Natural and Cultural Heritage List.
(c) Construction of Eco-tour Roads. The restoration of Gyeonggi and Donghae South-North railroad, the construction of Gyeonggi and Donghae inter-Korean road, and Mt. Keumkang road are expected to contribute significantly in south-north travel. However, more access roads are required for international eco-tourism.

Agriculture and Forestry

Agricultural activities are relatively low in the DMZ due to its border situation and lack of industrialization. But ecological agriculture takes place in some areas. Since the DMZ has valuable biological resources that may possibly contribute to the development of new medicines among other in-demand materials, but there should be an attempt to understand the impact of farming and forestry activities on biodiversity value as a priority over economic value. Based on research outcomes, there is a need to launch a joint ecosystem restoration project by the DPR Korea and the Republic of Korea in areas found to be necessarily used.

10.4.1.3 Transportation and Economic Cooperation

According to recent information, five inter-Korean roads and two railroads are discussed to be necessary. These roads and railroads need to be restored or built across the DMZ. Therefore, there is a need to examine the impact of such restoration and construction on the environment and ecosystems of the DMZ with the perspective for Transboundary Biosphere Reserve (TBR) arrangements before it is too late.

Principles to be considered in building new roads:

1. Examination of the suitability of road paths and the environment friendliness of techniques in the perspective of natural conservation of the very sensitive ecosystems in the DMZ.

Table 10.20 Potential sites for designation of TBR in East Asia (including Russia)

	ROK	DPRK	China	Russia	Monglia	Japan
ROK		DMZ	Not applicable	Not applicable	Not applicable	Not applicable
DPRK			Tumen River delta	Dalaihu, Tumen River delta	Not applicable	Not applicable
China				Daurksiy, Tumen River delta	Mongol Dornol	Not applicable
Russia					Uvs Nurr Depression	Not applicable
Monglia						Not applicable
Japan						

2. Impact on socioeconomic development of nearby communities including eco-tourism.

10.4.1.4 Necessity of South–North Korean Joint Efforts to Address International Cooperation and Strategy on Biodiversity Conservation and sustainable and peaceful

The Convention on Biological Diversity issued a declaration which was adopted at the Rio Earth Summit in 1992, and has 193 parties as of June 2013. Based on the purpose of the Convention, there is a need to develop a Biodiversity Action Plan viewing the entire DMZ as a single ecosystem using the ecosystem approach. In particular, there should be efforts to integrate biodiversity conservation in the national planning process of the two Koreas.

10.4.1.5 Necessity of Designating the DMZ TBR in Viewpoint of EABRN

As one of objectives of the North East Biosphere Reserve Network (EABRN), designation of transboundary biosphere reserve (TBR) was a major agenda for discussions at the 6th EABRN meeting. As shown in Table 10.20, potential sites in East Asia (including Russia) are under discussion, but nothing has been realized, which raises the necessity of designating a TBR in northeast Asia. The designation of the DMZ TBR will be the first in this area and may become a benchmark of other efforts.

10.4.1.6 Peace and Security

The DMZ is still considered as a politically sensitive area. The designation of DMZ TBR will facilitate open dialogue and cooperation on natural resource management and conservation through the creation of a South–North environment community. This will contribute to the peace and security of the Korean Peninsula as well as those of Northeast Asia

It is important to note that IUCN (1997) called on the international community to encourage the State to cooperate in the establishment and management of transfrontier conservation areas as a means of strengthening international cooperation, maximizing benefits, and fostering regional peace and stability.

10.4.2 Operational Strategy to Develop the DMZ TBR

This section describes procedures for the designation of DMZ TBR and key initiatives for implementation.

10.4.2.1 Procedures for DMZ TBR Designation

Designation of DMZ TBR may proceed in two options, which are shown in Figs. 10.57, 10.58.

(1) Option 1: Combine Southern and Northern DMZ and Establish at Once (Establishing TBR in "One Step")
(2) Option 2: DPRK and ROK Designate their DMZ as a Biosphere Reserve First and Integrate into a TBR

10.4.2.2 Establishing International Cooperation and Partnership to Promote Designation of DMZ TBR

Financing
The designation of DMZ TBR requires an extensive, systematic, and in-depth environment ecological survey and, based on its results, a natural ecological map and zoning map need to be developed. In particular, restoration of destroyed ecosystem will cost a significant amount of capital. Therefore, it would be desirable for the DPRK and the ROK to make joint efforts to leverage Global Environmental Facility (GEF) support and funding mechanism available at the Convention on Biological Diversity.

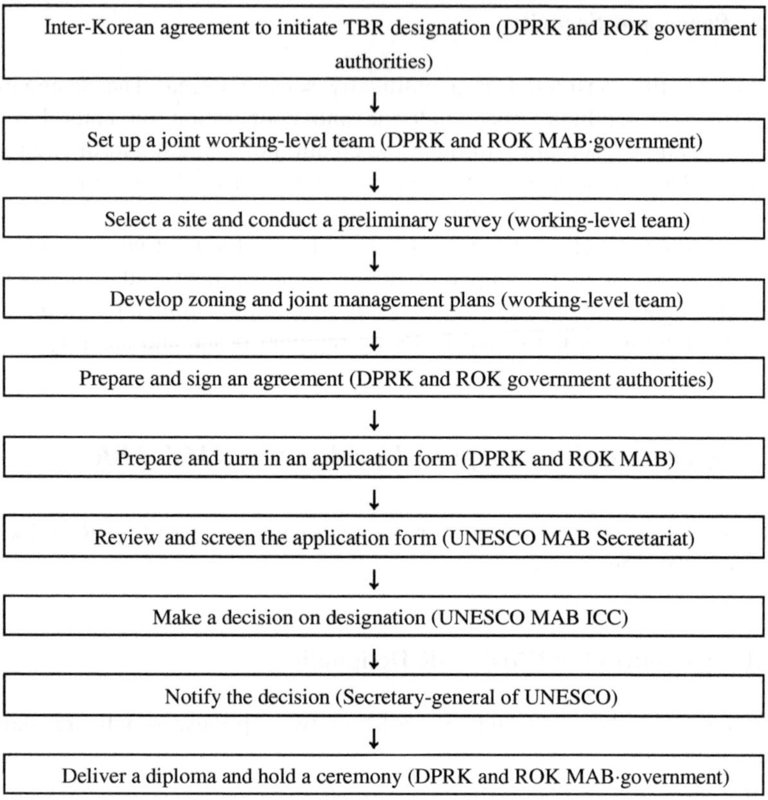

Fig. 10.57 Process when Southern and Northern DMZ is combined and establishment at once

Formulating an international partnership

Harmonizing environment protection and requirements of economic development is not a simple task. It may even cause conflict of interest between DPRK DMZ and ROK DMZ. Therefore, establishment of an international partnership for new thoughts on the DMZ, effective coordination and cooperation in transboundary areas, and active participation may be recommended.

The IUCN stated, in the Declaration of Principles, that there should be efforts to designate the DMZ on the Korean Peninsula as a Transfrontier Protected Area. This declaration had been adopted at the Parks for Peace Conference attended by 92 representatives from 32 nations,

Development of an integrated monitoring program

Without a formal monitoring mechanism that is applied to both the DPRK and the ROK, the sustainability of the DMZ may be dangerously determined through unscientific methods. Therefore, environment ecological indicators need to be

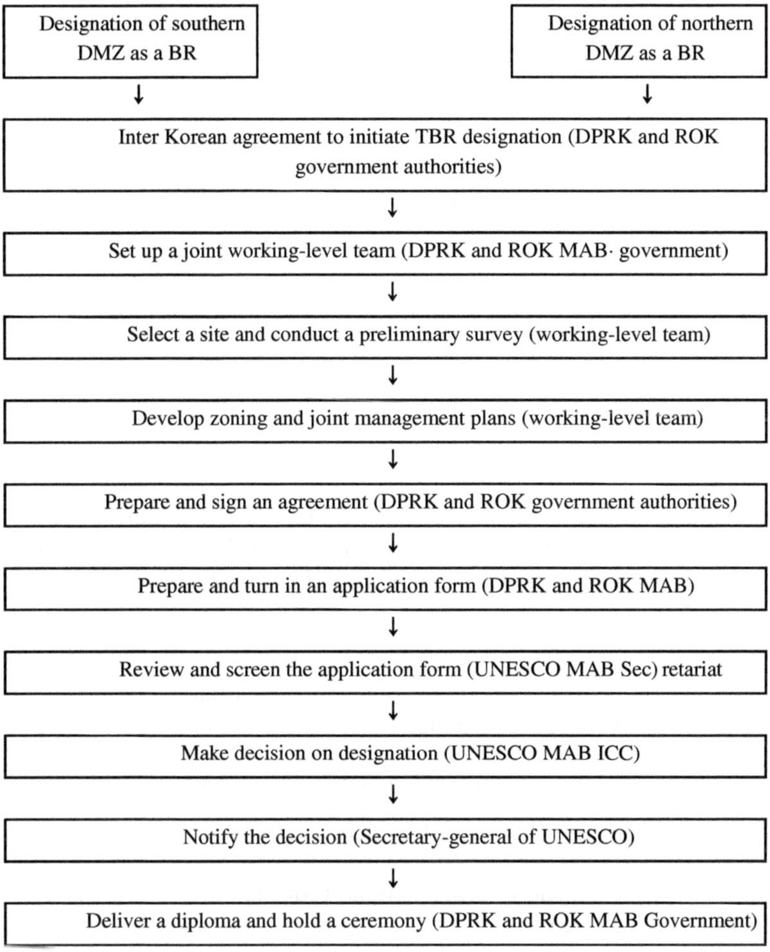

Fig. 10.58 Process when DPRK and ROK designate their DMZ as a biosphere reserve first and integrate the BRs into a TBR

developed with the cooperation of Biosphere Reserve Integrated Monitoring (BRIM) Programme or World Conservation Monitoring Centre (WCMC), which is one of UN organizations. In addition, comprehensive monitoring needs to be launched. This should include the survey and study of DMZ's ecological succession in the past 60 years.

Designation of a TBR is a good example for the conservation of nature in border areas through the cooperation of two or three nations. It offers opportunities to achieve common goals in economy, tourism, recreation, natural conservation, and culture through joint efforts. TBR is encouraged by international organizations.

If DMZ is designated as UNESCO TBR, it is expected that there will be significant opportunities for cooperation in the fields of economy, tourism, recreation, and nature conservation through inter-Korea joint efforts.

10.4.3 A Proposed Joint Project for North and South Korea

A Pilot Demonstration Project between North and South Korea: "CONSERVATION AND RESTORATION OF THE RIVERINE WETLANDS BETWEEN DPRK AND ROK" **proposed by the author**

10.4.3.1 Necessity of the Project

There are many transboundary rivers and streams in the middle of the Korean Peninsula that flow down from the north to the south of DMZ. Therefore, there is a need for integrated riverine wetland planning and management based on watershed concepts:

1. Prevention of the fragmentation of the riverine wetlands and riverine forests,
2. Maintenance of the health of stream-side corridors as an important habitat and,
3. Restoration of damaged ecosystems.

To implement the above-mentioned concepts, it is proposed that:

First, to undertake a pilot demonstration project in a specific area (experimental area),
Second, to gradually expand riverine wetland projects based on the data obtained from the pilot project area and,
Third, to designate the area as a transboundary Ramsar and UNESCO Transboundary Biosphere Reserve (TBR) Site.

10.4.3.2 Rationale of the Project

1. If not addressed, expected result:
   - Loss of "valuable" ecosystems and habitats, rare, and endangered species.
   - Loss of human dimensions due to flooding.
2. If addressed, expected result:
   - Avoidance of the definite loss of "valuable" ecosystems, habitats, and rare and endangered species.
   - Mitigation of climate impact and reduction of risks from disasters by sustaining a certain level of stable forest and wetlands.

10.4.3.3 Objectives and Outputs of the Proposed Project

Objective	Output
Objective 1 Joint development of models/methodology for conservation and restoration of riverine wetlands based on watershed concepts and approaches	Output 1.1 Model methodology for riverine wetland conservation and restoration jointly developed including: ① Site selection criteria, and ② Watershed definition criteria Output 1.2 Model/methodology for riverine wetland restoration jointly developed
Objective 2 Selection of a joint project site	Output 2 Description on the site jointly selected
Objective 3 Undergoing environmental and ecological survey and building-up of database (DB) of the selected site	Output 3.1 Report on environmental and ecological survey Output 3.2 Database for the selected site and its adjacent areas
Objective 4 Joint preparation of ecological map	Output 4.1 Map of species distribution Output 4.2 Map of habitat distribution Output 4.3 Ecological map
Objective 5 Joint identification of hotspots	Output 5.1 Criteria for hotspots defined in methodology Output 5.2 List and locations of hotspots
Objective 6 Restoration of the hotspots (riverine wetlands, and watershed of selected site) Sustainable joint monitoring and preservation Fully resourced and institutionalized mechanism in place	Output 6 Hotspots restored Medium- and long-term riverine wetlands/watershed restoration/conservation and monitoring plan A joint fully resourced restoration and monitoring institutional mechanism established
Objective 7 Joint identification of other potential sites for replication	Output 7 Criteria definition and Inventory of other potential sites Prioritization of sites for restoration

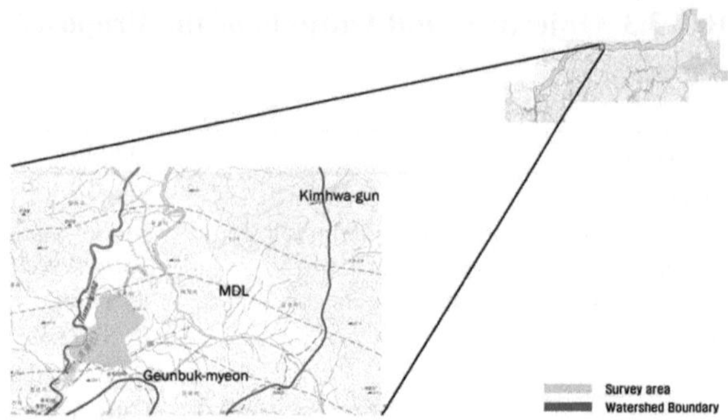

Fig. 10.59 Location of the site

10.4.4 Implementation Plan

To produce the outputs mentioned in the previous section logical framework with targeted milestones should be developed.

10.4.4.1 Project Beneficiaries

The beneficiaries of the project may include:

- Resource decision-makers, planners, and managers
- Local people
- Ecotourism visitors
- Future generations
- DPRK
- ROK
- UNESCO
- Ramsar Convention
- IUCN
- DMZ International Research Institute
- UN-HABITAT
- UNEP
- CBD

10.4.4.2 Potential Site for Demonstration Project Site

Selection of Demonstration Project Site

Mindeulle (dandelion) plain wetland area, in Hantan river, Cheorwon county, Gangwon province was selected as a demonstration project site for the south–north cooperation.

10.4.4.3 Overview of the Site

- The site belongs to the Hantan river watershed (Fig. 10.59).
- Hantan river is on the left and Mt. Oseong is located on the northern side.
- Many palustrine wetlands of diverse sizes and shapes are located.
- Assumption: farmland and village existed before.

Site Profile

The profile of the selected site is shown in Figs. 10.60, 10.61, 10.62, 10.63, 10.64, 10.65, 10.66, 10.67, 10.68 and 10.69).

Fig. 10.60 Overview of the site (Photographed on 9 Dec. 2009) (1)

Fig. 10.61 Overview of the site (Photographed on 9 Dec. 2009) (2)

Fig. 10.62 Site in the fall (Photographed on 17 Sep. 2009)

Fig. 10.63 Site in the winter (Photographed on 9 Dec. 2009) (1)

Fig. 10.64 Site in the winter (Photographed on 9 Dec. 2009) (2)

10.4.5 A Strategy for Regional Cooperation for Northeast Asian Migratory Species Sustainability: A Proposal for Northeast Asian ECONET

In the past, pollution and eventual environmental damage were problems to only few regions of the world. But since the late twentieth century, these problems became global threats. At present time, the scope and impact of environmental

Fig. 10.65 Water Deer drinking water in depression wetland of Mindeulle (dandelion) plain

Fig. 10.66 Large egret in depression wetland of Mindeulle (dandelion) plain

damage magnified in a global scale and is still mounting beyond comprehension. After recognizing the anticipated impending global catastrophe, the international community consistently strived for cooperative efforts among different nations (Jongman et al. 2004).

Among many other considerations, preservation and promotion of biodiversity are now crucial to the achievement of sustainable development of a sound and

Fig. 10.67 Salviniaceae

Fig. 10.68 *Aegypius monachus* flying over the Mindeulle (dandelion) plain

healthy environment (Kim et al. 2002). Beginning with Agenda 21, the importance of biodiversity has been emphasized in major stages such as MDGs and WEHAB (Fig. 10.70). Despite such efforts, the natural environment still faces various threats including habitat destruction, decrease in the number of species, population of each species, and even extinction of rare species.

In the past, many efforts targeting on affected regions had been done in order to solve these problems but, to no avail. The failure to achieve the initial objectives led to the reconsideration of the existing nature preservation policies and implementing strategies. Eventually, with the perspective that the entire global ecosystem is interconnected and spatially linked, global cooperation was realized. The shift in perspective for ecosystem preservation is called "Ecological Network" or more popularly known as "EcoNet."

EcoNet is a comprehensive ***concept*** encompassing the holistic conservation of the environment that is fundamentally premised on the ***principle*** of global cooperation and active participation. The immediate goal of EcoNet is the mitigation of the impact of the damaged ecology and the ultimate goal is the restoration of the

Fig. 10.69 Picture taken during the 11th East Asian Biosphere Reserve Network (EABRN) Workshop (Nov. 2009), in Wuyishan, China. From the *left*: *1*. North Korean Representative 2. North Korean Representative 3. Dr. Jayakumar Ramasamy, Programme Specialist UNESCO Office Beijing 4. Prof. Dr. Kwi-Gon Kim, Seoul National University, ROK 5. Chairperson, MAB National Committee of China 6. Dr. Son Kyong-Nam, President, Branch Academy of Biology, State Academy of Sciences, DPRK

world's natural environment. As a concept, it deals with Conservation Biology, Landscape Ecology, and Restoration Ecology. Conservation Biology deals with the concepts and principles regarding distribution and migration of species; Landscape ecology deals with the concepts and principles regarding land use and ecological patterns and; Restoration ecology deals with the concepts and principles regarding the repair and revival of damaged natural habitats.

EcoNet began to be acknowledged in Europe and in the U.S. The efforts for the establishment of the ecological network initially started at a national level and eventually expanded to border nations. A multinational council of European nations was organized for the cooperative management of biosphere reserves. A general direction was then provided for the systematic integration of the migration and distribution of identified species (Conservation Biology); for the strategic connectivity of the natural distribution of the elements and functions of the natural habitat through responsible land use and proper zoning (Landscape Ecology) and; for the scientific repair and revival of the man-made damage on the diversity and dynamics of native ecology.

Fig. 10.70 International efforts for the development of ESSD concept

10.4.5.1 North East Asian Ecological Network (NEA EcoNet)

The fast development and urbanization in the North East Asian region has led to surmounting negative impacts on the global environment. Significantly on biodiversity, the worst consequences were the enormous loss of species and habitats and, the destruction of the natural ecological network—now considered as the most threatening factor in regional sustainability (Min 2001). Leaders and advocates are now seriously contemplating the need for establishing an ecological network in north eastern regions of Asia, seeing the accomplishments of the European EcoNet. The possibility of establishing the north eastern Asian Ecological Network is becoming high (Korean National Commission for UNESCO 2004; Choi 2005). The north east Asian region includes Russia, Far East Asia, Mongolia, three (3) north eastern provinces of China, and North and South Korea (Fig. 10.71).

Kim et al. (2005) sought for a cooperative plan for the establishment of an ecological network in north east Asia aimed at securing environmental sustainability and creating partnership for sustainable development.

Pursuing this vision, the study examined the status and major issues in north east Asia regarding the sustainability of migratory species and derived future initiatives for NEAEcoNet in support to the currently implemented programs. Hence, a plan for the establishment of NEAEcoNet is proposed.

Due to the vast expanse of the identified regions in north east Asia, Advanced Very High Resolution (AVHRR) satellite images with 1 km × 1 km resolution were used. Based on the results, the Land Coverage map, DEM, and Habitat map of the examined regions were drafted.

Previous studies of ECNC (2002), and Darman et al. (2003) identified priority areas for specific species in the north east Asian regions. Based on such areas, this

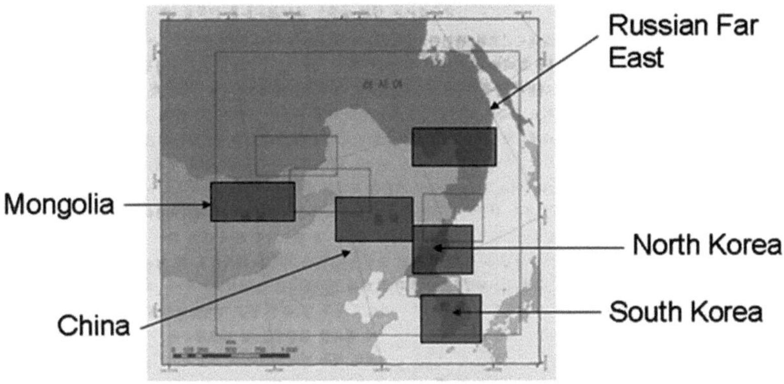

Fig. 10.71 Areas covered by north east Asia

study designated Core Areas with sizeable and sustainable habitat patterns of specific species as research locale. The animal migratory paths identified in previous studies were considered then the buffer and corridor areas in the core area were demarcated. Finally, the status map for the NEAEcoNET was developed.

10.4.5.2 Sustainability of Migratory Species in North East Asia

Status of Existing Habitats.

In north east Asia, various habitats exist depending on topographical features and climatic conditions of the region. Habitats may be classified in many ways but in this study, habitats were classified into 14 types based on the 94 categories suggested by Olson (1994) as shown in the Fig. 10.72. Northeast Asian regions have relatively well-developed river systems and wetlands everywhere. However, dry regions are expanding. The drying of regions resulted to the decrease of forested areas and the worst scenario is desertification. The loss of habitat affects the movement of migratory species.

Status of the Habitat of Migratory Species.

Ecosystem is not confined by human-made borderlines, but has close relationship with the living space in a larger scale. Responding to the increased concerns on biodiversity and interests on the conservation of natural environment, each country consistently put various efforts including the designation of reserve areas. However, there had been clear limitations in such efforts. While the international cooperative planning and implementation are increasing, many case studies and research results indicate that the approach used was not able to guarantee long-term survival of the species. In the past, the focus of preservation and management

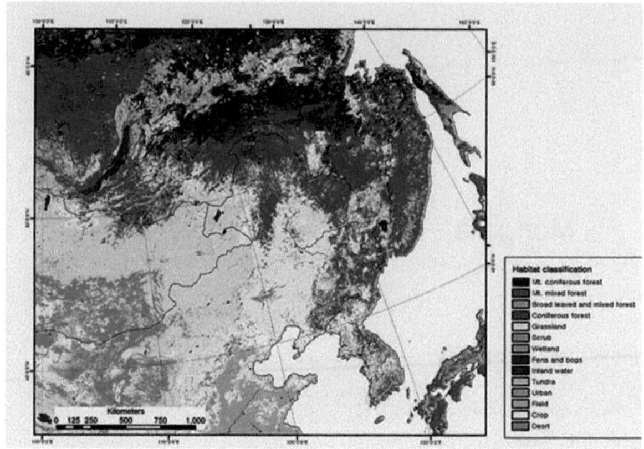

Fig. 10.72 Habitats in north east Asia

was on specific species. The preservation of each habitat posed clear limitations in the maintenance of diversity and movement of species.

The north east Asian region now faces the problem on habitat dispersion. The movement between habitats was found to be very crucial to the survival of species. Figure 10.73 shows the change in habitats and passages for selected species. There has been loss of habitats and reduction in the number of endangered species. Main problems include loss, degradation, fragmentation, and disconnection of habitats (Fig. 10.74).

The selection criteria for target species to be protected in the north east Asia will be explained later, but in principle, the species should have global or regional significance. Figure 10.75 shows the change in habitat of the Mongolian Gazelle. In Fig. 10.75, the pale green color code refers to historical habitats while dark green refers to current habitats.

Interconnected habitat network is necessary for the success of the dispersion and movement of wild animals. Previous experiences reveal that in most of cases, outcomes of programs independently implemented by individual countries do not affect regional and global significance. The effect of environmental destruction excuses no one, not even to the country that conducts its own conservation initiatives. Therefore, cooperative implementation between nations is imperative to achieve effective solutions.

Threatening Factors to the Status of Habitats

Threatening factors to habitats in the north east Asia includes large-scale energy resource development, road expansion, timber harvesting, and inappropriate expansion of agricultural lands among others.

Large-Scale Energy Resource Development:

Expansion of energy transport network in east Siberia and the Far East includes integration of power supply networks.

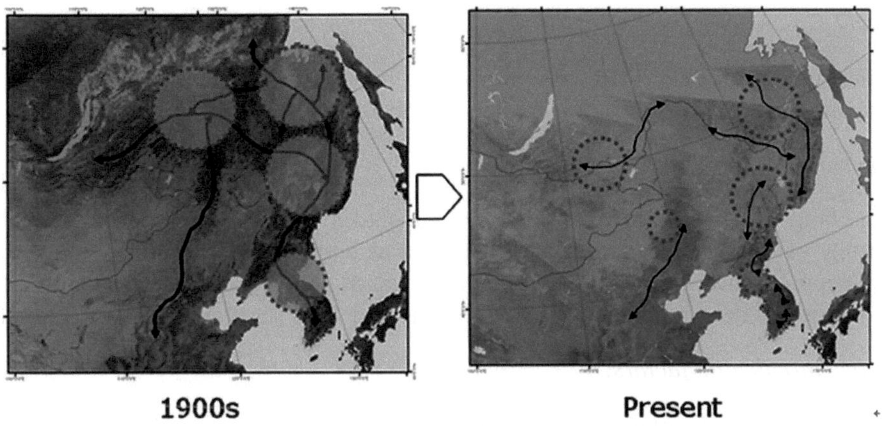

Fig. 10.73 Fragmentation of habitats in north east Asia

Fig. 10.74 Mongolian gazelle

Fig. 10.75 Decline of *M. gazelle* habitats

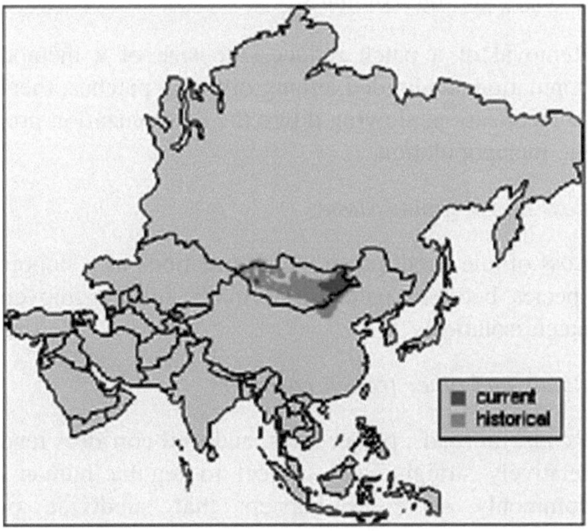

Road Expansion:

Construction of TSR, TKR, and major road expansion projects to cope with the growing quantity of goods transported are planned.

Timber Harvesting:

On-going large-scale logging in Siberia includes Tumen River basin and the cutting of trees by local residents to be used as lumber and firewood.

Inappropriate Expansion of Agricultural Lands: The expansion of agricultural lands causes destruction of wild life habitats and food shortage in North Korea.

Other threatening factors:

Expansion of industrialization (e.g. Hunchun), rapid urbanization, reopening of longtime-closed borders, rapid expansion of transit trade and traffic, growth of tourism, poaching and hunting, irresponsible human activities (fire, use of pesticides etc.), poor regulations, poor law compliance, poor political will, poor enforcement and coordination of conservation programs, use of CFCs and non-biodegradable materials, improper solid waste disposal, poor drainage and sewerage systems, natural catastrophes, climate change and global warming are some examples of other threatening factors.

Process of Change

Local extinction probability:

Larger patches normally have a larger population size of species than smaller patches making it less likely for the species to fluctuate in population size or to be extinct.

Metapopulation dynamics:

Removal of a patch reduces the size of a metapopulation, i.e., an interacting population subdivided among different patches, thereby increasing the probability of localization, slowing down the recolonization process and reducing stability of the metapopulation.

Loss of a stepping stone:

Loss of one small patch which functions as a stepping stone for the movement of species between patches, normally inhibits movement and thereby increasing patch isolation.

Roads and other trough corridors:

Roads, railroads, power lines, and trail corridors tend to be completely connected, relatively straight, and subject to regular human disturbance. Therefore, they commonly serve as barriers that subdivide populations of species into

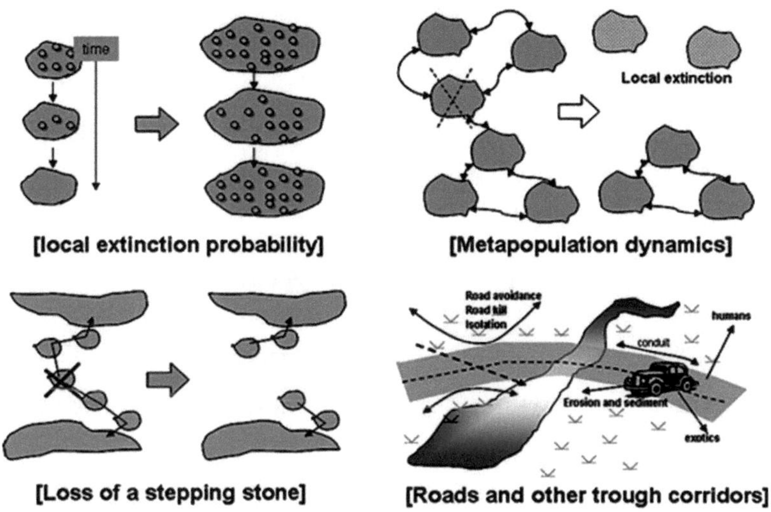

Fig. 10.76 Causes of change including loss of a stepping stone (Forman 1996)

metapopulations. The causes of change, process of change, and loss of stepping stone due to threats is shown in Fig. 10.76.

10.4.5.3 Key Issues: Sustainability of Migratory Species

The major issues related to the sustainability of migratory species in north east Asia can be indicated as follows:

Change in Core Area

Figure 10.77 shows how core habitat areas will be changed over time if proper measures will not be undertaken.

Habitat of Critical and Endangered Birds

In north east Asia, the decrease in species is triggered by various threatening factors. Figure 10.78 shows the status of birds under threat in north east Asia. It shows that the stable inhabitation of birds is being threatened extensively.

Evolution of Key Globally Significant Migratory Species–North East Asian Migratory Bird Flyway

There are 40 animal species listed in the Red Data Books. These are species that are endangered or disappearing and categorized as first-grade national protection category. Ten of which are distributed in China and 23 in Russia. Figure 10.79 shows the migration path of the cranes and black-faced spoon bill (*Grus vipio*),

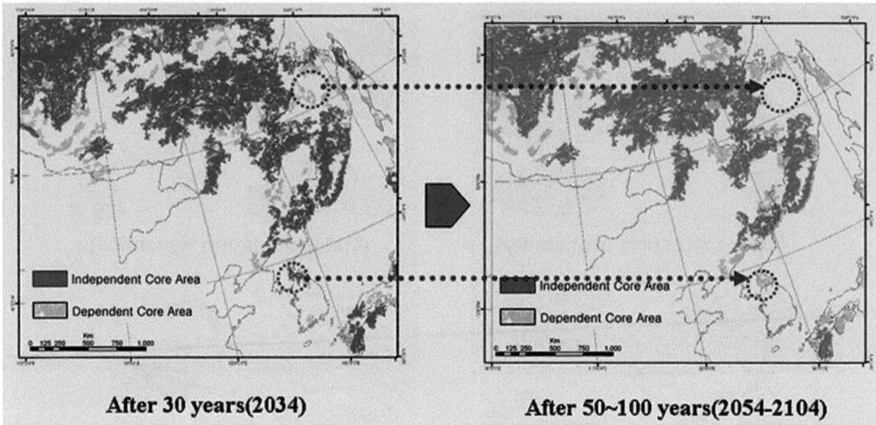

Fig. 10.77 Change in core area

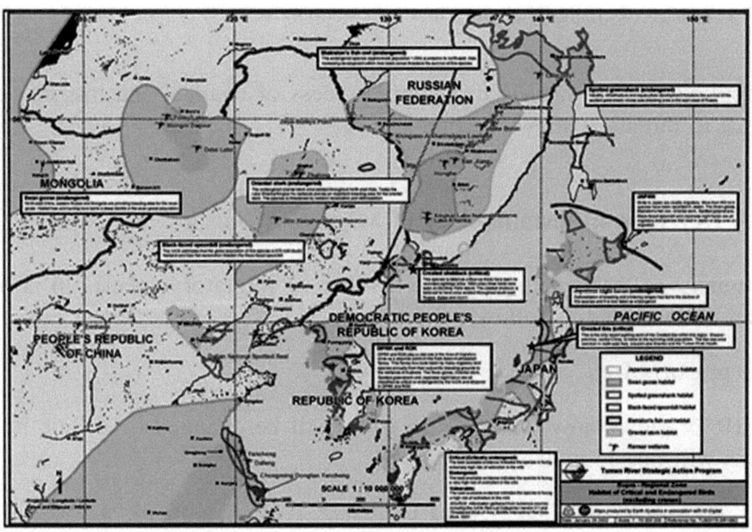

Fig. 10.78 Habitats of critical and endangered birds (UNDP, 2002)

which are of global significance. The number of cranes has been decreasing in Korea (Table 10.21).

Evolution of Key Globally Significant Migratory Species–North East Asian Land Migratory Animals

Tiger habitats have tragically decreased over the past 100 years as a result of habitat fragmentation caused by human settlement, logging, forest loss, and poaching. Previously covering the entire north east Asian area, wild tigers have

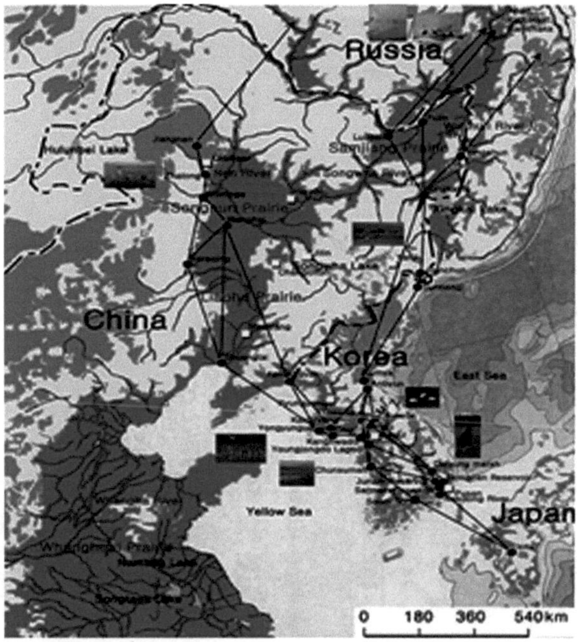

Fig. 10.79 Route of migrating birds

Table 10.21 Decrease of endangered birds in north east Asia

Time species	Past	Present
Crane	1,000 (before 1945)	200–300 (2000)
Black faced spoon bill	11	20

Source Darman et al. (2003)

now been confined to their last remnant habitat at the mixed coniferous and broad-leaved forest areas of Primorsky Territory, Russia including in Khasan District (Fig. 10.80), in Jilin Hunchun Nature Reserve and some neighboring areas of northern Hunchun and Wanfquing Country in the Yanbian Korean Autonomous Prefecture of China. A small isolated tiger population exists in the high northern mountains of neighboring DPR Korea (Table 10.22).

In north east Asia, diverse animal species are distributed. The Russian Far East is inhabited by 422 species excluding insects, out of which 23 are rare species. The Siberian tiger is a major species representing the area. About 2 million Gazelles are estimated to live in the Mongolian steppe, but the range of their inhabitation is significantly declining compared to the past. The Korean Peninsula is inhabited by 102 mammal species and 450 fish species.

Habitat Required for Survival

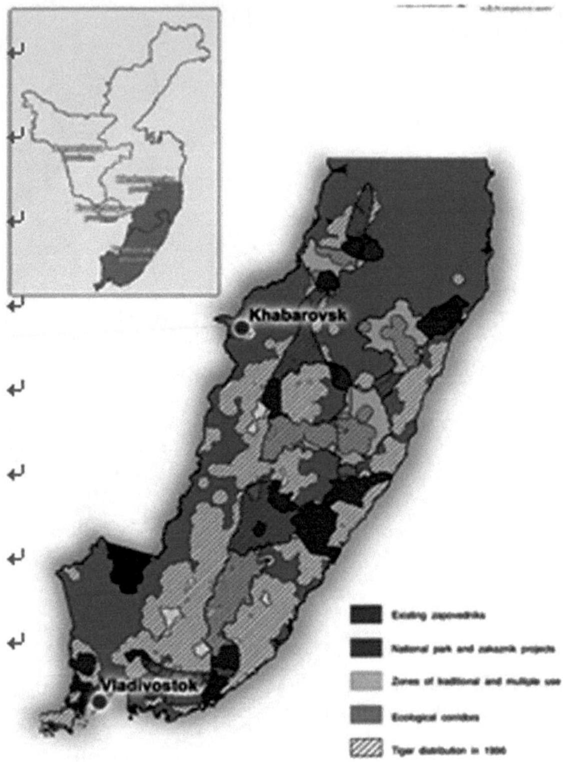

Fig. 10.80 Protection area and corridor for tiger in far East Russian area (Darman et al. 2002)

Table 10.22 Decrease of endangered mammals in north east Asia

Time species		Past	Present
Tiger (in Russia)		1,000 (before 1940)	415–476 (1995)
Leopard	China	38–46 (before 1973)	30–36 (1991)
	Russia	45	10

Table 10.23 shows the habitat type and area required for the survival of selected target species. The Red circle indicates loss of habitat and the blue circle for the decrease in connectivity as shown in Fig. 10.81.

The Arc-GIS was used on the priority species and habitat sizes of each habitat pattern. The changes in habitat patterns in the short-term dimension are 30 years from today, and in the long-term dimension is 50–100 years from today. In long-term perspective, the loss of a large area of habitat consequently disconnecting habitats can be shown in Fig. 10.81.

Table 10.23 Habitat type and area required for survival

	Fox		Lynx		Leopard		Tiger	
	Ideal	Medium	Ideal	Medium	Ideal	Medium	Ideal	Medium
Forest	>2,000	>400	>2,000	>400	>4,000	>800	>20,000	>4,000
Scrub			>400					
Peatland								
Tundra								

10.4.5.4 Present Mitigating Initiatives Based on Regional Cooperation

For the prevention of habitat change and the consequent decrease in species, the international cooperation in the north east Asian region have been organized as a multilateral cooperation of three countries and other border line nations with the participation of international organizations. The cooperation was formed during the early and middle of the 1990s and they have given so much effort to solve environmental problems in north east Asia.

Asia in Various Fields and Subjects.

Comparing the initial goal of the cooperation with that of current outcomes, the results did not work out as expected. The reasons behind the disappointing outcomes are: the leadership struggle of each nation, lack of long-term vision, and the lack of collaboration among regional nations. The conservation of the natural environment was not considered as prime concern in most of the multilateral cooperation.

Besides the need for the creation of NEAEcoNet, shortage of related materials and insufficiency of promotion systems, many more bottlenecks are yet to be toppled by NEAEcoNet before it can be at par with the European EECONET and PEEN (Pan European Ecological Network). Foremost is securing preliminary data and benchmarking of efficient management strategies reflective of the best practices and effective alternative actions utilized by successful countries.

In north east Asia, many areas were designated as Ramsar sites and many more are potentials (Fig. 10.82). In border areas, a number of habitats for rare animals and plants are also designated and managed as protected areas.

Some existing mitigating initiatives are introduced to cope with the threats to habitats which includes (Fig. 10.83).

- Designation of Ramsar sites,
- Designation of management of protected areas in border areas,
- Tumen Net Initiatives,
- UNESCO/MAB EABRN,
- WWF Initiatives, etc.

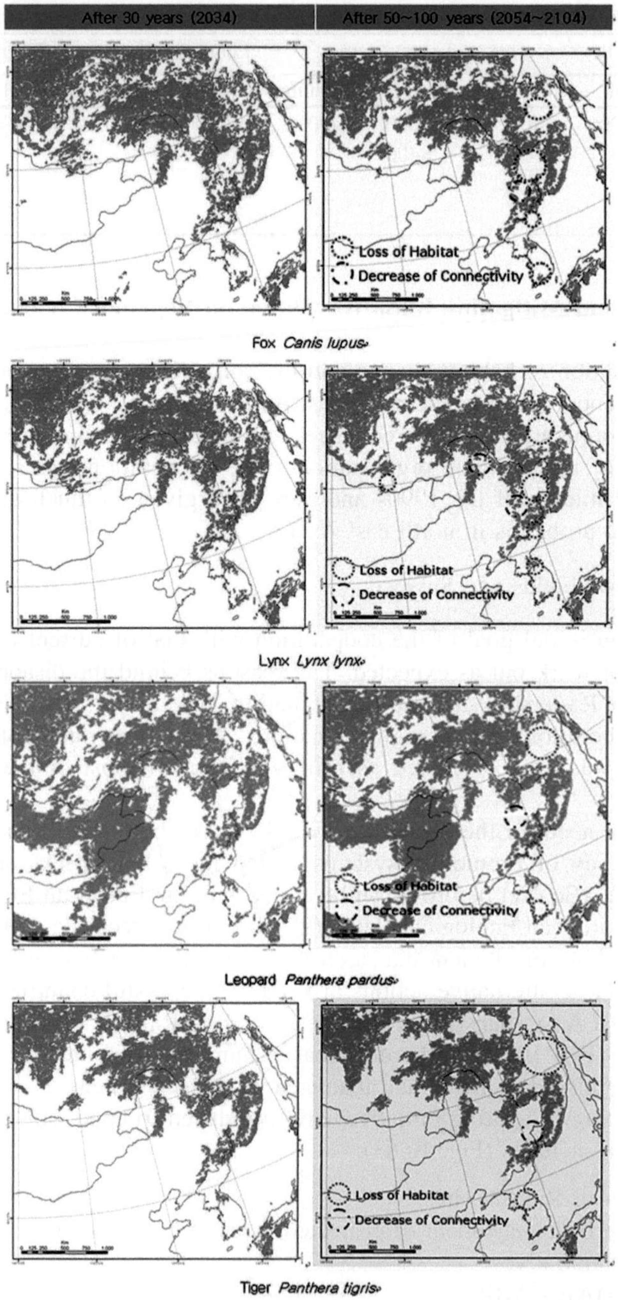

Fig. 10.81 Decline of sustainable habitats and connectivity for priority species in North East Asia

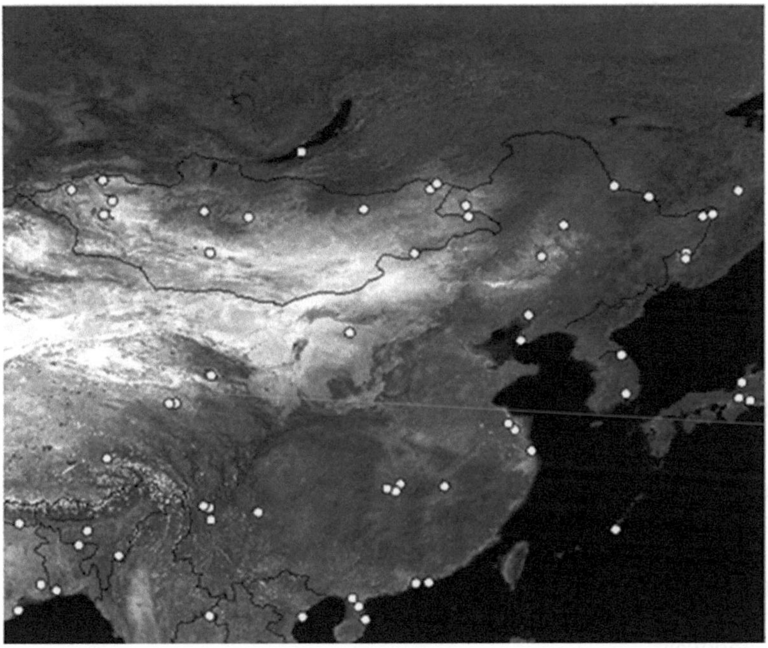

Fig. 10.82 Ramsar convention sites in north east Asia

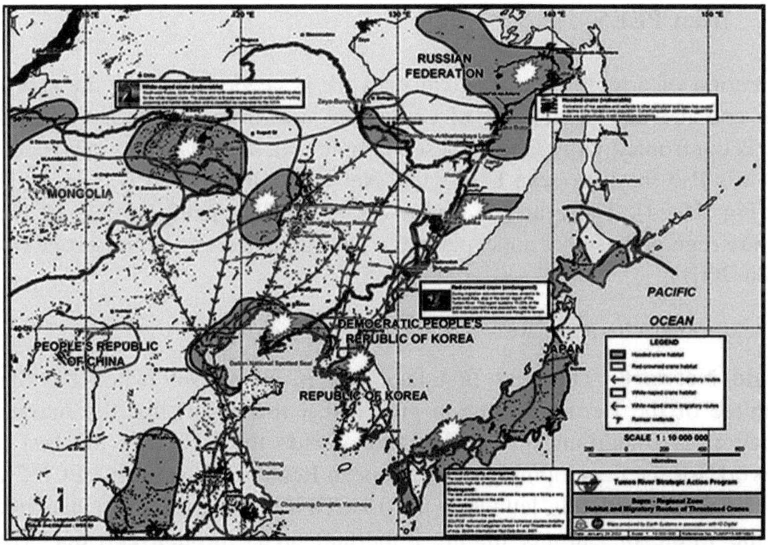

Fig. 10.83 Habitats and migratory routes of threatened bird species (UNDP 2002)

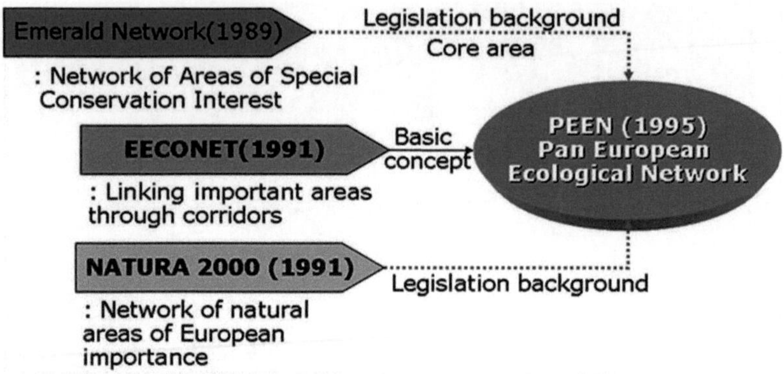

Fig. 10.84 Background of PEEN

Figure 10.83 was prepared as a part of Tumen NET Project, which is initiated in cooperation with UNDP and contributing governments. A study on migratory birds in the north east Asia was conducted in cooperation with the concerned country. The result shows that the direction of the flyways of major birds is south to north in accordance with the desired seasons and that the Korean Peninsula is a major stopover.

10.4.5.5 Present Mitigating Initiatives Based on Lessons Learned from PEEN

The creation of a regional ecological network is needed to solve the shortage of habitat and the disconnectivity of ecological networks in which the north east region is confronted with. One of best example of a successful regional ecological network is Pan the European Ecological Network (PEEN) that was established in 1995 (Fig. 10.84). There are 42 initiatives in the European Ecological Network which are regarded as core concepts in the natural conservation in Europe (Bennett and Wit 2001).

The background characteristics of PEEN are as follows:

Emerald Network: This was first launched in June 1989 when the Standing Committee of the Bern Convention provided a framework for the international designation of important sites inside and outside the EU. It is the first formal framework for the development of a European Ecological Network (ECNC 2003).
EECONET: Netherlands launched the EECONET, an ecological network covering the whole of Europe, linking important areas through corridors (Bennett 1991; Council of Europe 2000).
NATURA 2000: Focused primarily on the identification of core natural areas and did not specifically aim to designate corridors, buffer zones, or stepping stones (Julien et al. 2000; Liro 1999)

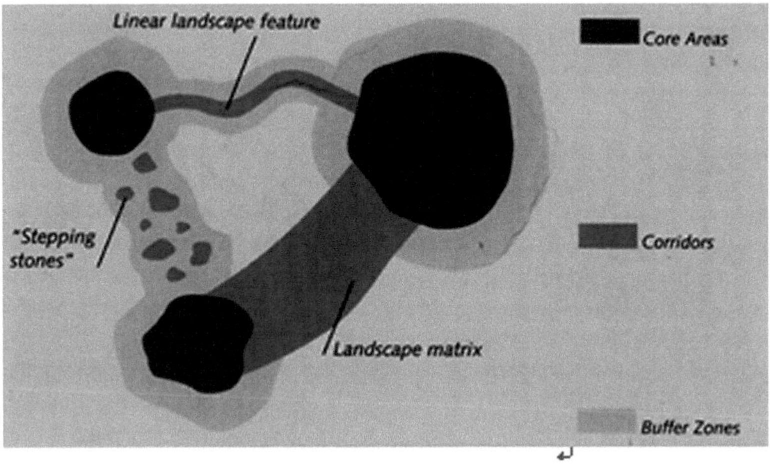

Fig. 10.85 Approach of PEEN (Council of Europe et al. 1996)

The goals of PEEN are as follows (Jongman et al. 1999; Kim 2003; Council of Europe et al. 1996) (Fig. 10.85).

- Habitats are large enough to place species in a favorable conservation status.
- Sufficient opportunities for the dispersal and migration of species.
- Damaged parts of the key environmental systems are restored.

The approach of PEEN is shown in Fig. 10.85. As part of their efforts, the PEEN produced an indicative map (Fig. 10.86).

The lessons learned from PEEN that are applicable to NEAEcoNet include common strategies, guidelines, and policy coordination mechanism applicable to all countries:

- Integrating and utilizing existing cooperation systems.
- Building and operating diverse partnerships.
- Policy coordination mechanism.
- EU parliament has lead role-political commitment.

From the Emerald Network in 1989 to EECONET and NATURA in 2000, the PEEN which is applied to the entire Europe, can be utilized for the creation of strategies for the establishment of a habitat network for the migratory species in Northeast Asia and for the designation of reserve areas.

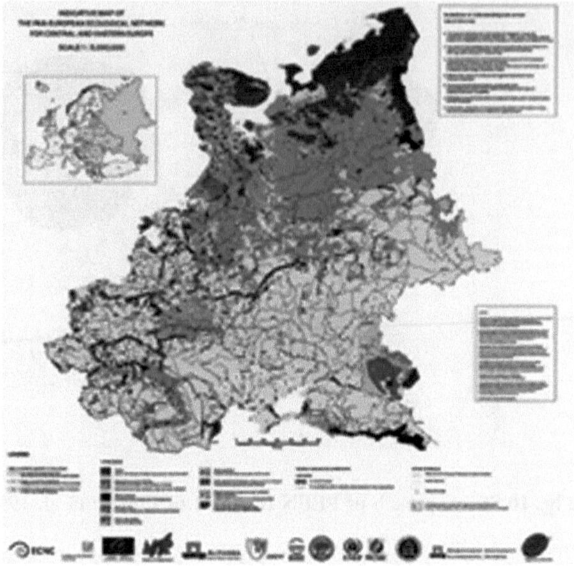

Fig. 10.86 Indicative Map of PEEN (ECNC 2002)

10.4.5.6 Future Initiatives for a North East Asian Ecological Network (NEAEcoNet)

Issues from Present Initiatives to Future Initiatives

The issues related to the North East Asian Ecological Network and the future initiatives were examined by Kim et al. (2005). Due to various causes, the north east Asian region currently is beset by the disconnectivity and the fast pace of the destruction of habitats that resulted in the dislocation and migration of wild animals. Despite the many efforts made and the diverse approaches and strategies adopted, the outcome is far from expected.

Consequently, it is urgent to arrange a strategy to establish a cooperative system for the creation of NEAEcoNET. The application of PEEN should be considered.

Future initiatives are as follows:

- Finding of implementation method for the creation of NEAEcoNET.
- Establishment of NEAEcoNET through an innovative approach.
- Arrangement of efficient and systemically linkable north east Asian cooperation.
- Management of strategy for the establishment of north east Asian cooperation and work programme.

Proposal for a NEAEcoNET

Conceptual Model

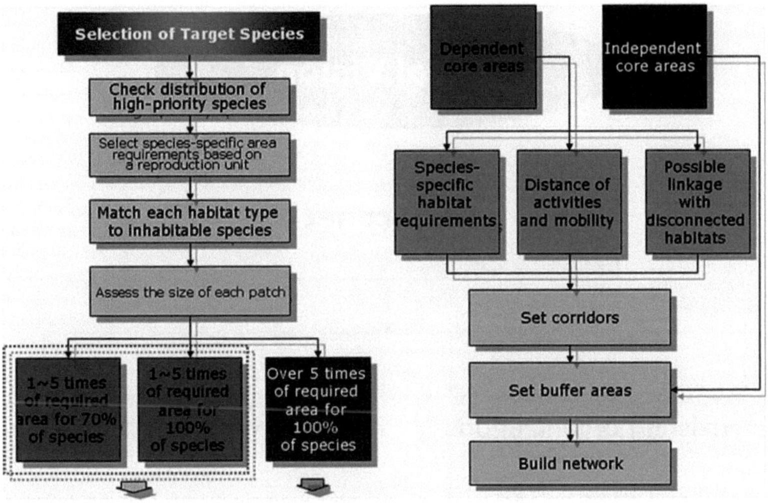

Fig. 10.87 Conceptual model for establishing an international ecological network

The most important stage in the creation of international ecological network is the proper selection of core, buffer, preservation, and corridors, and the development of its targeted function.

A conceptual model to build NEAEcoNet is shown in Fig. 10.87. Here, core areas indicate areas which can provide stable habitats to the selected high-priority species. On the other hand, dependent core areas indicate the areas that need to be interconnected.

Innovative Aspects of NEAEcoNET

The establishment of NEAEcoNET requires a more innovative approach and implementation strategies to achieve future goals. The innovative aspects of NEAEcoNet proposed include holistic and systems/chain approach.

Main Objectives

- Monitor and preserve existing habitat along the migratory paths.
- Restore and relink/recreate habitat along the migratory paths.
- Enhance livelihood opportunities/empower local people.
- Enhance dialog and cooperation between NEA countries.
- Foster greater sense of connection between NEA countries.

Approaches

The approach suggested for the NEAEcoNET includes:

- Building on existing initiatives.

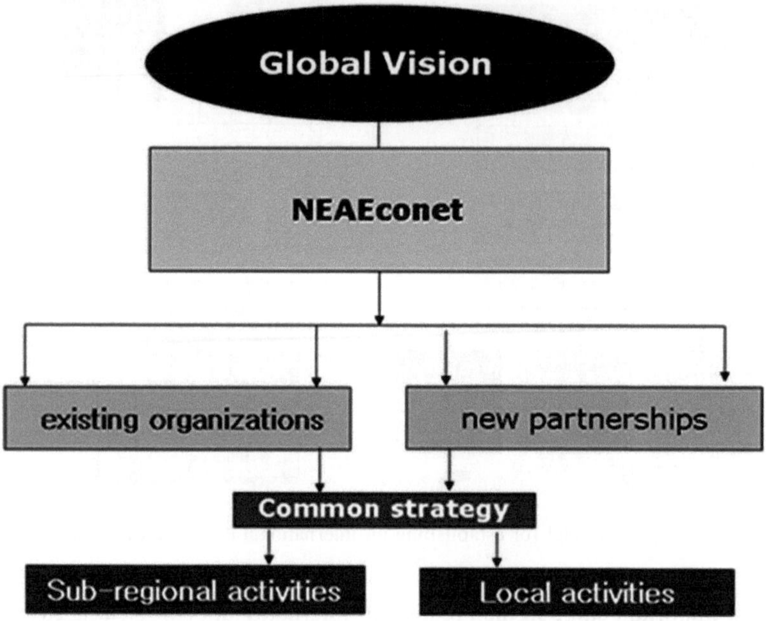

Fig. 10.88 The Role of NEAEcoNET

- Broad partnership between central governments, NGOs, private sector, and local populations.
- Use of modern technology.
- Apply principles of multifunctional landscapes/multisectoral approach.

Supra Nature of the NEAEcoNET

Figure 10.88 presents the supra nature of the NEAEcoNET that embrace existing organizations and new partnerships. Based on global vision, the existing organization and new partnership, NEAEconet should create a general strategy to be executed in sub-regional and local level (Fig. 10.88).

Organizational Options

Based on the supra nature of NEAEcoNET, two organizational options are suggested for the proposed EcoNET. Strengths and weaknesses of each option are described as Fig. 10.89.

Future Work Programme

The future work programme includes development of a sustainable development strategy focusing on partnership, advocacy, policy, restoration, and monitoring strategies including database. This is presented in Fig. 10.90.

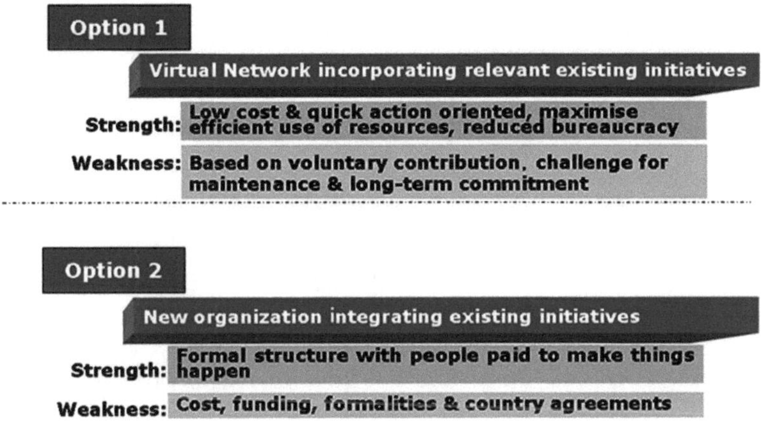

Fig. 10.89 Organization options for NEAEcoNET

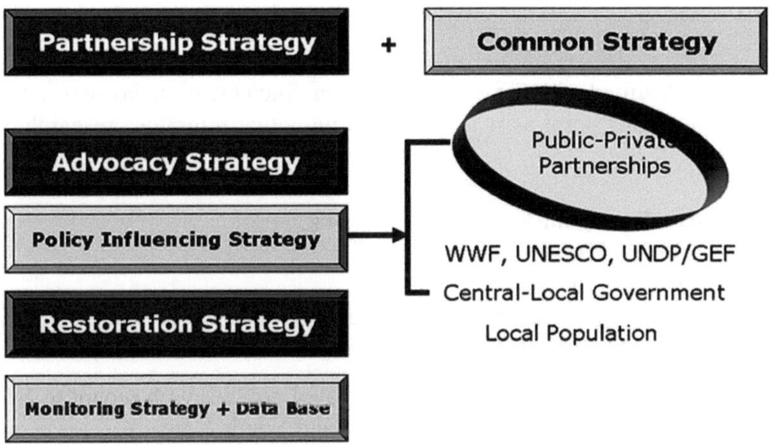

Fig. 10.90 Sustainable development strategy

Mapping ideal migratory paths

Ideal migratory paths for the north east can be mapped in terms of possible core areas and corridors. Figure 10.91 shows the migratory paths interconnecting the core regions in NEAEcoNET and migratory paths reflecting the proposal of UNDP (2002) on tiger movement path, and ecological migratory path to Tumen river watershed.

In case of the created migratory paths, the path of the cranes as revealed by past investigations overlaps considerable areas, so that the corridor linkages on core

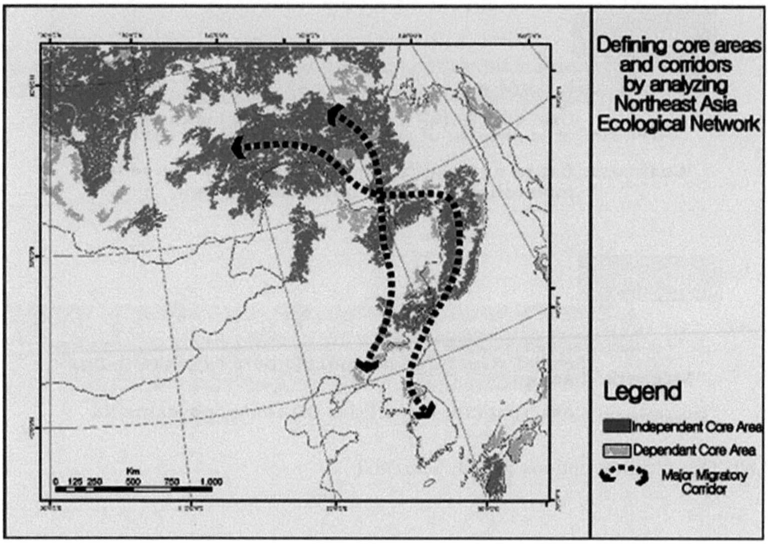

Fig. 10.91 Defining core areas and corridors by analyzing North East Asia Ecological Network

regions are well suited with the current situation. Such creation, however, indicates the limitation at this point as it is not based on actual migratory research data.

10.4.5.7 North East Asia Ecological Network

Sample Priority Actions
The prerequired procedures for the application of a conceptual model in the process of the establishment of NEAEcoNET are as follows:

- Review the ecological environment and related policies with regard to EcoNET in Mongolia, Russia, China, North Korea, and South Korea.
- Survey the stance and current status of north east Asian countries on setting up the EcoNET.
- Classify major habitats in north east Asia and conduct a threat analysis.
- Designate biota core areas and corridors to set up the NEAEcoNET (including areas to be restored).
- Define the scope of cooperation and roles in setting up the NEAEcoNET.

In 1995, the Third Ministerial Environment Conference for Europe has approved the establishment of PEEN. However, after more than 10 years, the cooperative system, strategy, and action plan for the establishment of North East Asian Ecological Network have not yet been systematically structured and still beset with loss of habitats and disconnectivity. The elements threatening the habitats are spread all over north east Asia and the plan for the establishment of

NEAEcoNET to protect the habitat from such threatening elements is the ultimate objective of this research.

Based on the status, issues, and the current initiative in the north east Asia region, the future initiative was conceptualized and the approach, direction, and supra nature for the NEAEcoNET were described.

Prior to the implementation of the conceptual model of this study, as mentioned previously

1. Review the ecological environment and related policies as regard to EcoNET in Mongolia, Russia, China, North Korea, and South Korea.
2. Survey the stance and current status of north east Asian countries on setting up the EcoNET.
3. Classify major habitats in north east Asia and conduct a threat analysis.
4. Designate biota core areas and corridors to set up the NEAEcoNET, including areas to be restored.
5. Define the scope of cooperation and roles in setting up the NEAEcoNET, specific programs should be conducted.

Politically and ecologically, this is the right time to bring countries together for greater accomplishment of existing initiatives. Through NEAEcoNET, the day will come when the Siberian tiger would show up at Jiri mountain which is located on the southern part of the Korean Peninsula.

References

Arkansas Multi-Agency Wetland Planning Team (2001) http://www.mawpt.org
Barbier EB, Acreman MC, Knowler D (1996) Economic valuation of wetlands: a guide for policy makers and planners. Ramsar Convention Bureau, Gland, pp 3–17
Bennett G (1991) Toward a European ecological network. IEEP, Arnhem
Bennett G, Wit P (2001) The development and application of ecological network: a review of proposals, plans and programme. ADD Environment, Amsterdam, p 131
Best EPH, Baker JP (1993) The Netherlands—wetland. Kluwer, Dordrecht, pp 306–310
Caims J Jr (ed) (1995) Rehabilitating damaged ecosystems, 2nd edn. p 425
Choi J (2005) A study on the international environmental cooperation for establishment of North East Asia ecological network. Ph.D. Dissertation, Seoul National University
Council of Europe (2000) General guidelines for the development of the Pan-European ecological network. Nature and environment, vol 107. Council of Europe, Strasbourg
Council of Europe, UNEP and ECNC (1996) The Pan-European biological and landscape diversity strategy: a vision for Europe's natural heritage. Mart Spriijt bv, Amsterdam
Cowardin LM et al (1976) Interim classification of wetland and aquatic habitats of the United States. US Fish and Wildlife Service, Washington, pp 5–16
Cowardin LM et al (1977) Classification of wetland and aquatic habitats of the United States. US Fish and Wildlife Service, Washington, pp 4–16
Cowardin LM et al (1977) Classification of wetland deep water habitats of the United States. US Fish and Wildlife Service, Washington, pp 4–18
Cowardin LM et al (1979) Classification of wetland and deepwater habitats of the United States. US Fish and Wildlife Service, FWS/OBS-79/31, Washington, pp 4–20
Daniel Smith DR et al (1995) An approach for assessing wetland function using hydrogeomorphic classification, reference wetlands, and functional indices. US Army Corps of Engineers, Engineer Research and Development Center, pp 11–22
Dargle T, Daragie M, Dornoch DT (1994) English nature research report, vol 7(149). Lowland wet grassland resource survey, module 2
Darman Y, Karakin V, Martynenko A, Williamsm L (2003) Conservation action plan for the Russian Far Ecoregion complex: Part 1. Vladiostok, Russia WWF Russia
ECNC (2002) Indicative map of the Pan-European ecological network for central and Eastern Europe. Drukkerij Groels, Tlburg, the Netherlands
ECNC (2003) Support for ecological network in European nature conservation. Drukkerij Groels, Tlburg, the Netherlands
fuer Naturschutz B (1995) Systematik der Biotoptypenund Nutzungstypenkartierung (kartieranleitung). Schriftenreihe fuer Landschaftspflege und Naturschutz 45
Gallimard O (2000) Biosphere reserves: domains for people and nature. MAB France, P 6
Hauer FR et al (2002) A regional guidebook for applying the hydrogeomorphic approach to assessing wetland functions of intermontane prairie pothole wetlands in the northern Rocky Mountains. US Army Corps of Engineers, Engineer Research and Development Center, pp 3–7

K.-G. Kim, *The Demilitarized Zone (DMZ) of Korea*,
DOI: 10.1007/978-3-642-38463-9,

Hyohaemi L (2000) Korean wetland classifications. Master's Thesis, Inha University Report, pp 20–72 (in Korea with English abstract)
IUCN (1977) Declaration of principles, PARKS-Parka for peace, vol 7(3). p 4
Jongman RHG, Küövik M, Kristiansen I (2004) European ecological network and greenways. Landscape Urban Plann 68(2–3):305–319
Jongman RHG, Kamphost D(1999) Ecological corridors in land use planning and development policies, national approaches for ecological corridors of countries implementing the Pan-European biological and landscape diversity strategy. Council of Europe, p 51
Julien B, Lammertz M, Barbie JM, Sylvicuteurs F (2000) Voicing interest and concerns: NATURA 2000—an ecological network in conflict with people. Forest Policy Econ 1:357–366
Koo BH (2002) A study on the classification and mapping method of wetland in Korea. A Thesis for doctor's degree, Seoul National University, pp 44–70, 150 (in Korea with English abstract)
Koo BH, Kim KG (2001) A study on Wetland classification in the ROK: focusing on Wetlands at the DMZ and the Han River Riparian Zone. Korean Society of Environment Restoration Technology, Spring Study Presentation Thesis for 2001, pp 72–77
Karr JR, Dudley DR (1981) Ecological perspective on water quality goals. Environ Manage 5:55–68
Kilmas CV et al (2005) A regional guidebook for applying the hydrogeomorphic approach to assessing wetland functions of forested wetland in the west gulf costal plain region of Arkansas. US Army Corps of Engineers, pp 3–10
Korea Environment Institute (2002) Rainwater run-off management for mitigating non-point source of water pollution
Kim KG et al (2001) A study on management plan and assessment methods the characteristics of wetlands inland. Ministry of Environment Report, pp 37–95 (in Korea)
Kim KG et al (2005) Establishment of type classification for national wetland, Republic of Korea. In: Ministry of Environment·UNDP·GEF·UNDP/GEF national wetland project management unit report, pp 4–62 (published in English)
Kim KG, Cho DG (2004) Introduction of natural environment and ecology restoration. Academybook, Seoul, p 44
Kim KG (2003) The application of the biosphere reserve concept to urban areas: the case of green rooftops for habitat network in SEOUL. In: Urban biosphere and society conference. Partnership of cities conference, pp 28–31
Kim KG (1999) The evaluation of the East Asian biosphere reserve network. Research report submitted to UNESCO Jakarta Office, presented at the EABRN-6 meeting, China, 14–20 Sep 1999
Korean Nature Conservation Institute (1997) Second environmental field survey guide in the whole country. The Ministry of Environment Report, pp 1–35 (in Korea)
Korean Wetlands Society (2000) Environmental field survey guide in whole country report. The Ministry of Environment Report, pp 15–58 (in Korea)
Korean National Commission for UNESCO (2004) UNDP project final report: feasibility study on the establishment of the lower. Tumen river area trans-boundary biosphere reserve (ROK/02/ 2004), pp 2–3
Liro A (1999), The linkages between Natura (2000) ECONET/PEEN and rural development plans. In Natural 2000, rural development and ecological network. IUCN Office for Central Europe, Warsaw
Lockwood JL, Hoopes MF, Marchetti MP (2007) Invasion Ecology. Blackwell, pp 1–9
Molles MC Jr (2010) Ecology: concepts and applications. Mc Graw-Hill International Edition, Boston, pp 466, 474 and 484
Ministry of Environment (2001) Proposal on detailed plan on designation of DMZ TBR, steering committee meeting document. Kwachon Community Hall Seminar Room, 27 Mar 2001 (modified by the author)

Martin AC, Hotchkiss N, Uhler FM, Bourn WS (1953) Classification of wetland of the United States. Special Science Report, Wildlife no. 20. US Fish and Wildlife Service, Washington, pp 4–22
Ministry of the Environment (2006) http://www.env.go.jp/en and http://www.sizenken.biodic.go.jp/wetland/
Min BS (2001) Regional cooperation for control of transboundary air pollution in East Asia. J Asian Econ 12(2001):137–153
Mitsch WJ, Gosselink JG (1993) Wetlands, 2nd edn. Van Nostrand Reinhold, New York, pp 21–29
Mitsch WJ, Gosselink JG (2000) Wetlands, 3rd edn. Wiley, New York, pp 726–766
Olson JS (1994) Global ecosystem framework-definitions: USGS EROS data center internal report, Sioux Falls, pp 37
Richardson JL, Vepraskas MJ (2001) Wetland soils: genesis, hydrology, landscape and classification. Lewis, Boca Raton, pp 88–115
Riecken U, Ries U, Ssymank A (1994) Rote Liste der gefaehrdeten Biotoptypen der Bundersrepublik Deutschland. Schriftenreihe fuer Landschftspflege und Naturschutz, vol 41
Succow M, Joosten H (2001) Landschaftsoekologische Moorkunde, E. Schweizerbart'sche Verlasbuchhandlung Stuttgart
The Korea Forest Service (2004) Survey, conservation and management of forested wetland ecosystem report (in Korea)
The national UNESCO-MAB Committee of Poland (2000) Biosphere reserves on borders. Warsaw, pp 78, 94, and 108
The Ramsar Conservation on Wetlands (2006) http://www.ramsar.org/
Tiner RW (1999) Wetland indicators—a guide to wetland identification, delineation, classification, and mapping, levels. Boca Raton, pp 264–275
UNESCO/MAB (1995) Standing working group of the biosphere reserves in Germany, guidelines for protection, maintenance and development of the biosphere reserves in Germany. Bonn, PP 16–18
UNESCO (2000) Solving the Puzzle: the ecosystem approach and biosphere reserve. Paris, pp 4–6
US Army Corps of Engineers (1987) Corps of engineers wetlands delineation manual. Technical report Y-87-1, US Army Corps of Engineers Waterways Experiment Station, Vicksburg, pp 88–100
US Army Corps of Engineers (1996a) National action plan to develop the hydrogeomorphic approach for assessing wetland functions. Regulatory program of the US Army Corps of Engineers, pp 6–25
US Army Corps of Engineers (1996b) Wetlands: water, wildlife, plants and people. Poster series, Middle school, pp 5–20
US Army Corps of Engineers (1998) Recognizing wetlands. US Army Corps of Engineers
US Fish and Wildlife Service (US FWS) (2000) National list of vascular plant species that occur in wetland: 1996

Index

A
Adaptive management, 16
Advocacy movements, 69
Aerial photographs, 15
Agenda 21, 550
Agricultural pesticides, 477
Alliance for zero extinction, 311
Alluvial wetland, 236
Amazon, 19
Armistice pact, 1
Artificial ecological waterway, 396
Artificial stream, 379
Artificial village ecosystem, 84
Artificial waterway, 394

B
Baekhak mountain, 40
Balance, 128
Bass, 472
Beating, 152
Bio-assessment, 429
Biodiversity, 6
Biodiversity comprehensive strategy, 310
Biodiversity hotspots, 311
Biodiversity index, 158
Biodiversity institution, 308
Biodiversity management agreement, 128
Biodiversity Management Contract Program, 303
Biodiversity value, 312
Biological environment, 96
Biological evolutions, 69
Biological integrity, 429
Biological resources, 539
Bioregion, 33
Bioregional approach, 480
Bioregional areas, 33
Biosphere Reserve, 55
Biosphere Reserve Integrated Monitoring (BRIM), 543
Biosphere, 11
Biota, 88
Biota core areas, 570, 571
Biotope, 82
Bird feeding program, 497
Birdlife International, 311
Birds
- azure-winged Magpie (*Cyanopica cyana*), 286
- black-capped kingfisher (*Halcyon pileata*), 259
- black-faced spoonbill (*Platalea minor*), 257
- black-tailed gull (*Larus crassirostris*), 281
- broad-billed Roller (*Eurystomus orientalis*), 261
- cormorant (*Phalacrocorax capillatus*), 266
- eagle owl (*Bubo bubo kiautschensis*), 285
- einereous vulture (*Aegypius monachus*), 272
- gray wagtail (*Motacilla cinerea*), 264
- hen harrier (*Circus cyaneus*), 274
- kestrel (*Falco tinnunculus interstinctus*), 280
- large and medium egret (*Egretta alba modesta*), 262
- mandarin duck (*Aix galericulata*), 283
- *otus scops*, 252
- parrotbill (*Paradoxomis webbiana*), 277
- red-crowned Crane, White-naped Crane (*Grus japonensis*, *Grus vipio*), 267
- spot-billed duck (*Anas poecilorhyncha*), 279

K.-G. Kim, *The Demilitarized Zone (DMZ) of Korea*,
DOI: 10.1007/978-3-642-38463-9, © Springer-Verlag Berlin Heidelberg 2013

Birds (*cont.*)
swan Goose (*Anser cygnoides*), 269
white-fronted Goose (*Anser albifrons*), 275
white-tailed sea eagle (*Haliaeetus albicilla*), 270
Bogs, 119
Border nations, 551
Border situation, 539
Borders, 510
Boundaries, 30
Brackish water, 504
Breeding place, 327
Buffer, 47
Buffer forest, 456
Burning, 450
Bush grove ecosystems, 84
Bushfires, 367

C
Capacity building, 17
Carrying capacity of ecosystems, 5
Case studies, 553
Cease-fire line, 37
Chemical fertilizers, 444, 477
Civic and people cooperative systems, 309
Civilian controlled zone (CCZ), 1
Civilian wetland monitoring system, 444
Classification system, 74
Climate impact, 544
Coastal dunes, 93
Coastal ecosystem, 84
Coastal habitats, 85
Coastal line, 506
Coding, 81
Community, 367
Community issues, 444
Community-based technologies, 17
Compensatory mitigation policies, 71
Conceptual model, 571
Connection, 567
Connectivity, 447
Conservation biology, 551
Conservation International, 311
Conservation value, 11
Contaminants, 108
Cooperative management, 551
Core areas, 553
Core habitat areas, 557
Corridors, 570
Crisis Ecoregions, 312
Critical ecosystem partnership fund, 311
Cropland ecosystems, 84
Cultivation suspension management contract, 304
Cultural asset, 506
Cultural landscapes, 16

D
Damage, 452
Database, 443, 494
Degradation, 554
Demilitarized Zone (DMZ), 1
Depressions, 243
Desertification, 553
Development, 567
Dialogue, 541
Diversity index, 158
DMZ TBR, 540
Dominance, 396
Dominance index, 158
Dominant, 352
Donghae line, 57
DPRK, 541

E
Ecological bridge, 6
Ecological certificate program, 308
Ecological corridor, 309
Ecological engineering, 16
Ecological network, 51
Ecological pond, 532
Ecological process, 15
Ecological research, 464
Eco-region, 399
Eco-research park, 492
Eco-residential area, 492
Eco-tourism, 10
Ecological resources, 55
Ecological restoration, 74
Ecological route, 506
Ecological security, 18
Ecological study, 496
Ecological succession, 70, 378, 522
Ecological survey methods, 56
Ecological sustainability, 537
Ecological value, 404, 449
EcoNet, 550
Economic cooperation, 447
Ecoscience, 29
Ecosystem disturbance, 464
Ecosystem modeling, 16
Ecosystem services, 479
Ecotourism, 496

Ecotourism plan, 512
Ecotourism resources, 510
Edge, 30, 455
EECONET, 564
Emerald network, 564
Endangered specie, 272
Endangered wild animals, 315
Endemic bird areas (EBAs), 311
Environment and ecology planning, 56
Environmental assessment, 17
Environmental carrying capacity and suitability analysis, 480
Environmental Research Group, 295
Environmental security, 6
Environmental sustainability, 8
Environmentally sound and sustainable development (ESSD), 5
Estuarine ecosystems, 84
Estuary, 9
Ethical-legal responsibility, 96
EU parliament, 565
Evaluation, 440
Evenness index, 158
Evolutionary processes, 68
Evolving process, 24, 446
Exotic animals, 469
Exotic plants, 455

F
Farmland, 2
Fauna survey, 147
Fauna, 97
Feeding grounds, 128
Fens, 119
Fertilizers, 416
Field surveys, 152
Fish
  Coreoperca herzi, 231
  Hemibarbus mylodon, 221
  Iksookimia koreensis, 228
  Koreocobitis rotundicaudata, 225
  Lampetra japonica, 234
  Microphysogobio jeoni, 224
  Microphysogobio longidorsalis, 221
  Pungitius kaibarae, 236
  Silurus microdorsalis, 230
  Siniperca scherzeri, 233
Flora, 97, 190
  bladderwort, 198
  bur reed, 190
  duckweed, 198
  round-leaved sundew, 200
  rush, 195
  sweet-flag, 195
  turkestan rose, 201
  waternut, 196
Flora survey, 145
Flyway, 12
Food capacity, 73
Food chain, 118, 477
Food resources, 13
Forest axis, 41
Forest ecosystems, 84
Forest fires, 15, 453
Forest matrix, 43
Fragility, 485
Fragmentation, 446
Framework, 546
Freshwater, 93
Functional assessment, 401
Functions, 426

G
Genetic diversity, 15
Geographical features, 352
Geomorphic setting, 414
Ginseng fields, 476
Global biodiversity information system (GBIS), 16
Global calamities, 445
Global climate change, 18, 287, 479
Global cooperation, 550
Global ecosystem, 550
Global threats, 548
Global vision, 568
Global warming, 556
Geographic information system (GIS), 16
Globalization Initiative Committee, 494
Governance, 446
Grassland ecosystems, 84
Gyeonggi line, 57
Gyoenggi railroad, 10, 12

H
Habitat corridors, 520
Habitat dispersion, 554
Habitat mapping, 80
Habitat patch network, 509
Habitat restoration, 6
Habitats, 565
Habitat survey and analysis, 74
Herbivores, 112
HGM approach, 399
High-biodiversity wilderness areas, 312
Human disturbance, 556

Human security, 6
Hybrid approach, 399
Hydraulic and hydrologic analysis, 329
Hydric soils, 405
Hydrogeomorphic method, 400
Hydrologic process, 102
Hydrologic structures, 14
Hydrological development, 115
Hydrological disturbance, 121

I
Impact, 71
Inbreeding, 298
Indicator, 432, 433
Indigenes, 474
Industrialization, 539
Initiatives, 561
Inland ecosystem, 362
Inland freshwater marshes, 134
Inland wetland, 422
Insects, 214
  Clossiana selene, 216
  Coenagrion lanceolatun, 220
  Lycaena dispar, 214
  Nannophya pygmaea, 219
  Prosopocoilus inclinatus, 217
Integrated model, 480
Inter-Korean Cooperation Economic development, 396
Inter-Korean reconciliation, 56
Interconnected ecological unit, 535
Interconnected habitat network, 554
Intergovernmental Science-Policy Platform on Biodiversity and Ecosystem Services (IPBES), 17
International community, 549
International partnership, 542
International Peat Society criteria, 55, 67
Intrinsic appeal, 485
Intrinsic character, 489
Intrinsic resources, 488
Invaders, 462
Invasion ecology, 17
Invasion theory, 15
Invasive process model, 463
Investigation, 347
Isolation, 520
IUCN, 541

J
Jiri mountain, 571
Knowledge generation, 17
Korean war, 1
Lacustrine wetlands, 443
Lagoon, 21
Lake and reservoir ecosystems, 84
Land suitability and carrying capacity analysis, 492
Land-use strategies, 7
Landform, 14
Landing spots, 13
Landmines, 16
Landscape district, 33
Landscape ecological models, 30
Landscape ecological pattern, 37–47
Landscape ecological principles, 32
Landscape ecology, 17, 29, 551
Landslides, 114
Leadership, 561
Lease management contract, 304
Light trap, 152
Linear zonation, 496
Living laboratory, 16
Longtime-closed borders, 556

M
Maintenance, 8
Mammals, 287
  chinese water deer (*Hydropotes inermis*), 288
  European otter (*Lutra lutra*), 294
  goral (*Nemorhaedus goral raddeanus*), 295
  leopard cat (*Felis bengalensis manchurica*), 288
  raccoon dog (*Nyctereutes procyonoides*), 293
  wild boar/Wild pig (*Sus scrofa creanus*), 290
Management information system, 308
Matrix, 44
MDGs, 550
Measure, 460, 557
Metamorphosis, 22
Metapopulation, 556
Migratory birds, 19
Migratory species, 553
Military activities, 450
Military demarcation line (MDL), 56
Military operations, 24
Mitigating initiatives, 564
Mole trap, 156
Monitoring, 455
Mosaic habitats, 15
Mosaics
  pattern and scale, 30

Movement of animals, 520
Mud baths, 291
Mud flats, 88
Multinational council, 551

N
National Geographic Society, 311
National Institute of Environmental Research, 160
National Land Use Management Law, 496
National wetlands inventory, 400
Native frequency, 387
Native species, 462
NATURA 2000, 564
Natural areas, 312, 483
Natural catastrophes, 556
Natural ecological best practice area, 309
Natural ecosystem, 51
Natural environment, 37
Natural habitat, 429
Natural land classification survey, 496
Natural monuments, 388
Natural pests, 264
Natural state, 7
Natural stream, 387
Natural wonders, 496
Naturalization, 108
Naturalness, 484
Nature environment audit model, 480, 481
Nesting, 307
Non-regulatory approach, 444
Nonpoint pollution source, 476
North East Biosphere Reserve Network (EA-BRN), 540
Northeast Asia flyways, 538

O
Oil storage, 475
Optimum habitat, 378
Ordinal scale, 430
Organic soil, 221

P
Palustrine wetlands, 443
Pan the European Ecological Network (PEEN), 564
Panmun field, 10
Panoramic seascape, 489
Paradise, 453
Parks for Peace Conference, 542
Partnership, 7, 568
Passages, 554
Patch, 15, 30
Peatlands, 67, 134
Peninsula, 571
Phenomenon, 459
Pit-fall trap, 153
Plain, 48
Planned design, 512
Poaching, 290
Policy, 568
Policy support, 17
Ponds, 219
Potential, 485
Predator group, 442
Predators, 293
Preservation, 447
Pressure, 396
Primary consumer, 442
Prime biodiversity area (PBA), 55, 506, 509
Protection activity management contract, 304
Prototype wetland, 522
Ramsar, 544
Ramsar convention, 426
Ramsar sites, 66
Rarity, 484
Recolonization process, 556
Recreational opportunities, 490
Red data book, 64, 557
Refuge, 446
Regional cooperation, 548, 561
Regional sustainability, 552
Reiteration analysis, 146
Relationships, 487
Replacement ecosystem, 16
Reptiles and amphibians
  Agkistrodon brevicaudus, 249
  Bufo stejnegeri, 238
  Dybowski's brown frog, 244
  Elaphe schrenckii, 250
  Kaloula borealis, 247
  Rana amurensis corneana, 243
  Rana nigromaculata, 243
  Salamander, 238
  Takydromus wolteri, 248
Reservoirs, 219
Responsible development, 448
Restoration, 6
Restoration ecology, 551
Restoration management, 497
Richness index, 159
Riparian buffer zones, 465
Riparian wetlands, 134
Risks, 544
River basin, 230

River levee, 456
Riverbed, 367
Riverine wetland planning, 544
Riverine wetlands, 443
Road Corridor, 40

S
Sacheon river, 12
Salinity, 92
Salt water, 93
Satellite images, 15
Scenic resources, 489
Science-based assessments, 17
Seaside basin, 461
Seasonal changes, 530
Sediment, 121
Segmentation, 488
Seoul national university, 551
Silt loam, 357
Soil contaminants, 475
Soil contamination, 474
Space, 73
Spatial scope, 409
Species, 89
Spread, 463
Standard biodiversity, 302
Stepping stone, 557
Stopover, 564
Stream ecosystems, 84
Streams, 449
Subregion, 33
Succession of plant species, 18
Succession theory, 15
Suitability analysis, 480
Suitability, 539
Supra nature of NEAEcoNET, 568
Survey location, 203
Sustainability, 552
Sustainable development, 552
Sustainable use, 55
Sweeping, 152
SWOT analysis, 446
Systematic analysis, 74

T
Threat analysis, 571
Threatening factors, 554
Timber harvesting, 556
Topographic maps, 14
Total assessment, 494
Tourism zone, 506
Traces, 461
Trail corridors, 556
Transboundary Biosphere Reserve (TBR), 69, 539, 540
Transboundary ecosystems, 536
Transfrontier cooperation, 536
Transition zone, 442
Treasure, 64
Tree planting, 288
Tributary, 38
Tumen NET project, 564
Tunnel, 498

U
UNEP GEO-5, 17
UNESCO (2000) Natural Heritage program, 64
UNESCO/MAB model, 480
Unification Observation Deck, 60
Urban ecological environment, 79

V
Vegetation, 203
  Asiatica, 13
  Aster scaber, 167, 169, 206
  Carex humilis var. nana, 207
  Commelina communis, 359
  Diarrhena japonica Fr. et Sav., 207
  Disporum smilacinum, 206
  Epimedium koreanum Nakai, 169, 206
  Hepatica asiatica Nakai, 206
  Miscanthus sacchariflorus var. gracilis Y.N.Lee, 314
Village, 42, 547
Visual experiences, 489

W
War monument, 502
Water canals, 465
Water contamination, 476
Water fringe wetland, 197
Water purification, 522
Water quality, 295
Water system, 33
Watershed, 544
Waterways, 32, 37
Wetland delineation, 407
Wetland ecosystems, 84
Wetland fringe area, 442
Wetland habitat, 370
Wetland Identification, 405
Wetland integrity, 429

Wetland mapping, 419
Wetland protection zone, 441, 442
Wetland survey, 407
Wetland vegetation map, 209
Wetland-type classification, 399
Wildlife habitat, 2
Willow communities, 115
Woodland patches, 41
Woody plants, 121
World Conservation Monitoring Centre
(WCMC), 543
World Natural Heritage site, 55
World peace, 447

Z
Zonation, 494

MIX
Papier aus verantwortungsvollen Quellen
Paper from responsible sources
FSC® C105338

If you have any concerns about our products,
you can contact us on
ProductSafety@springernature.com

In case Publisher is established outside the EU,
the EU authorized representative is:
Springer Nature Customer Service Center GmbH
Europaplatz 3, 69115 Heidelberg, Germany

Printed by Libri Plureos GmbH
in Hamburg, Germany